Symbiosis as a Source of Evolutionary Innovation

The participants in the conference *Symbiosis as a Source of Evolutionary Innovation*. Back row, from left to right: Mary Beth Saffo, Sorin Sonea, Peter Atsatt, David Lewis, Margaret J. McFall-Ngai, Silvano Scannerini, Jan Sapp, Russell Vetter, John Maynard Smith, Kris Pirozynski, Richard Law, Gregory Hinkle, Werner Schwemmler, Toomas Tiivel. Front row, from left to right: René Fester, Lynda J. Goff, Kenneth H. Nealson, Lynn Margulis, Kwang W. Jeon, Rosmarie Honegger, Paul Nardon.

Symbiosis as a Source of Evolutionary Innovation

Speciation and Morphogenesis

edited by
Lynn Margulis
and
René Fester

The MIT Press
Cambridge, Massachusetts
London, England

This book was set in Palatino by Asco Trade Typesetting Ltd.

Library of Congress Cataloging-in-Publication Data

Symbiosis as a source of evolutionary innovation : speciation and morphogenesis / edited by Lynn Margulis and René Fester.
p. cm.
Based on a conference held in Bellagio, Italy, June 25–30, 1989.
Includes bibliographical references and index.
ISBN 978-0-262-13269-5 (hc.: alk. paper)
ISBN 978-0-262-51990-8 (pb.)
1. Symbiosis—Congresses. 2. Species—Congresses. I. Margulis, Lynn, 1938–. II. Fester, René, 1965–.
QH548.S93 1991 90-20439
575.1′3—dc20 CIP

The MIT Press is pleased to keep this title available in print by manufacturing single copies, on demand, via digital printing technology.

Contents

Foreword

Without symbiosis, the nature of life on Earth would be unrecognizable from that which is found today. This is not just because symbiosis was crucial to the evolution of eukaryotes from their prokaryote ancestors. It is also that most modern terrestrial ecosystems are critically dependent on symbiosis: 90 percent of land plants in nature are mycorrhizal, and virtually all mammalian and insect herbivores would starve without their cellulose-digesting symbionts.

Yet, despite its all-embracing importance as a phenomenon, symbiosis usually receives no more than lip service from most mainstream biologists, whose ideas about it are usually naive and a good half-century out of date. This was nowhere better reflected than by the sometimes ridiculous hostility which greeted the Serial Endosymbiosis Theory when it was first advocated in its modern form in the late 1960s by Lynn (Sagan) Margulis. The later acceptance of the theory by cell biologists has often been accompanied by a reluctant unwillingness to acquaint themselves with modern advances in the study of symbiosis.

Nevertheless, biologists whose primary concern has been research into symbiosis have seen the subject advance by leaps and bounds in the last 30 years. Those who once confined themselves to the narrow study of a single type of association, such as coral reefs, insects, legumes, or lichens, now draw avidly on ideas and concepts from other associations, giving the whole subject much-needed cohesiveness.

The Bellagio Conference, at which the papers in this volume were presented, addressed another key aspect of symbiosis which had not previously had the major attention it deserved, namely the importance of symbiosis in the origin of evolutionary and morphological novelty. This is a topic of major importance, and the organizers of the conference are to be congratulated on assembling such a distinguished array of talent.

Reading the book reinforced my sense of great sadness at being unable to attend the conference. Many of the papers raise novel and provocative ideas—not all of which I agree with, but it would have been such fun to have been at Bellagio to challenge some of them, and then marvel at the stimulating originality of others of my colleagues. It was also gratifying that some biologists outside the field of symbiosis attended, and this was clearly to everyone's mutual advantage. The overwhelming and fascinating conclusion to be drawn from reading this book is that symbiosis is a very powerful source of innovation, and that it will be a matter of great and lasting excitement to understand and explore why this should be so.

David C. Smith

Preface

The chapters in this book reflect oral presentations given at the conference on Symbiosis as a Source of Evolutionary Innovation, held June 25–30, 1989, at the Bellagio Study and Conference Center in Bellagio, Italy. Organized by Lynn Margulis and Kenneth H. Nealson and funded by the Rockefeller Foundation, the meeting provided a unique opportunity for evolutionary theorists and symbiosis biologists to cross the boundaries of their respective disciplines and share ideas, addressing the adequacy of the prevailing neo-Darwinian concept of evolution and the relative importance of symbiosis in the origin of morphological and evolutionary novelty.

Twenty scientists from nine countries gathered to discuss these issues in an idyllic setting overlooking Lake Como amid the foothills of the Italian Alps. Everyone was accommodated on the shore of the lake, and each morning we would walk up the hill through fragrant terrace gardens to meet in the elegant Villa Serbelloni. To increase interaction among the scholars, short presentations by the individual participants were followed by lively round-table discussions. Fascinating, amusing, and convivial, the meeting was successful in fostering communication between academic disciplines: symbiosis biologists were made cognizant of the need for the publication of their work in the mainstream evolution literature, and evolutionary theorists were encouraged to more profoundly consider symbiosis in evolutionary models. Addressing a need for inclusion of mechanisms besides gradualism in the prevailing neo-Darwinian dogma, we hope to enhance awareness among evolutionists from all disciplines of the significance of the persistent novelties generated by symbiosis, many wonderful illustrations of which are presented by our authors.

Some invited speakers (Bermudes, Haynes, Guerrero, Kendrick, Price, Smith, and Trench) were unable to attend the conference.

Rather than being entirely denied their insights, we include their written contributions here.

Application materials for international conferences and for a residence program for individual scholars or artists are available from the Bellagio Center Office, Rockefeller Foundation, 1133 Avenue of the Americas, New York, NY 10036.

René Fester

Acknowledgments

We are grateful to the Rockefeller Foundation, whose generosity made possible the conference on which this volume is based. We would also like to thank all of the helpful staff at the Villa Serbelloni, especially Roberto and Gianna Celli and Antonella Acanfora. Susan Garfield in the Bellagio Center Office in New York answered many queries. Valued assistance with manuscript preparation was given by J. Christopher Brown, Kathryn Delisle, Laura Nault, Amanda Ferro, Dorion Sagan, Lorraine Olendzenski, Matthew Farmer, Stephanie Hiebert, Christie Lyons, Eileen Crist, and Thomas Lang. We gratefully acknowledge that some travel and publication costs were borne by the Richard Lounsbery Foundation, the NASA Life Sciences office (including the Planetary Biology Internship Program), and the Department of Botany of the University of Massachusetts at Amherst.

I

Individuality and Evolution

1 Symbiogenesis and Symbionticism

Lynn Margulis

The historian of biology L. N. Khakhina works for the Soviet Academy of Sciences. She emphasizes that "symbiogenesis," the evolutionary origin of new morphologies and physiologies by symbiosis, has been in the forefront of Russian concepts of evolution since the last century (Khakhina 1979). The seminal contributions of the leading "symbiologists," Famintsin (1835–1918), Mereschkovskii (1855–1921), and Kozo-Polianski (1890–1957), are little known in Western Europe and the United States; eclipsed by the influence of genetics and molecular biology even in the Soviet Union, the writings of these old "symbiologists" are only now coming to light in a fashion that makes them accessible to English-language readers (Margulis and McMenamin, in press). In retrospect, however, these early Russian biologists, introduced to the West by E. B. Wilson (1928), are clearly the founders of "evolutionary symbiology."

The conference represented by this volume was designed and carried out with very little recognition of our debt to our illustrious Russian predecessors and their early colleagues, among them Portier (1918), Wallin (1927), Pierantoni (1948), and Buchner (1965). Mechanisms for generating "geologically sudden" evolutionary change were already detailed by these scientists; this we see now only in retrospect. As Sapp points out in chapter 2 of this volume, the participants in this conference are more inspired by the Anglo-Saxon tradition of "non-Mendelian heredity" or "cytoplasmic heredity" than by Russian symbiogenesis. In this introductory chapter, I explore the roots of the recent surge of interest of symbiosis in an evolutionary context.

The term *symbiosis* was defined by the German mycologist Anton De Bary (1879) as meaning the "living together" of "dissimilar" or "differently named" organisms. Symbiosis, in most current biological literature, is taken to mean "mutualistic biotrophic associations"

Table 1
Definitions of terms.

partners[1] *(bionts)* two or more organisms, members of different species

symbiosis (holobiont) association[2] throughout a significant portion of the life history[3]

Spatial relationships[4]

obligate One partner requires physical contact with the other throughout most or all of its life history. In "phoresy"[5] one partner physically "carries" the other; in "mutualism" both partners "benefit."

facultative One partner can complete its life history in the absence of the other partner. In "commensalism" nutrient sources are shared; in "phoresy" one partner is borne or carried by another; in "mutualism" one partner "benefits" another.

Temporal relationships

allelochemical Chemical compounds produced by one partner evoke a behavioral or growth response in the other partner ("mutualism").

behavioral The behavior of each partner is required for the establishment or maintenance of the association ("mutualism").

cyclical Physical association between partners is periodically established and disestablished.

permanent Physical association between partners is required throughout the life history of each (hereditary symbioses, "mutualism").

Metabolic relationships

metabolite A product of metabolism (e.g., an amino acid, a carbohydrate, or a nucleotide derivative) of one partner becomes a semiochemical or a component of a semiotic product for other partner.[6]

biotrophy One partner requires carbon, nitrogen, or some other nutrient that is a metabolic product of the other partner.

symbiotrophy (s), necrotrophy (n) One partner's nutritional needs are entirely supplied by the other partner, which (s) remains alive during the association ("mutualism," "parasitism") or (n) which is weakened or killed by the association ("parasitism," "pathogenesis").

Genetic relationships

gene-product transfer (protein, RNA) Protein(s) or RNA(s) synthesized off the genome of one partner is used in the metabolism of the other partner ("mutualism," "parasitism").

gene transfer Gene(s) of the partner are transferred to the genome of the other partner ("mutualism," "parasitism").

---

1. *partners*: definitions with respect to only one partner. *biont*: individual organism. *holobiont*: symbiont compound of recognizable bionts.
2. *association*: physical contact between organisms that are members of different species.
3. *life history*: events throughout the development of an individual organism correlating environment with changes in external morphology, formation of propagules, and other observable aspects. This refers to, but is distinguish-

(Schiff and Lyman 1982). In reference books, it is typically defined with respect to outcome—for example, "An internal partnership between two organisms (symbionts) in which the mutual advantages normally outweigh the disadvantages" (Collocott 1972).

Clear and consistent definitions of *symbiosis* are not in general use, and the relationship between symbiosis and evolution is underexplored (Margulis 1981). Even the word *evolution* has various meanings among biologists; hence it is not surprising that confusion and controversy surround the two terms *symbiosis* and *evolution* and their relation to each other. The lack of consensus about first principles of symbiosis and evolution has serious consequences for both the teaching and the practice of evolutionary biology. Attempting to alleviate some of this malaise, I define here terms related to *symbiosis* which acknowledge the potential for behavioral, allelochemical, anatomical, developmental, metabolic, and genetic relationships between organisms of different species (table 1).

Botanists, who at some time in their education study lichens, tend to know about the prevalence of symbiosis in nature. The algal and fungal components of all lichens are nearly equal in mass. The late-19th-century controversy which surrounded "Schwendenerisme," the concept of the symbiotic or "dual" nature of these "plants," has left its legacy in the botanical literature. Indeed, De Bary himself was the first to recognize the composite nature of lichens, five years before Schwendener's comprehensive work (Abbayes 1954). By the opening years of the 20th century, "Schwendenerisme" was unambiguously victorious: the polygenomic ("dual") nature of lichens was firmly

---

able from, *life cycle*: events throughout the development of an individual organism correlating environment and morphology with genetic and cytological observations, e.g. ploidy of the nuclei, fertilization, meiosis, karyokinesis, cytokinesis (Margulis et al. 1990).

4. See Lewis 1973a,b, Starr 1975, and Margulis 1976 for discussions of these issues. Ecological relations (e.g., "parasitism," "pathogenicity," "mutualism") are given in quotation marks because only the outcome with respect to the relative growth rates of the partners can determine whether each term is appropriate in any given case.

5. Traditional terms, given in quotation marks, may correspond to the relationships tabulated here.

6. *semiochemical*: chemical substance acting as signal (sense), i.e., capable of involving biological response. Allelochemicals, hormones, and pheromones are all examples of semiochemicals. *semiotic*: meaningful, or the making of meaning. Chemical, verbal or other exchanges of signal or signs (Sebeok et al. 1989).

established as biological fact. Yet, as was noted by David Smith and Angela Douglas in their graduate text on the experimental analysis of symbiosis, "symbiosis was once regarded as a curiosity, remote from the mainstream of biology" (Smith and Douglas 1987).

P. J. Van Beneden (1845–1910), a microscopist and one of the discoverers of fertilization and centrioles, first wrote in 1873: "There is mutual aid in many species, with services being repaid with good behaviour or in kind, and mutualism can well take its place besides commensalism." From the late 19th century until the present, from the Belgian Van Beneden to recent American biology textbooks, the term *symbiosis* has had economic and financial overtones: primarily it has meant a relation between different organisms in which one lives "at the expense of the other" or which "benefits" each partner. In neo-Darwinist jargon, symbiosis has been "quantitated." The symbionts are assumed to be more "fit" if they leave more offspring when they are associated than when they live independently. Much contention in the literature derives from the difficulty of actually measuring the value of any symbiosis by comparing the number of offspring of the holobionts (the symbiotic complex) with that of the unassociated partners. In any case, to paraphrase Smith and Douglas, symbiosis, whatever its definition and in spite of a profound literature to the contrary, is *still* regarded as a curiosity, remote from the mainstream of evolution.

Today, when they contemplate the role of symbiosis in evolution, cognitive dissonance must envelope the scholars who control the funding for "evolutionary biology," at least in the United States. Indeed, if a molecular, genetic, and ecological understanding of symbiosis were more widespread among academic biologists, current theories of "population ecology" and "evolutionary biology" (i.e., Newtonian ecology, neo-Darwinism) would have to be denounced as illogical and internally inconsistent. In his excellent review of the use of the term *mutualism*, Boucher (1985) details its historical connection with politics, ethics, religion, philosophy, and other nonscientific endeavors. From their beginnings, both De Bary's *symbiosis* and Van Beneden's *mutualism* and *commensalism* have been imbued with meanings far transcending their limited use in biological science.

In this book, *symbiosis* refers only to a set of ecological interactions between nonhuman organisms. In defining symbiosis as the protracted physical association of one or more members of different species, I return to De Bary's original intention. Symbiotic partnerships may be loosely or exceedingly tightly integrated on the behavioral, the metabolic (trophic), the gene-product, or the genic level (Margulis 1976). None of these levels excludes the others. Symbioses in nature differ in outcome; the number of offspring left by the holobiont, relative to unassociated bionts (partners), depends on environmental conditions. Reviewing the terminological morass surrounding these labels, Starr (1975) also resuscitates De Bary's meaning, with the proviso that, in order to warrant the appellation *symbiosis,* associations between partners must not be casual; rather, they must be significant to the well-being or the "unwell-being" of one or both of the participants. In the two recent extensive texts on symbiosis, Ahmadjian and Paracer (1986) and Smith and Douglas (1987) reiterate De Bary's use of the term. Like those authors, I accept De Bary's *symbiosis,* meaning "living together," which always implies some degree of permanence of the physical association of the bionts.

Ivan Wallin (1927), in a book with a remarkably modern outlook, thoroughly recognized the central creative role played by symbiosis in evolution. He claimed that three major features characterize Darwinian (organic) evolution: the origin of new forms; the tendency, through time, toward complexification and specialization; and the retention or destruction of individuals and species. The latter is "natural selection." "In *Natural Selection,*" writes Wallin, "Darwin has established one of the cardinal principles that [are] operative." Natural selection, by "controlling the retention or destruction" of organisms as members of populations, determines the *presence* of living forms at any given period. But Wallin, the zoologist and critic of Darwin P.-P. Grassé, and Charles Darwin himself recognized the insufficiency of natural selection alone (which is simply the failure of biotic potential to be reached) to produce either new species, other new taxa, or increased morphological complexity in the living world. Grassé (1977) wrote:

> Darwinians rarely and discretely mention the possibility of finding in the allele pool of an unspecific population the gene capable of satisfying a given need of the species. They know, of course, that this possibility is very low, but this does not bother them since, they say, millions of present species managed to get through in spite of their narrow chances of success. In any event, this extremely low chance of success certainly contrasts with the constant success of the species and should puzzle the least curious among us. What gambler would be crazy enough to play roulette with random evolution? The probability of dust carried by the wind reproducing Dürer's "Melancholia" is less infinitesimal than the probability of copy errors in the DNA molecule leading to the formation of the eye. . . .

Lamarck had developed the idea of "inheritance of acquired characteristics," an idea which Darwin effectively used in his postulate of "gemmules" and which the later-day Darwinians have rigorously rejected. Ignoring their critics, neo-Darwinians insist on the accumulation of random mutations by natural selection as the major source of innovation in evolution. To forge an uneasy alliance between Mendel's entirely fixed and stable inherited factors (later dubbed *genes* by Johannsen [1911]) and Darwin's ever-changing natural inheritance, the "new synthesizers" (Hardy, Weinberg, Wright, Mayr, and Simpson; see Futuyma 1986 for references) invoked the magic of mathematization. Population geneticists claimed to have "solved" Darwin's fundamental problem of the origin of organic change—including that of the origin of species—by defining evolution as "changes, in natural populations, of gene frequencies." Today's "mechanist" (as opposed to "organicist") outlook, called "Newtonian ecology" by Boucher (1985), then ripened to maturity by incorporating Lotka-Volterra population biology and other models. In these mathematic machinations of evolution, the number of individuals in a population or a species is taken as the basic measure. The usual interactions that are explored between individuals (or species) are competition (for resources, space, etc.) and predation. A simple reversing of the signs of the interaction coefficient in the Lotka-Volterra model turns a "competition model" into one of "mutualism" (Boucher 1985). The incorporation of "cost-benefit analysis" methods borrowed from insurance practices has led to the biologically puerile numerology that systematically ignores chemistry, biochemistry, molecular biology, and geology (sedimentology, paleontology, and oceanography). Nevertheless, such aseptic language dominates current evolutionary theory.

With the notable exceptions of Wallin, K. S. Mereschkovskii, and Paul Portier, all of whom emphasized the crucial role played by hereditary symbiosis in the evolution of biological novelty, the literature of evolutionary biology developed independently from profound and important experimental and observational analyses of symbiosis. Indeed, the foremost symbiosis biologist of his day, the German Paul Buchner (1886–1969), who defined [endo]symbiosis as "well-regulated and essentially undisturbed co-operative living between two differently constituted partners," was able to write:

> Symbiosis research was hampered . . . by theoretical misjudgments. . . . Again and again there have been authors who insist that endosymbiosis is an elementary principle of all organisms and that structures usually regarded as

specific to the plant or animal are in reality symbionts which have attained a high degree of adaptation to the host plasma (1965, p. 69). . . . There were also scientists who mistakenly traced the chromatophores and leucoplasts of plant life to symbiotic algae. . . . The course of research has been uninfluenced by these bold hypotheses expanding endosymbiosis into a fundamental principle cleaving the uniformity of cells, both plant and animal. Independent of such extravagant concepts, in clear-headed, unassuming work, the science of endosymbiosis has laid stone upon stone to build the structure which is to be described in the following pages and which even in this less grandiose form embodies much that is miraculous. For us who have remained aloof from such speculations, endosymbiosis between animals and plant microorganisms [bacteria] represents a widespread, though always supplementary, device, enhancing the vital possibilities of the host animals in a multiplicity of ways. (p. 7A)

Wallin (1927), who mistakenly claimed to have grown mitochondria in nutrient broth, was a major proclaimant of "such extravagant concepts." Unlike his contemporaries, he recognized that bacteria "are neither plant nor animal" (p. 7) and that these "lowliest forms of life," which "exhibit the greatest degree of variability," respond more readily to the environment than more complex forms of life. Wallin developed the principle of *"prototaxis,"* "the innate tendency of one organism or cell to react in a definite manner [positive or negative] to another organism or cell" (p. 58). Symbionticism, according to Wallin, is the end result of positive prototaxis. Wallin insisted that symbionticism (by which he meant obligate microbial symbiosis) is the "fundamental factor or the cardinal principle involved in the origin of species" (p. 8). Indeed, Wallin defined symbionticism as "the establishment of intimate microsymbiotic complexes." Not only did he argue the details of the origins of mitochondria and plastids from bacteria, but he held that both ontological development (morphogenesis) and phylogenetic development (speciation) resulted from symbionticism. The comments of Buchner quoted above give us a clear idea of the response of mainstream biologists to Wallin and his coupling of the term symbiosis with evolution. Indeed, it was exactly the speculations of Mereschkovskii (1909), Wallin (1927), and Portier (1918) on the importance of symbiosis in evolution from which Buchner was remaining aloof.

As Caullery (1952) and Lewis (1973a,b) both made clear, *symbiosis, mutualism, commensalism, parasitism,* and other value-laden words for relations between organisms that are members of different species are essentially ecological terms which refer to the outcomes of associations. Caullery (1952) wrote: "Parasitism may be defined as the condition of life which is normal and necessary for an organism nourishing itself at the expense of another—called the host—without destroying it as the predator does the prey."

Scott, who looked for physical contact between participating organisms and to whom the symbiotic association is a "permanent feature of the life cycle of organisms," showed that in many cases one of the symbionts is removed from direct physical contact with substrate or environment. He defines symbiosis as "a state of equilibrated physiological interdependence of two or more organisms involving no permanent stimulation of defensive reaction mechanisms" (Scott 1969).

Recent authors have tended more and more to recognize that physical associations between members of different species (i.e., symbioses) can be integrated in many ways. Examples of regulating mechanisms controlling symbiotic interactions were detailed by Trager (1970). Specific examples abound of pathogens becoming benign, symbionts becoming pathogenic, and parasites becoming required organelles (Jeon 1983). Such observations of the changing nature of association depending on time and environment have confirmed the contention of Caullery (1952) and Lewis (1973b) that these relationships form a continuity of categories. The continuity of associations from casual through total genetic integration has been emphasized by Goff (1983), who cites the importance of algal symbioses as measured by their prevalence in aquatic environments. Algae, with their distinctive colors and chemistries, are very useful for experimental analyses of cell-cell recognition, trophic relations, and integration of genetic systems of the partners. Symbiosis analysis elucidates the following: (1) chemical and behavior recognition of organisms of different species, (2) initial contacts between prospective symbiotic partners, (3) selection pressures leading to the establishment and disestablishment of associations, and (4) genetic, metabolic, and behavioral aspects of partnership integration (Cook et al. 1980; Margulis 1976).

Probably the most important lesson symbiosis analysis has to offer evolutionary theory comes from the example of the transition of cell symbiosis theory to the new subfield of biology called "endocytobiology" (Schwemmler and Schenk 1980; Schenk and Schwemmler 1983). As Taylor (1987) notes, "serial endosymbiosis theory has been trans-

Table 2
Sources of evolutionary innovation.

Mutations ("micro" hereditary alterations)	**Karyotypic alterations** ("macro" hereditary alterations)	**Genomic Acquisitions** ("mega" hereditary alterations)[1]
Base pair changes (e.g., AT → GC)	Polyploidy ($2N = 4N$)	Transformation (e.g., DNA uptake)
Deletions (e.g., ACTG → ATG)	Polyteny ($2N = 2N$)	Transduction (phage, virus, replicon acquisition)
Duplications (e.g., ATCG → ATCGATCGT	Polyenergids ($2N \rightarrow xN$)	Bacterial conjugations
Transpositions (e.g., GCGCCATG → GCGATCCG)	Robertsonian fusions ($2N = 2N - 1$)	Meiotic sex
	Karyotypic fissions ($2N = 2N'$)	Symbioses

1. See page 46 of the present volume.

formed from an amusing ingenuity into a respectable alternative and is now a preferred explanation for the origins of plastids and mitochondria."

The close analogies between DNA-containing eukaryotic cell organelles and microbial symbionts require revision of classic cell theory, wrote Schwemmler and Schenk (1980) on introducing the field of endocytobiology. Cells are no longer units; they are environments, not only for other organisms, but for DNA from a wide variety of sources (Richmond and Smith 1979).

The advances of molecular biology, molecular genetics, electron microscopy, and other fields of modern biology suggest that symbiosis was a major mechanism in the establishment of the first eukaryotes from which the ancestral protoctists evolved (Margulis et al. 1990). We also know that members of the other three kingdoms—fungi, plants, and animals—have protoctist ancestors. We know too that, (at least in the microcosm) genes cross taxonomic boundaries rampantly, because DNA travels easily in the form of small replicons: plasmids, viruses, transposons, and so forth (Sonea and Panisset 1983). Thus, many mechanisms besides random mutation cause change in the hereditary endowments of organisms, including animals and plants (table 2).

Ironically, these lessons from symbiosis research and from molecular biology directly contradict the assumptions of mathematical evolutionary biology (and its stepchild sociobiology) as it is fashionably practiced. The "individuals" handled as unities in the population equations are themselves symbiotic complexes involving uncounted numbers of live entities integrated in diverse ways in an unstudied fashion. In representations of standard evolutionary theory, branches on "family trees" (phylogenies) are allowed only to bifurcate. Yet symbiosis analyses reveals that branches on evolutionary trees are bushy and must anastomose; indeed, every eukaryote, like every lichen, has more than a single type of ancestor. Such analyses also reveal rampant polyphyly (e.g., more than eight independent origins of parasitism in dicotyledonous plants [Kuijt 1969]) and thousands of lichens (holobionts) which evolved from independent associations between mycobionts (the fungal partner) and phycobionts (the algal partner). The fact that "individuals"—as the countable unities of population genetics—do not exist wreaks havoc with "cladistics," a science in which common ancestors of composite beings are supposedly rigorously determined. Failure to acknowledge the composite nature of the organisms studied invalidates entire "fields" of study. Symbiosis analysis as outlined above must be applied to the organisms under study. Let a single example suffice: Paleontologists are quick to recognize stromatolites as trace fossils of communities of microorganisms. Yet these fossils, named by binomials (*Genus species*), are used as stratigraphic markers; they have morphologies that can be traced in an evolutionary sequence through time. Examples include *Conophyton rigida* and *Domonomo cloudiana* (Cloud 1989). My claim is that all animal and plant fossils are analogous: they too are composite products of coevolved microbial communities and in this sense are "form-taxa." That they bear genus and species names, as do the fossil stromatolites, does not make them unities. Cladists be forewarned!

In most texts and treatises, according to present-day neo-Darwinian evolutionary theory, the only source of novelty is claimed to be by incorporation of random mutations, by recombination, gene duplication, and other DNA arrangements. As is emphasized by those using the term *symbiogenesis*, symbiosis analysis contradicts these assertions by revealing "Lamarckian" cases of the inheritance of acquired genomes (Nardon and Grenier, this volume). Neo-Darwinian evolutionary theorists claim that "individuals" behave to increase their inclusive fitness, the number of offspring left by them and relatives that share their genes. Analysts of symbioses retort that no individuals exist—with the exception of the unstudiable single bacterium. In spite of sociobiological dicta to the contrary, organisms behave to increase the fitness of symbionts with which they have *very few genes* in common (e.g., the cow licking her calf ensures the continuity of her entodiniomorph rumen ciliates). The "standard" neo-Darwinian evolutionary theory claims that cows evolved by "gradual accumulation of favorable mutations" while it ignores the cellulytic activities of cow symbionts. The standard textbooks on evolution catechize all species and higher taxa (genera, families, phyla) as having evolved in the same way: by gradual accumulation of favorable mutations. *Yet not a single example of the origin of such lower taxa (species) exists in the literature.* Rather, the highest taxa (kingdoms and phyla) have evolved by acquisition of symbionts that have become hereditary (Bermudes and Margulis 1987, table 1.1). Various contributors to this volume (including Jeon, Bermudes, and Back) suspect lower taxa to have originated in the same way. Whether the newly evolved taxon is higher (more inclusive) or lower (less inclusive) has to do with how different the symbiotic partners are from each other (Margulis and Bermudes 1985; Bermudes and Margulis 1987). Extraordinary differences (e.g., sedentary phototrophs associated with motile heterotrophs) potentially lead to more innovative higher taxa than closely related associates (e.g., two heterotrophs). We only request here that those biologists concerned with mechanisms of speciation inform themselves of the symbiosis literature, just as we students of symbiosis are attempting in this volume to inform ourselves of the evolution literature.

Cell physiology must become isomorphic with microbial community analysis; developmental biology must become a specialized branch of microbial ecology by this same reasoning; current practices of population biology and genetics must be obliterated by their own false assumptions. In contrasting these current practices of evolutionary biologists with the revelations of recent symbiosis analyses (e.g., Scannerini et al. 1988), we are dealing with the clashes between established religious doctrine and its reformation. As the brilliantly original sociologist of science Ludwik Fleck claimed in 1936 (see Fleck 1979), "words become battle cries." Although symbiosis, in the evolutionary context, has been just such a word, the concept of "symbiogenesis," amply documented in this book, deserves to be revitalized.

References

Abbayes, H. des. 1954. Lichénologie. In: De Virville, Davy. Histoire de la Botanique en France. Société d'Édition d'Enseignement Supérieur, Paris.

Ahmadjian, V., and Paracer, S. 1986. *Symbiosis: An Introduction to Biological Associations*. University Press of New England.

Bermudes, D., and Margulis, L. 1987. Symbiont acquistion as neoseme: Origin of species and higher taxa. *Symbiosis* 4: 379–397.

Boucher, D. H. 1985. *The Biology of Mutualism: Ecology and Evolution*. Oxford University Press.

Buchner, P. 1965. *Endosymbiosis of Animals with Plant Microorganisms*. Interscience.

Caullery, M. 1952. *Parasitism and Symbiosis*. Sidgwick and Jackson, London.

Cloud, P. E., Jr. 1989. *Oasis in Space*. Norton.

Collocott, T. C., ed. 1972. *Dictionary of Science and Technology*. Barnes and Noble.

Cook, C. B., Pappas, P. W., and Rudolph, E. D., eds. 1980. *Cellular Interactions in Symbiosis and Parasitism*. Ohio State University Press.

De Bary, H. A. 1879. *Die Erscheinung der Symbiose*. Vortrag, gehalten auf der Versammlung Deutscher Naturforscher und Aerzte zu Cassel. R. J. Trübner, Strassburg.

Fleck, L. 1979. *Genesis and Development of a Scientific Fact*. University of Chicago Press.

Futuyma, D. J. 1986. *Evolutionary Biology*, second edition. Sinauer.

Goff, L. J. 1983. *Algal Symbiosis: A Continuum of Interaction Strategies*. Cambridge University Press.

Grassé, P.–P. 1977. *Evolution of Living Organisms: Evidence for a New Theory of Transformation*. Academic Press.

Jeon, K. W. 1983. Intracellular symbiosis (Supplement 14). In: Bourne, G. H., Danielli, J. F., and Jeon, K. W., eds., *International Review of Cytology*. Academic Press. See also Jeon, this volume.

Johannsen, W. 1911. The genotype conception of heredity. *American Naturalist* 45: 129–159.

Khakhina, L. N. 1979. *Concepts of Symbiogenesis*. Akademie NAUK, USSR.

Kuijt, J. 1969. *Biology of Parasitic Flowering Plants*. University of California Press.

Lewis, D. H. 1973a. The relevance of symbiosis to taxonomy and ecology with particular reference to mutualistic symbiosis and the exploitation of marginal habitats. In: Heywood, V. H., ed., *Taxonomy and Ecology*. Academic Press.

Lewis, D. H. 1973b. Concepts in fungal nutrition and the origin of biotrophy. *Biological Reviews of the Cambridge Philosophical Society* 48: 261–278.

Margulis, L. 1976. A review: Genetic and evolutionary consequences of symbiosis. *Experimental Parasitology* 39: 277–349.

Margulis, L. 1981. *Symbiosis in Cell Evolution*. Freeman.

Margulis, L. 1990. Words as battlecries—Symbiogenesis and the new field of endocytobiology. *BioScience* 40: 673–677.

Margulis, L., and Bermudes, D. 1985. Symbiosis as a mechanism of evolution: Status of cell symbiosis theory. *Symbiosis* 1: 101–124.

Margulis, L., Corliss, J. O., Melkonian, M., Chapman, D. J., eds. 1990. *Handbook of Protoctista: The Structure, Cultivation, Habitats and Life Histories of the Eukaryotic Microoganisms and Their Descendants Exclusive of Animals, Plants and Fungi*. Jones and Bartlett, Boston.

Margulis, L., and McMenamin, M., eds., *Concepts of Symbiosis*. Yale University Press. In press.

Mereschkovskii, K. C. 1909. *Theory of Two Plasms and the Basis of Symbiogenesis, New Studies About the Origin of Organisms*. Kazan, USSR.

Pierantoni, U. 1948. *Trattato di Biologia e Zoologia Generale*. Casa Editorial Humus, Naples.

Portier, P. 1918. *Les Symbiotes*. Masson.

Richmond, M. H., and Smith, D. C., eds. 1979 *The Cell as a Habitat*. Royal Society, London.

Scannerini, S., Smith, D. C., Bonfante-Fasolo, P., and Gianinazzi-Pearson, V., eds. 1988. *Cell-to-Cell Signals in Plant, Animal and Microbial Symbioses*. Springer-Verlag.

Schenk, H., and Schwemmler, W. 1983. *Endocytobiology II: Intracellular Space as Oligogenetic System*. Walter de Gruyter.

Schiff, J. A., and Lyman, H. 1982. *On the Origins of Chloroplasts*. Elsevier/North-Holland.

Schwemmler, W., and Schenk, H. E. A., eds. 1980. *Endocytobiology: Endosymbiosis and Cell Biology. A Synthesis of Recent Research* (Proceedings of the International Colloquium on Endosymbiosis and Cell Research, Tübingen, 1980). Walter de Gruyter.

Scott, G. D. 1969. *Plant Symbiosis*. Edward Arnold, London.

Sebeok, T. A., Lamb, S. M. and Regan, J. O. 1988. Semiotics in education: A dialogue. In: *Issues in Communication*. Claremont Graduate School, Claremont, California.

Smith, D. C., and Douglas, A. E. 1987. *The Biology of Symbiosis*. Edward Arnold, London.

Sonea, S., and Panisset, M. 1983. *A New Bacteriology*. Jones and Bartlett.

Starr, M. P. 1975. A generalized scheme for classifying organismic associations. In: Jennings, D. H., and Lee, D. L., eds. *Symbiosis: Symposia of the Society for Experimental Biology*. Cambridge University Press.

Taylor, F. J. R. 1987. An overview of the status of evolutlonary cell symbiosis theories. In: Lee, J. L., and Fredrick, J. F., eds., *Endocytobiology III*, Annals of the New York Academy of Sciences, Volume 503.

Trager, W. 1970. *Symbiosis*. Van Nostrand Reinhold.

Van Beneden, P. J. 1873. Un mot sur la vie sociale des Animaux Inferieurs. *Bulletin de l'Academie Royale de Belgique*, serie 2, 28: 621–648.

Wallin, I. E. 1927. *Symbionticism and the Origin of Species*. Williams and Wilkins.

Wilson, E. B. 1928. *The Cell in Development and Heredity*. Macmillan.

2 Living Together: Symbiosis and Cytoplasmic Inheritance

Jan Sapp

Suggestions that the eukaryotic cell is made up of collections of primordial organisms, and that symbiosis is an important source of evolutionary innovation, have been made throughout the 20th century. Yet only recently have these suggestions had any significant impact on biological thought (Margulis 1981). To understand the general reluctance of biologists in general to recognize the evolutionary importance of hereditary symbiosis, one has to place the development of these ideas in the context of the doctrines and methods of genetic research programs, and also in the context of other debates, especially ones about the scope and significance of cytoplasmic inheritance (Sapp 1987). The history of the research and ideas pertaining to hereditary symbiosis is complex. What I offer here is only a rough outline of some of the main ideas and the research traditions associated with them.

Some of the most prominent suggestions that "higher organisms" were made up of collections of symbiotic microorganisms emerged from cytological studies of mitochondria during the first three decades of this century. Advances in staining techniques enabled cytologists and histologists to study these cytoplasmic granules in detail and identify them in both eggs and sperm. Many leading cytologists ascribed to them the power of independent growth and division and considered them to represent a mechanism of heredity comparable to chromosomes. Mitochondria were often regarded to be fundamentally important both in the chemical activities of the cell and for tissue development and differentiation. Mitochondria were even thought to be the source of many other cell components, including chloroplasts. Most researchers attributed the initial interest in mitochondria to R. Altmann's treatise *The Elementary Organisms and Their Relationship to the Cell*, published in 1890. Altmann referred to "elementary organisms," later identified as mitochondria, as the ultimate units of living matter which have entered into intimate association within the cell.

In 1918, Paul Portier published his treatise *Les Symbiotes* and advanced evidence to support his theory that mitochondria are bacterial organisms symbiotically combined with cells of all "higher organisms." To prove the essential point, the existence within the cell of autonomous organisms said to be mitochondria, Portier attempted to extract mitochondria from the cell and to culture them. His experiments caused a great deal of excitement in the French biological community and were subjected to vigorous criticism. His techniques were severely criticized and his results were deemed unbelievable. For example, he reported that the mitochondria/bacteria retained life after being kept in absolute alcohol for long periods. Portier's work lent some notoriety to the idea that symbiosis was a primordial characteristic of the cell. The next year, A. Lumière published *Le Myth des Symbiotes*.

However, the view that higher organisms were symbiotic complexes continued to be suggested by biologists outside of France. Between 1905 and 1920, K. S. Mereschkovskii advanced his hypothesis that chloroplasts were microsymbionts genetically related to the blue-green algae and that all green plants are symbiotic complexes. Mereschkovskii's hypothesis of the origin of chloroplasts was certainly not accepted by all biologists, nor was it accepted even by those who, like him, championed microsymbiotic theories during the 1920s. The predominant opinion among cytologists of his day was that chloroplasts were derivatives of mitochondria. This view was advanced in the United States by Ivan E. Wallin, a professor of anatomy in the School of Medicine at the University of Colorado.

Wallin was one of the leading proponents of the importance of symbiosis in evolution. His research and theorizing in this area culminated in 1927 with his book *Symbionticism and the Origin of Species*. Like Portier, Wallin attempted to carry out what he regarded to be definitive experiments, trying to culture and then subculture mitochondria to prove their bacterial nature. However, he tried to avoid the errors that had contaminated Portier's attempts. Basing his arguments on his demonstration of the bacterial nature of mitochondria and the responses of organisms to microbic invasions, Wallin advanced his theory that symbionticism was the fundamental factor in the origin of species. Bacteria were the "building stones," or "primordial stuff," from which all higher organisms had been constructed and modified. "Just as reproduction ensures the perpetuation of existing species," Wallin argued, "symbionticism insures the origin of new species" (1927, p. 64). Wallin offered his theory as a solution to one of the major problems discussed by geneticists during the first decades of the century: the origin of new genes.

Wallin argued that the acquisition of a new symbiont in the cells of an organism may be the means by which new genes are added to those already present in the germ cell. He suggested that such new genes might not only control their own hereditary characters but also might modify other characters. Wallin (1927, p. 127) further speculated that part or all of the chromatin of a microsymbiont could be given up to the nucleus of the host cell and germ plasm during the development of symbiosis, leaving the remains of the microsymbiont in the cytoplasm.

Whatever merit we may grant these ideas today, they were not taken seriously by biologists of Wallin's day, nor by geneticists in particular. There were several major reasons for this. One, suggested by Wallin himself, was that, to the popular mind, bacteria represented disease: "It is a rather startling proposal that bacteria, the organisms which are popularly associated with disease, may represent the fundamental causative factor in the origin of species. Evidence of the constructive activities of bacteria has been at hand for many years, but popular conceptions of bacteria have been colored chiefly by their destructive activities as represented in disease." (1927, p.8) It would be naive to underestimate the extent to which such attitudes affected biological opinion about ideas of symbionticism. Moreover, the view of the complex organism as colonies of smaller organisms ran against the holism of the 1920s, which characterized much thinking about the organism. In the third edition of his famous book *The Cell in Development and Heredity*, the leading cell biologist E. B. Wilson reviewed theories about the cell in terms of a symbiosis between the nucleus and the cytoplasmic components. He referred to the writings of Mereschkovskii as "entertaining fantasy." He also mentioned Theodor Boveri's idea that chromosomes themselves may have been smaller organisms which established themselves within a larger one. After mentioning Wallin's idea that mitochondria may be regarded as symbiotic bacteria, Wilson (1928, p. 730) remarked: "To many, no doubt such speculations may appear too fantastic for present mention in polite biological society; nevertheless it is within the range of possibility that they may some day call for more serious attention."

In the meantime, Wallin's ideas about the symbiotic origin of new genes were "zapped" by a wave of radiation genetics. In 1927, the same year Wallin published his book, H. J. Muller published his first report that the use of heavy doses of x rays could increase the frequency of gene mutations in *Drosophila* by some 1500 times. The next year, L. J. Stadler published the results of similar studies on barley. The types of phenotypic changes that occurred were the same as those known to occur spontaneously. Geneticists considered such gene mutations to be the source of new alleles and thus the principal source of evolutionary variation. As Muller (1929) argued, the ability of genes to vary (mutate) and to reproduce themselves in new form conferred on these cell elements the properties of the building blocks required by evolution. The artificial production of mutations gave genetics a new burst of life during the late 1920s and the 1930s. The evolutionary synthesis of the 1930s and the 1940s was based on a marriage of Darwinian natural selection to gene mutations and recombination. Symbiosis as a source of genetic innovation was not considered.

A comprehensive account of symbiosis as a mechanism of evolutionary innovation would also have to take into consideration that the neo-Darwinian synthesis was based on competition as a major force in natural selection. It virtually ignored cooperation. Insomuch as symbiosis promotes cooperation as a major factor in selection, it conflicted with one of the fundamental assumptions of the principal architects of the "modern synthesis" and many other biologists involved in evolutionary studies. A detailed study of these conflicting assumptions would, no doubt, go far in helping to explain the vehement rejection of symbiosis theory.

But there is still another crucial reason why symbiosis as an important source of evolutionary innovation was not taken seriously by biologists and by geneticists in particular. The claim that symbionticism was a fundamental factor in the origin of species relied heavily on the exogenous origin of cytoplasmic organelles: mitochondria and chloroplasts. That claim, in turn, relied on the assumption that these cytoplasmic bodies were endowed with genetic continuity and represented vehicles of inheritance comparable to the nucleus. Even this seemingly less daring suggestion—that there existed more than one seat of hereditary determination in the cell—was the subject of heated debate throughout most of the 20th century.

The idea of an essential and general cytoplasmic inheritance competed with a long tradition upholding the exclusive role of the nucleus in heredity. The belief that all essential cytoplasmic bodies were derived from the action of the nucleus had been prevalent in biological thought since it was championed in the late 19th century in the writings of August Weismann. In his celebrated book *The Germ-plasm*, Weismann (1893) had offered a series of reasons why the nucleus was the sole seat of heredity. Many of these reasons were repeated by leading geneticists throughout most of the 20th century. The first argument might be called "the issue of the common denominator." In higher organisms, the egg cell is many hundred times larger than the sperm cell. Yet, Weismann claimed, "we know that the father's capacity for transmission is as great as the mother's." The nucleus provided a place for equal transmission of hereditary substance from both parents. Second, studies of fertilization seemed to indicate that the essential part of this process consisted of the union of the nuclei of the egg and sperm cells. Third, observations of cell division showed that the nuclear complex possessed a wonderfully exact apparatus for dividing the chromatin substance in a fixed and regular manner. Finally, Weismann (1893, p. 29) appealed to the "economy of Nature" against a view that there could be more than one location for the hereditary material: "this substance can hardly be stored up in two different places, seeing that a very complicated apparatus is required for its distribution: a double apparatus would certainly not have been formed by nature if a single one suffices for the purpose."

Certainly not all biologists came to be committed to this view. Indeed, the history of genetics research is characterized by a continuing debate over the relative importance of the nucleus and the cytoplasm in heredity. The theoretical limits of this dispute were framed by two extreme positions. At one extreme was the view that determiners in the cytoplasm were responsible for the large characters of the organism—that is, those characteristics which distinguished higher taxonomic groups: phylum, class, order, family, genus. According to this view, Mendelian genes "topped off" the more fundamental organismic features. They controlled only traits that distinguished individuals, varieties, and species. This suggestion was championed by many embryologists and European geneticists who found it difficult to account for ontogenetic development within the confines of the chromosome theory. It was also maintained by those naturalists who found it difficult to explain macro-evolution in terms of gene mutations (Sapp 1987). At the other extreme, many American geneticists upheld the dominant if not exclusive role of Mendelian genes in heredity. They proclaimed genes to be the "governing elements," largely immune from the rest of the cell and dictating its activities. T. H. Morgan (1926, p. 491) expressed this view succinctly: "In a word the cytoplasm may be ignored genetically."

The efforts of those geneticists who disagreed with Morgan's view were directed to providing definitive genetic evidence for the existence of a universal cytoplasmic inheritance. Between the two world wars many German botanists set out to challenge what they called "the nuclear monopoly" of the cell, detailing various cases of non-Mendelian inheritance. Cases of the transmission of infectious agents were often discussed in reference to the evidence for cytoplasmic inheritance. But the onus on these researchers was to demonstrate that cytoplasmic genetic particles constituted an essential part of the genetic system of all organisms.

In such debates the inheritance of characters due to the transmission of symbionts was a source of error and criticism for both sides. Indeed, the evidence presented by German investigators for the existence of essential cytoplasmic genetic systems was often criticized by defenders of the "nuclear monopoly." Some cases of non-Mendelian inheritance were dismissed on the basis that the cytoplasmic traits in question might be due to symbionts or parasites of exogenous origin and playing no normal or essential role in heredity and evolution. Other cases were reinterpreted formally in terms of the effects of nuclear genes; still others which seemed to involve chloroplasts in plants were accepted by American geneticists as rare occurrences (Sapp 1987).

Discussion of cytoplasmic inheritance came to occupy a prominent place in genetic discourse in the 1940s and the 1950s when microorganisms were domesticated for use in genetics. In the United States, research on cytoplasmic inheritance was led by T. M. Sonneborn and a few others who brought forth evidence of non-Mendelian inheritance in microorganisms. Based on his genetic investigations in *Paramecium,* Sonneborn (1950, p. 31) developed the concept of plasmagenes: "self-duplicating, mutable, cytoplasmic particles . . . which depend on the nucleus for their maintenance and normal functioning but not for their origin or for their specificity."

After World War II, Boris Ephrussi, Andre Lwoff, Jean Brachet, and Philippe L'Heritier in France and C. D. Darlington in England joined Sonneborn in challenging the nuclear monopoly of the cell. Their arguments for the essential genetic role of various cytoplasmic bodies, including chloroplasts, mitochondria, centrioles, and kinetosomes, drew on embryological, cytological, and genetic investigations (Lwoff 1950; Ephrussi 1953). The conflict over the relative importance of the nucleus and the cytoplasm became more heated. The idea that the cytoplasm could be in control of fundamental traits of the organism was rehabilitated, and the evidence for extranuclear inheritance became entangled in the Cold War dispute surrounding the Lysenko affair (Sapp 1987).

The evidence for cytoplasmic inheritance was criticized by several leading American geneticists, including E. Altenberg (1946), G. H. Beadle (1948), and H. J. Muller (1951). There were three arguments against the conception that cytoplasmic genes or gene complexes formed an essential part of the genetic constitution of all organisms. First, there was the issue of the continued rarity of genetically demonstrated cases of cytoplasmic inheritance, in contrast to the thousands of Mendelian differences studied. Second, there was the question of a reliable mechanism comparable to mitosis and meiosis, to transmit cytoplasmic genes and gene complexes safely from one generation to the next. Third, there was the question of the infectious nature of some of the cytoplasmic particles. Kappa in *Paramecium* and Sigma in *Drosophila* could be transmitted by infection under certain laboratory conditions. The lack of a fundamental basis for distinguishing between these and cases of undoubtedly parasitic or symbiotic microorganisms or viruses of exogenous derivation merely gave nucleocentric geneticists of the 1940s and the 1950s more arguments to marshal against the idea that cytoplasmic genetic particles constituted an essential part of the genetic constitution of all organisms.

At the same time, however, genetic investigations of bacteria and their viruses led to attempts to broaden the concept of heredity itself. Geneticists had restricted their investigations to the sexual transmission of differences between individuals. In so doing they had constructed a restricted notion of heredity to suit their practice. However, in bacteria there are other means besides sex for transmitting genetic material from one generation to the next. Viruses can act as vehicles of inheritance. The geneticists C. D. Darlington (1951, 1958) and J. Lederberg (1951, 1952) attempted to construct a "unified theory" of heredity that included concepts of "infective heredity." Basing their claim on contemporary investigations of bacteria and their viruses, they argued that it would broaden genetic and evolutionary insights if a hard and fast distinction were not made between a symbiont and an integrated genetic particle. As Lederberg (1951, p. 286) put it, extrachromosomal genetic agents represented a continuum between deleterious parasitic viruses at one extreme and integrated cytoplasmic genes such as chloroplasts at the other. He proposed *plasmid* as a generic term for any extrachromosomal hereditary determinant.

During the 1960s the evidence for the existence of cytoplasmic genes and gene complexes in all organisms became overwhelming to geneticists when cytoplasmic DNAs associated with mitochondria and chloroplasts were identified. With the recognition that chloroplasts and mitochondria possessed all the essential equipment for "life" (DNA, transcription enzymes for making RNA, and a full protein-synthesis apparatus), the endosymbiotic theory reemerged, championed most prominently by Lynn Margulis (1970, 1981).

However, the idea that these organelles originated as symbionts certainly did not meet immediate acceptance during the 1960s and the 1970s. Classical symbiosis researchers, including Paul Buchner (1965, pp. 67–74) continued to discuss the ideas concerning the symbiotic origin of the cellular organelles under the heading "wrong paths in symbiosis research." This perspective was reinforced somewhat in subsequent years. During the 1960s and the 1970s, geneticists had shown cytoplasmic gene complexes to be well integrated into the genetic systems of all higher organisms and to control essential traits concerning photosynthesis and respiration (Sager 1972; Gilham 1978). The ships had been placed in their bottles, and it seemed inconceivable to some that they could have ever originated from outside. Klein and Cronquist (1967, p. 167), for example, referred to the notion that chloroplasts are endosymbionts as a "bad penny" that "has been circulating for a long time." A great deal of effort was also exerted in an attempt to contest the idea that mitochondria originated as a symbiont (Raff and Mahler 1972, 1975).

In response to their critics, leading symbiosis theorists constructed evidence from a broad range of disciplines and specialties to provide convincing arguments for the symbiotic origin of mitochondria and chloroplasts and to discuss its implications (Taylor 1974; Margulis 1981). Symbiotic origins have been postulated for other organelles: cilia (Margulis 1970, 1981) and the nucleus (Pickett-Heaps 1974). Though neither of these additional suggestions was met with general acceptance (Cavalier-Smith 1975, 1987), the importance of symbiosis in the origin of the eukaryotic cell was placed beyond doubt by cell biologists. However, the scope and significance of symbiosis and the mechanisms underlying the phenomena remained a closed book in evolutionary theory generally, one which evolutionists showed neither the means nor the desire to open.

Over the past 20 years, then, ideas about symbiosis as a source of evolutionary innovation have come a long way, from the days when they were considered X-rated science fiction, not to be mentioned in "polite biological society." Today, these ideas represent a source of scientific innovation, with societies and journals devoted to their study, and offer the basis for a grand synthesis of once disparate scientific initiatives. It is not surprising, therefore, that evolutionists and biologists of various persuasions are challenged to consider symbiosis as a major source of evolutionary innovation.

References

Altenburg, E. 1946. The symbiont theory in explanation of the apparent cytoplasmic inheritance in *Paramecium*. *American Naturalist* 80: 661–662.

Altmann, R. 1890. *Die Elementarorganismen und ihre Beziehungen zu den Zellen*. Viet, Leipzig.

Beadle, G. H. 1948. Genes and biological enigmas. In: Baitsell, G. A., ed., *Science in Progress*, 6th Series. Yale University Press.

Buchner, P. 1965. *Endosymbiosis of Animals with Plant Microorganisms*, revised English edition. Wiley.

Cavalier-Smith, T. 1975. The origin of nuclei and of eukaryotic cells. *Nature* 256: 463–468.

Cavalier-Smith, T. 1987. The simultaneous symbiotic origin of mitochondria, chloroplasts and microbodies. *Annals of the New York Academy of Science* 503: 55–71.

Darlington, C. D. 1951. Mendel and the determinants. In: Dunn, L. C., ed., *Genetics in the Twentieth Century*. Macmillan.

Darlington, C. D. 1958. *The Evolution of Genetic Systems*, 2nd edition. Basic Books.

Ephrussi, B. 1953. *Nucleo-Cytoplasmic Relations in Microorganisms*. Clarendon.

Gilham, N. W. 1978. *Organelle Heredity*. Raven.

Klein, R., and Cronquist, A. 1967. A consideration of the evolutionary and taxonomic significance of some biochemical, micromorphological and physiological characteristics in the Thallophytes. *Quarterly Review of Biology* 42: 105–296.

Lederberg, J. 1951. Genetic systems in bacteria. In: Dunn, L. C., ed., *Genetics in the Twentieth Century*. Macmillan.

Lederberg, J. 1952. Cell genetics and hereditary symbiosis. *Physiological Reviews* 32: 403–430.

Lumière, A. 1919. *Le Myth des Symbiotes*. Masson.

Lwoff, A. 1950. *Problems of Morphogenesis in Ciliates*. Wiley.

Margulis, L. 1970. *Origin of Eukaryotic Cells*. Yale University Press.

Margulis, L. 1981. *Symbiosis in Cell Evolution*. Freeman.

Mereschkovskii, K. S. 1905. Uber Natur und Ursprung der Chromatophoren im Pflanzenreich. *Biologische Centralblatt* 25: 593–604.

Mereschkovskii, K. S. 1910. Theorie der zwei Plasmaarten als Grundlage der Symbiogenesis, einer neuen Lehre von der Entstehung der Organismen. *Biologische Centralblatt* 30: 278, 321, 351.

Mereschkovskii, K. S. 1920. La plante considerée comme un complexe symbiotique. *Bull. Soc. Sci. Nat. Ouest France* 6: 17–98.

Morgan, T. H. 1926. Genetics and the physiology of development. *American Naturalist* 60: 489–515.

Muller, H. J. 1927. Artificial transmutation of the gene. *Science* 46: 84–87.

Muller, H. J. 1929. The gene as the basis of life. *Proceedings of the International Congress of Plant Sciences* 1: 897–921.

Muller, H. J. 1951. The development of the gene theory. In: Dunn, L. C., ed., *Genetics in the Twentieth Century*. Macmillan.

Pickett-Heaps, J. 1974. Evolution of mitosis and the eukaryotic condition. *Biosystems* 6: 37–48.

Portier, P. 1918. *Les symbiotes*. Masson.

Raff, R. A., and Mahler, H. R. 1972. The non-symbiotic origin of mitochondria. *Science* 177: 575–582.

Raff, R. A., and Mahler, H. R. 1975. The symbiont that never was: An inquiry into the evolutionary origin of the mitochondrion. In: Jennings, D. H., and Lee, D. L., eds., *Symbiosis*. Symposia for the Society for Experimental Biology. Cambridge University Press.

Sager, R. 1972. *Cytoplasmic Genes and Organelles*. Academic Press.

Sapp, J. 1987. *Beyond the Gene*. Oxford University Press.

Sonneborn, T. M. 1960. The cytoplasm in heredity. *Heredity* 4: 11–36.

Stadler, L. J. 1928. Mutations in barley induced by x-rays and radium. *Science* 68: 168–187.

Taylor, F. J. R. 1974. Implications and extensions of the serial endosymbiosis theory of the origin of eukaryotes. *Taxon* 23: 229–258.

Wallin, I. E. 1927. *Symbionticism and the Origin of Species*. Williams and Wilkins.

Weismann, A. 1893. *The Germ-plasm*. Translated by Parker, W. F., and Rofeldt. H. Walter Scott Ltd., London.

Wilson, E. B. 1928. *The Cell in Development and Heredity*. Macmillan.

3 A Darwinian View of Symbiosis

John Maynard Smith

There are several reasons why an evolutionary biologist should be interested in symbiosis. The first concerns the idea of "evolutionary progress." There is no empirical evidence, nor any theoretical reason, to lead us to suppose that natural selection will, as a general rule, lead to anything that could sensibly be called progress. However, I have argued (1988) that we can recognize a series of stages during evolution, each characterized by the way in which the genetic material is organized and transmitted between generations. A major problem is to explain the transitions between these stages. In one of them—the transition from prokaryote to eukaryote—symbiosis played a crucial role. It is therefore natural to ask whether it may have played a role in the other transitions also. A study of contemporary symbioses may help to answer this question. In particular, it may help to answer the following question: If a transition involved the joining together in a single evolutionary unit of previously independent entities, how did it come about that selection between these lower-level entities did not disrupt integration at the higher level?

A second and related reason to consider symbiosis is that it has been seen as a major source of evolutionary novelty. Some authors believe it constitutes a challenge to the neo-Darwinian view of evolution. Whether this is so depends, of course, on what one understands by "neo-Darwinism." I interpret it to mean the hypothesis that mutation (change in the genetic material) is not, except occasionally and by accident, adaptive to its causative agent, and therefore, insofar as organisms are adapted to their ways of life, that adaptation must have arisen by natural selection acting on originally nonadaptive genetic variation. In this sense I am a neo-Darwinist. However, the ways in which genetic material from different ancestors is united in a single descendant has profound effects. Usually, population geneticists think of this as coming about in eukaryotes through meiotic sex, and in prokaryotes through conjugation, transformation, and vector-mediated transduction. The relevance of symbiosis is that it affords a mechanism whereby genetic material from very distantly related organisms can be brought together in a single descendant.

The structure of this paper is as follows. In the next section, I formulate Darwin's theory of evolution by natural selection, in order to bring out the point that it is a theory that applies only to populations of "units" that have the properties of multiplication, variation, and heredity. In the third section, I argue that there have been, in the course of evolution, a number of changes in the nature of these units. The fourth section proposes three processes—duplication and divergence, symbiosis, and epigenesis—whereby these changes may have come about. The fifth section describes a particular, and rather abstract, model of how a set of interacting but independently evolving species might change into a single unit of evolution. The sixth section reviews existing symbioses, and argues that their evolution depends critically on how genes are transmitted between generations. In the final section, I speculate about the role of symbiosis in evolution.

The Darwinian Theory and Units of Evolution

Darwin's theory can be summarized as follows: Suppose there is a population of entities with the three properties of multiplication (one can give rise to two), variation (not all entities are alike), and heredity (like usually begets like in the multiplication process), and suppose also that some of the differences between entities influence their likelihood of surviving and reproducing (i.e., their "fitness"). Such a population will change in time—it will "evolve." Further, the individual entities will come to possess traits that increase their likelihood of survival and reproduction—i.e., "adaptations." This statement, I think, is not a testable scientific theory but follows necessarily from the original assumptions (including the assumption that there is a continual supply of new variations, some of which increase fitness). In its weak form, Darwin's theory asserts that the resulting selective process is one of the causes of evolutionary change. Few people, I suspect, would disagree with this. Darwin himself went further, and argued that the selective process is the major cause of evolutionary change; I share this view, but I do not think that it is the only cause.

Our present problem is to identify the "entities" in Darwin's theory. Are they individual organisms (as Darwin supposed), or are they genes, or organs, or groups of organisms, or populations? I have suggested (1987) that we use the term *unit of evolution* for an entity that has the necessary three properties, and which therefore can be expected to evolve adaptations ensuring its survival. I prefer this term to *unit of selection*, since many entities are selected but do not reproduce and hence do not evolve.

The Major Events in Evolution

We can identify a number of changes in the course of evolution in the way in which the genetic material is organized and transmitted, and hence in the nature of the units of evolution. Very briefly, I postulate the following stages:

(i) Replicating molecules, dependent on complementary base pairing.

(ii) Populations of replicating molecules within "compartments."

(iii) The prokaryotic stage. Sets of genes arranged in a circular chromoneme (and therefore replicated together).

(iv) The eukaryotic stage. Linear chromosomes, contained within a membrane-bounded nucleus (separating replication and transcription from translation), multiple origins of replication (enabling a larger genome to be replicated), mitosis (ensuring equational division), and organelles, some of which have separate prokaryote-like genomes.

(v) Multicellular eukaryotes, with epigenetic differentiation (i.e., cells differ, but, in general, their genomes do not[1]).

(vi) Animal societies, with genetically different individuals. In human societies, this depends on a second inheritance system, mediated by language instead of nucleic acids.

The picture is further complicated by processes that make possible the transfer of genetic material between individuals (e.g., conjugation and transformation in prokaryotes, meiotic sex in eukaryotes). In particular, a sexual species is a set of individuals that can exchange genes among themselves but only exceptionally with outsiders; this leads to the idea that species should be seen as units of evolution. In prokaryotes, it seems more natural to see individual genes, or small groups of genes, as units.

In understanding how the transitions took place between the above stages, the major difficulty lies in explaining why selection between the lower-level entities (e.g., genes within a chromosome, organelles in a cell, cells in an individual, individuals in a colony) does not disrupt integration at the higher level. (Why is meiosis fair? Why aren't mitochondrial cancers the rule? Why are certain castes sterile?)

I suggest that there are two main reasons:

- At some stage of the life cycle, the number of copies of the genome is reduced to one, or to very few, so that all the potential competitors are at least genetically similar. The simple answer to the problem (Buss 1987) of why there is not competition between somatic cells for the privilege of becoming gametes is that the organism (typically) develops from a single cell, so that (barring somatic mutation) all somatic cells are genetically identical. This is an extreme case of Hamilton's (1964) explanation of social behavior in terms of genetic relatedness. It makes particularly interesting those cases (multi-queen colonies, vegetatively reproducing plants) in which the "few-copy" rule is broken.
- After the higher-level entity has existed for a long time, its components may no longer be capable of independent existence. A malignant liver cell or an uncontrolled mitochondrion may multiply in the short run, but has no long-term future. A worker honeybee cannot start a new colony on its own (though it can, and sometimes does, lay male eggs). There is, then, a kind of irreversibility associated with the transitions. This may help to explain the long-term stability of higher levels, but cannot help to explain their origins.

Models of Transitions

To understand these transitions further, we need more precise models. Figures 1–3 show three alternative scenarios.

Duplication and Divergence (Figure 1)

A single genetic unit is first duplicated, giving rise to two or more identical copies, which then diverge. This is potentially a gradual process, because only a single gene can be duplicated at a time and because the subsequent divergence is likely to be slow. Even when a whole genome is duplicated (polyploidy), the phenotypic effects are

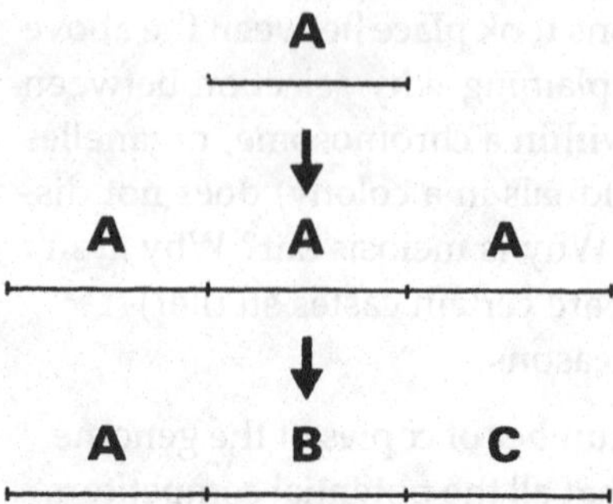

Figure 1
Increase in complexity by duplication, followed by divergence.

usually relatively slight. This process is important in genome evolution within eukaryotes (for example, the classic case of hemoglobin), but probably not in the major transitions listed above. We do not know whether the first chromosomes arose from the stringing together of several identical genes or of qualitatively different genes.

Symbiosis (Figure 2)

Two or more initially distinct genomes combine. The figure assumes three such entities: A, B, and C. It suggests three stages in the process: (1) the stable ecological coexistence of three distinct species; (2) the enclosure of small numbers of A's, B's, and C's within a "compartment," so that a particular A interacts only with one or a few B's and C's, and so on; (3) the evolution of a mechanism ensuring the synchronous and coordinated replication of A's, B's, and C's. In the figure, I have supposed that A, B, and C are genes, and that coordinated replication is ensured by stringing the genes together. However, as the symbiotic origin of eukaryotic cells indicates, physical linkage is not necessary for coordinated replication. The logic of the process is discussed further in the next section. It is now generally accepted that at least some eukaryotic organelles originated by this route. I am attracted by the idea that chromosomes originated symbiotically rather than by duplication, but this is more speculative. I do not (at present) think that symbiosis was involved in the origin of multicellular animals or plants, or of animal societies.

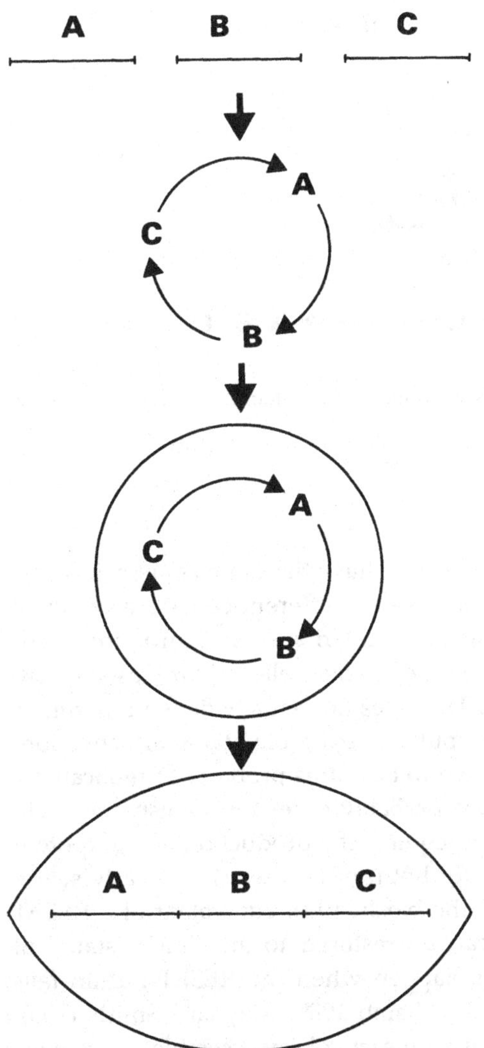

Figure 2
Increase in complexity by compartmentalization, followed by synchronized replication.

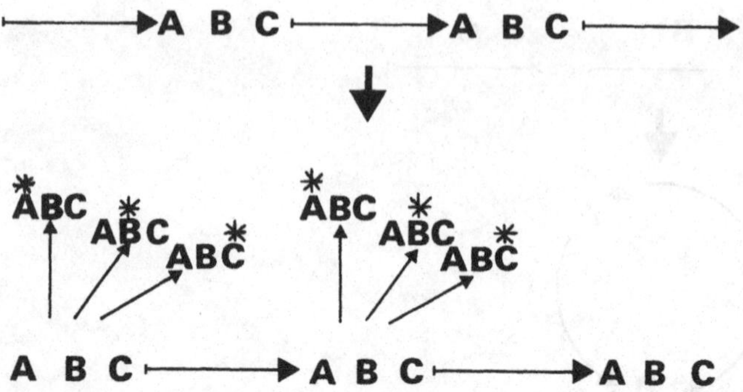

Figure 3
Increase in complexity by epigenesis, as exemplified by multicellular organisms and by insect societies.

Epigenesis (Figure 3)

All the cells of a plant or animal body have the same genetic information, but cells are differentiated because different genes are switched on in different cells. An epigenetic inheritance system is involved, because cells of a given type (e.g., epithelial cells or fibroblasts) usually reproduce their own kind. This does not usually depend on differences in the DNA sequence, but on replicable states of activation. Methylation patterns are known to have this property of replicability (Holliday 1987), and there are probably other mechanisms as well. Reproduction of the organism requires the production of a totipotent egg. This is achieved in one (or both) of two ways: the early segregation of the germ line, and the biochemical equivalent of a RESET button whereby all genes can be restored to an "initial state" of activation. Interesting things happen when the RESET button fails to work perfectly (Jablonka and Lamb 1989; Maynard Smith 1990). Formally, the development of an insect colony resembles epigenetic development.

Natural Selection and Symbiosis

Figure 4 represents an imaginary stable ecosystem. It has the structure of a hypercycle (Eigen and Schuster 1979), in that each of the three entities is a replicating unit of evolution, but the rate at which each entity replicates is accelerated by the numbers of individuals in the preceding state. That is, lions breed faster if there are many ante-

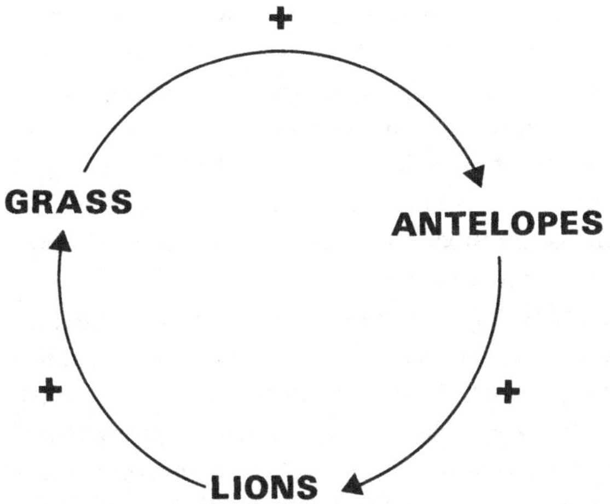

Figure 4
An imaginary hypercycle.

lopes, and antelopes breed faster if there is much grass; the imaginary feature is the assumption that grass grows faster if there are many lions to fertilize it. Such a system is stable. But how will it evolve? Imagine two mutations in the antelopes. Mutant 1 makes antelopes better at eating grass, and hence better at reproducing. Mutant 2 makes antelopes easier for lions to catch, and is therefore better at stimulating the reproduction of lions. Obviously, mutant 1 will spread, but mutant 2 will not. Now, in order for the system as a whole to evolve adaptions ensuring its survival as a system, in competition with other systems (e.g., herbs, grasshoppers, and rollers), both types of mutants are needed. The system in the figure is not a unit of evolution, and cannot be expected to evolve adaptions "for the good of the system."

Now suppose that a system of the kind shown in the figure is enclosed in a compartment—and now it is better to think of the entities as replicating molecules rather than animals and plants. Each compartment contains relatively few entities of each kind. An individual "antelope" with a mutant 2 will get eaten (as before), but now the increased growth of lions, and hence of grass, and hence of antelopes, will accrue to that antelope's offspring, which also carry mutant 2. Szathmary and Demeter (1987) have treated this model mathematically. They show that, given compartments with only a few entities of each kind, and given that a compartment (or a cell) multiplies if the number of entities within it increases, the individual entities can evolve characteristics which favor the survival of the system as a whole. The moral is that if the component entities of a symbiotic complex are to evolve mutually supportive characteristics, there must be effective compartmentalisation, and relatively few entities of each kind per compartment, at least once in the life cycle. Also required, I suspect, is little or no mixing of entities from different compartments at the time of reproduction.

For a Darwinian, the following questions must be answered before the evolution of a supposed symbiotic partnership can be understood: Can either of the partners exist without the other? At reproduction, are the partners always (or usually) transmitted together? How many individuals of each kind are there? Is one, or both, of the partners sexual? Only when the answers to such questions are known is it possible to predict whether selection will favor the evolution of the cooperative traits.

Symbiosis Today

There is a wide range in the extent to which endosymbionts can exist independent of their hosts. At one extreme, exemplified by the green algae symbiotic in aquatic invertebrates and by luminescent bacteria in fish (McFall-Ngai, this volume), the endosymbiont is able to live and reproduce in the absence of the host. Next in the series come endosymbionts that possess their own protein-synthesis apparatus and synthesize most or all of their proteins, but are incapable of reproducing outside their host: the intracellular bacteria of plant-sucking insects (Tiivel, this volume; Schwemmler, this volume) and weevils (Nardon and Grenier, this volume) are examples. Some intracellular symbiotic bacteria have substantially reduced genomes; for example, the cyanobacterial symbionts of *Cyanophora paradoxa* have 10% the DNA content of nonsymbiotic cyanobacteria (Herdman and Stanier 1977). These cyanobacteria may approach the stage reached by the organelles—mitochondria and chloroplasts—of eukaryotes, which retain their protein-synthesis apparatus but many of whose proteins are coded for by nuclear genes. Still more dependent are those units of evolution whose ancestors, in all probability, were never able to reproduce outside the host cell. Viruses can survive, but not metabolize, outside the host. Plasmids can replicate, but because they lack a protein coat they must be transmitted directly from cell to cell.

Viruses and plasmids do at least code for some of the proteins needed for their own replication; "defective interfering viruses" have lost even that ability, and depend for their replication on proteins coded for by nondefective viruses, from which they are descended. They, and the analogous defective transposable elements, are the ultimate parasites. (For a review of molecular parasites, see Nee and Maynard Smith 1990.)

Parallel with this range in the degree of independence of symbionts, there is variation in the mode of transmission between hosts. In a surprising number of cases, contact between host and symbiont must be reestablished in every generation. Larval vestimentiferan worms must ingest the sulfur bacteria on which they will depend as adults (Vetter, this volume). Young leiognathid fishes must swallow their luminous bacteria. Young termites acquire their gut symbionts, usually by feeding at the rear end of adult termites. Seedling plants reestablish contact with mycorrhiza in each generation (Lewis, this volume). These cases contrast with the direct transmission in the egg of the bacterial symbionts of plant-sucking bugs and weevils.

The relevance of the mode of transmission is obvious. With direct transmission, the genes of the symbiont will leave descendants only to the extent that the host survives and reproduces. In general, therefore, mutations in the genes of the symbiont will be established by selection only if they increase the fitness of the host. Thus, one would not expect directly transmitted symbionts to evolve into parasites. Even with direct transmission, however, there can be a conflict of interest between the genes of the host and the symbionts. For example, in some hermaphrodite plants, there are mutant mitochondrial genes that cause male sterility (Gouyon and Couvet 1987). Mitochondria are transmitted only in the seed, and it is therefore in the interest of mitochondrial genes that the plant should not waste resources on male functions. The effects of these male-sterile genes are counteracted by "restorer" genes in the nucleus. Hence, although for most purposes we can treat a eukaryotic organism as a unit of evolution, all of whose genes (nuclear or cytoplasmic) will be selected to maximize the inclusive fitness of their carrier, there are contexts in which that assumption would be misleading.

With indirect transmission, the likelihood that a symbiont will evolve toward parasitism is far greater. May and Anderson (1983) have argued that the myxoma virus in Australian rabbits has evolved to an intermediate degree of virulence, maximizing the number of new rabbits that will be infected by viruses from a single infected rabbit before it dies. If the virus is too virulent, it kills the rabbit before many new rabbits have been infected; if the virus strain is insufficiently virulent, the host rabbit may survive for a long time but the infectivity is low, again limiting the infection of new rabbits. Note that the argument assumes that, in most cases, a rabbit is infected by only one strain of virus. A strain of intermediate virulence would gain no advantage if the host rabbit was simultaneously infected by a more virulent strain; the latter would kill the rabbit. There is no point in striving to keep alive the goose that lays the golden eggs if someone else is going to kill and eat it.

When transmission is indirect, it is difficult to explain the evolution of mutualism. Douglas and Smith (1989) also point out that it may be hard to demonstrate mutualism in general. That is, it may be straightforward to show that the host grows and reproduces better in the presence of the symbiont, but it is harder, for practical reasons, to show a direct benefit to the symbiont: perhaps the latter is merely being exploited. In some cases, however, mutual benefit seems likely. Consider, for example, the symbiosis between fish and luminescent bacteria. It seems clear that the fish are benefiting, but what of the bacteria? I do not know, but I wonder why, if the bacteria are getting nothing out of the relationship, they continue to be luminous. The luminescence may be an unselected consequence of something else, but I doubt it.

It may be helpful to compare cooperation between members of different species versus cooperation within a species. In the latter case, two mechanisms have been proposed: kin selection (the interacting individuals share genes) and synergistic selection (both partners benefit, and it would not pay either to renege). In most real cases, both mechanisms are probably operating. If there is cooperation between members of different species, kin selection is replaced by the following: if an individual of species A enables an individual of species B to produce more descendants, those additional descendants will help the descendants of individual A. This will be true in cases of direct transmission, and in cases in which neither the host nor the symbiont disperses; the latter probably holds for termites and their gut parasites. In many cases, direct synergistic effects on the fitness of individuals are probably involved. For example, it may usually be the case that both host plant and mycorrhizal fungus benefit. However, either may evolve into a parasite: ghost orchids are parasitic on their associated fungi, and mycorrhizal fungi sometimes become parasitic. Particularly illuminating are studies of the fungi associated with insectivorous plants and their noninsectivorous relatives in nutrient-poor soils. The fungi of insectivorous plants are more often parasitic. This makes sense in terms of the likely costs and benefits: it only pays to tax the rich.

A final point about the coevolution of host and endosymbiont is worth making. Law and Lewis (1983) point out that in mutualistic interactions, a taxonomically restricted range of symbionts may associate with a much wider range of hosts. The contrast can be striking. Only two genera of dinoflagellates are found in marine invertebrates, but the latter belong to three phyla: Cnidaria, Mollusca, and Platyhelminthes. All ericaceous plants associate with a single species of mycorrhiza (Lewis, this volume). This uniformity in mutualistic endosymbionts contrasts with the diversity of parasites. If many host species benefit from an association, they will converge in those characteristics that facilitate symbiosis with the relatively few potential endosymbionts in the environment; if the latter benefit, there is no evolutionary pressure on them to diversify. In contrast, parasitic species evolve special mechanisms to generate diversity, even within a species; for example, recombination devices for altering the genes determining surface antigens are known in both protists (Borst et al. 1983) and bacteria (Seifert et al. 1988).

Conclusion

Symbiosis may give rise rather suddenly to evolutionary novelty; it is therefore seen as presenting a challenge to Darwinian gradualism. I think this is to misunderstand the reason why Darwin was a gradualist: essentially, it was because the origin of a complex adaptation would be miraculous. To give a modern example, luminescent bacteria require the presence of five genes (Nealson, this volume). The probability that the necessary DNA sequences would arise suddenly, without selection of a series of gradually improving intermediates, is so small as to be negligible. A fish could acquire luminescent bacteria suddenly, but the elaboration of the structures that enable the fish to use the bacteria in signaling, camouflage, and vision would again depend on a gradual process of mutation and selection.

There is, therefore, no contradiction between Darwin's belief that complex adaptations arise by the natural selection of numerous intermediates, and the possibility that new evolutionary potentialities may arise suddenly if genetic material that has been programmed by selection in different ancestral lineages is brought together by symbiosis. So long as transmission is indirect, we should still regard host and endosymiont as separate units of evolution, even if the interactions between them are mutualistic. If transmission is direct, the system has taken the first step on the road to becoming a single unit of evolution. When the symbiont has been reduced to an organelle, it is reasonable to consider the association as a single unit: mitochondria can be expected to behave in ways that ensure the survival of the organism. But, as the example of male sterility in plants shows, so long as different parts of the genome are transmitted in different ways, there is the possibility of conflict: ultimately, genes are the only replicators.

Note

1. Unsubstantiated assumptions that mitosis leads to identical genomes in offspring cells (heterocystous cyanobacteria, *Saccharomyces* mating type, trypanosoma variable antigens, mammalian immunoglobin-producing cells and erythrocytes) argue to the contrary.—*Editors*

References

Borst, P., et al. 1983. In: Chater, K. G., ed., *Genetic Rearrangement*. Croom Helm.

Buss, L. 1987. *The Evolution of Individuality*. Princeton University Press.

Douglas, A. E., and Smith, D. C. 1989. Are endosymbioses mutualistic? *TREE* 4: 350–352.

Eigen, M., and Schuster, P. 1979. *The Hypercycle*. Springer-Verlag.

Gouyon, P. H., and Couvet, D. 1987. The evolution of sex and its consequences. In: Stearns, S. C., ed., *The Evolution of Sex and Its Consequences*. Birkhauser.

Hamilton, W. D. 1964. The genetical evolution of social behavior. *Journal of Theoretical Biology* 7: 1–52.

Herdman, M., and Stanier, R. Y. 1977. The cyanelle: Chloroplast or endosymbiotic prokaryote? *FEMS Lett*. 1: 7–12.

Holliday, R. 1987. The inheritance of epigenetic defects. *Science* 238: 163–170.

Jablonka, E., and Lamb, M. J. 1989. The inheritance of acquired epigenetic variations. *Journal of Theoretical Biology* 139: 69–83.

Law, R., and Lewis, D. H. 1983. Biotic environments and the maintenance of sex: Some evidence from mutualistic symbioses. *Biological Journal of the Linnean Society* 27: 249–276.

May, R. M., and Anderson, R. M. 1983. Epidemiology and genetics in the coevolution of parasites and hosts. *Proceedings of the Royal Society of London* B219: 281–313.

Maynard Smith, J. 1987. How to model evolution. In: Dupré, J., ed., *The Latest on the Best*. MIT Press.

Maynard Smith, J. 1988. Evolutionary progress and levels of selection. In: Nitecki, M. H., ed., *Evolutionary Progress*. University of Chicago Press.

Maynard Smith, J. 1990. Models of a dual inheritance system. *Journal of Theoretical Biology* 143: 41–53.

Nee, S., and Maynard Smith, J. 1990. The evolution of molecular parasites. In: Keymer, A. E., and Read, A. F., eds., *The Evolutionary Biology of Parasitism*. Cambridge University Press.

Seifert, H. S., Ajioka, R. S., Marchal, C., Sparling, P. F., and So, M. 1988. DNA transformation leads to pilin antigenic variation in *Neisseria gonorrhoea*. *Nature* 336: 392–395.

Szathmary, E., and Demeter, L. 1987. Group selection of early replicators and the origin of life. *Journal of Theoretical Biology* 128: 463–486.

4 Modes of Mutation and Repair in Evolutionary Rhythms

Robert H. Haynes

Theodosius Dobzhansky often said that "in biology, nothing makes sense except in the light of evolution." My contention here is that Dobzhansky's aphorism can be extended to the further assertion that in evolution nothing makes sense except in light of the molecular and cellular basis of genetic stability and change. In particular, I suggest that hereditary symbioses of preadapted genomes should be regarded as *megamutations* in the evolution of life.

The best that can be achieved in any historical study, biological or otherwise, is the construction of "plausible stories" which attempt to bring some degree of consistency and coherence to those fragmentary data and observations which are available in the present. More than one such story usually can be told since one person's plausibility may be another's improbability. Thus, apart from puerile attacks by creationists and other connoisseurs of the irrational, evolutionary theory continues to be afflicted with much controversy, and even rancorous debate, among the heirs and disciples of Darwin. The inferences that I draw in this paper on the roles of various modes of mutation and repair in evolution therefore constitute nothing more than *my* attempts at constructing some plausible stories on the basis of what is known today. Some of these views are held by others who work in mutation research, though I hardly expect them to be shared by all biologists. *Ergo, caveat emptor!*

Charles Darwin argued that natural selection is the immediate, or proximal, cause of evolutionary change. However, selection is not a force of nature analogous to gravity or electromagnetism. The ultimate sources of evolutionary innovation must be sought in the mechanisms involved in the production of heritable variation among organisms. In the absence of variation, selection can do nothing. Furthermore, it has been shown, both experimentally and theoretically, that if certain physicochemical conditions prevail in simple, noncellular, replicating systems, selection inevitably ensues (Eigen and Schuster 1979; Küppers 1985).

Heredity is a conservative process. It is a manifestation of the remarkable *stability* of genes, and of their accurate replication, utilization, and transmission, from one generation to the next. On the other hand, variation is a revolutionary process. It is a manifestation of many different sorts of genetic *change*. At the DNA level it has been found that there exists in cells a surprisingly intimate relation between these superficially conflicting phenomena. This relation is rooted in the structure of the genetic material and the biochemical mechanisms for its replication, repair, and recombination (Sargentini and Smith 1985).

"Change" can be explained mechanistically only if one knows the structure(s) of whatever it is that is observed to change. However, any "thing" that undergoes change, transformation, or transmutation must possess some degree of temporal stability if it is to be recognized as an entity: that which has no lifetime has no "life." Understanding stability is logically prior to understanding change. Thus, one must enquire first into the structures and processes that maintain genetic stability if one is to comprehend genetic change.

I begin, therefore, with an overview of the mechanisms that maintain the stability of the genetic information in contemporary cells. This (quasi–)stability is manifest in two main ways: in the astonishingly high fidelity of DNA replication despite the low intrinsic accuracy of template-directed but nonenzymic nucleic-acid polymerization, and in the ability of cells to preserve their viability in the presence of many diverse sources of potentially lethal DNA damage and other less specific metabolic disturbances. I point out some of the ways in which the discovery of these mechanisms has influenced current thinking on mutagenesis, recombinagenesis, sex, and speciation, and then show how mutations, might reasonably be classified into three categories at the levels of the DNA sequence, the chromosome, and genome. I conclude by extending the concept of symbiosis metaphorically to the macromolecular level in precellular evolution.

Genetic Stability and Change at the Molecular Level

In 1935 Max Delbrück suggested that both the stability and the mutability of genes could be explained in terms of the Polanyi-Wigner theory of molecular fluctuations. In this theory, spontaneous mutations were considered to arise physically from quantum-statistical fluctuations in the genetic molecules. The rarity of detectable mutations and the anomalously high temperature coefficient (Q_{10}) for spontaneous mutagenesis in *Drosophila* implied that these molecules must have unusually stable physical structures. Induced mutagenesis also was thought to be a purely physiochemical process associated with the interactions of radiations having high quantum energies with the genetic material, as envisioned in the classical target theory (Schrödinger 1944; Haynes 1985).

We now know, however, that the genetic material is composed of ordinary molecular subunits and is not endowed with any peculiar kind of physicochemical stability. DNA is subject to many types of spontaneous structural degradation, as would be expected in warm, aqueous intracellular environments (Saul and Ames 1986). In addition, cells in nature are exposed to many mutagenic chemicals of both endogenous and exogenous origin (Ames 1983). Also, the potential error rate of nonenzymic DNA synthesis is high (on the order of 10^{-2} per base pair replicated), whereas the observed error rates in normal replication are remarkably low (about 10^{-10} to 10^{-8} per base pair replicated) (Reanny 1987). To get some feeling for this level of accuracy, consider that the most accurate keypunch operators average about one mistake per 5000 symbols typed.

If the various ambient sources of DNA structural decay, damage, and replication error had free rein, neither the informational integrity of DNA nor the viability of cells could be maintained. The well-regulated metabolism of living cells would collapse from what might be called "genetic meltdown." That this does not occur arises from the fact that cells possess an amazing battery of coordinated *biochemical* processes that actively promote the stability of genes and the accuracy of transcription, translation, and protein synthesis throughout the cell cycle (Kirkwood et al. 1986). The stability of the genetic material itself arises from at least three sources:

mechanisms that ensure high levels of replicational fidelity during normal, semi-conservative DNA synthesis, such as 3′-exonucleolytic proofreading by DNA polymerases and methylation-instructed mismatch correction (Modrich 1987; Kunkel 1988)

complex enzymic systems that repair or bypass potentially lethal or mutagenic damage in DNA (Friedberg and Hanawalt 1988) processes that protect DNA by neutralizing or detoxifying mutagenic molecules of both endogenous and exogenous origin (Ames 1983).

Regulation of deoxyribonucleotide precursor pools also contributes to the maintenance of cell viability and the stability of genes and chromosomes (MacPhee et al. 1988). Recent studies have shown that changes in the relative concentrations of these pools can provoke the entire range of genetic effects normally associated with exposure of cells to physical and chemical mutagens. These effects include point mutations, mitotic recombination, chromosome aberrations, aneuploidy, sister-chromatid exchanges, DNA-strand breakage, loss of centromeric plasmids, and neoplastic transformation (Kunz 1982; Haynes 1985). *In vitro* studies with various DNA polymerases have revealed that the fidelity of DNA synthesis is related to the relative concentrations of the deoxyribonucleotides in the reaction mixture (Das et al. 1985).

Consistent with these observations is the fact that mutations in genetic loci which code for enzymes involved in pyrimidine nucleotide biosynthesis can have the effect of increasing spontaneous mutation rates at other loci in both mammalian cells and yeast (Meuth 1984). Most surprising, perhaps, mutations in loci known to control various modes of DNA repair have pleiotropic effects on cellular responses to imbalances in deoxyribonucleotide pools (Kunz and Haynes 1982).

On thermodynamic grounds, if none other, no macromolecular system can function with perfect accuracy. Even though many mutations are deleterious, mutation rates can never be driven selectively to zero and the continuing production of genetic variation thereby eliminated. In molecular evolution, the opposing processes of genetic stabilization and genetic change are complementary, rather than antagonistic phenomena.

Biological Significance of DNA Repair and Fidelity

Enzymic mechanisms for the repair of damaged DNA and the maintenance of replicational fidelity have been found wherever they have been sought, in viral systems and in the simplest and the most complex organisms. Their importance is well attested by the extraordinary sensitivity to ultraviolet light of cells deficient in DNA repair: the 37% survival dose for yeast mutants lacking all major modes of repair is fantastically low—it corresponds to the formation of only one or two pyrimidine dimers per haploid genome (Haynes and Kunz 1981). Clearly, organisms devoid of repair could not survive in nature. Indeed, the potential for efficient repair inherent in the informational redundancy of complementary base pairing may account for the ubiquity of double-stranded nucleic acid as the genetic material of contemporary cells (Hanawalt and Haynes 1967).

Living cells provide the most amazing known examples of highly reliable, dynamic systems built from vulnerable and unreliable parts. Many different genetic loci are involved in the biochemical stabilization of the genetic material. The number of such loci in any organism is not known. In yeast, about 100 are involved in the repair of radiation damage and presumably many more exist (Haynes and Kunz 1981). The large number of these loci is consistent with a well-known principle of design engineering: if great fidelity is to be achieved in the operation of a complex system constructed from components of poor intrinsic precision, many quality assurance devices must somehow be built into the system. For optimum economy, the energy cost of these devices should be just sufficient to reduce overall error rates to a tolerable level (Dancoff and Quastler 1953). This "optimality principle" of maximum error is exemplified in the genetic machinery of cells.

The quasi-stability of long genetic messages depends on the fact that they encode extensive instructions for their own correction. On the optimality principle of maximum error and the mathematical analysis of Eigen and Schuster (1979), the evolution of such self-correcting programs must have entailed the stepwise development of ever-more-efficient quality-control mechanisms as a prerequisite for genomic expansion in evolution. How this process was initiated chemically in the earliest, highly inaccurate replication systems, with their extremely limited coding capacities is not known. This may be one of the most intractable "chicken or egg" problems for all theories of the origin of life (Küppers 1985).

In contemporary cells, it would appear that all major aspects of DNA metabolism are regulated and coordinated in such a way as to counteract the deleterious effects of "genetic noise" and to minimize mortality and mutability. This may be called the "3M" optimality principle of metabolic design (Haynes and Kunz 1988). However, in view of the existence of inducible "error-prone" processes which simultaneously promote viability in cells exposed to ultraviolet light and other mutagens and generate mutations, it would appear that viability takes precedence over genetic fidelity in the economy of cells (Witkin 1969). Cells possess both "error-free" and "error-prone" modes of repair, which contribute to viability by removing or bypassing, respectively, *potentially* lethal lesions in DNA induced by mutagens. Error-prone repair allows replication to proceed past "bulky" lesions in DNA (such as UV-induced pyrimidine dimers) that otherwise would block replication and cause cell death. However, this "bypass" mode of replication occurs with reduced fidelity. The resulting increase in mutagenesis can be regarded as the "price" paid for survival, insofar as new mutations often are deleterious to the cell. This has been called the "better red than dead" theory of mutagenesis.

The immediate benefit of the inducible, error-prone, "SOS response" in *Escherichia coli* (Walker 1985), like that of the damage-inducible responses to alkylation and oxidation, is to ameliorate the toxic effects of mutagens, rather than to expand genetic variation (Demple 1987). Nonetheless, it is reasonable to think that, in bacteria at least, should appropriate new mutations be produced by induction of the SOS response, these could serve to reduce the probability of population extinction in noxious environments (Radman 1980; reviewed in Echols 1982). On the other hand, it seems unlikely that modern sexually reproducing organisms would increase their *general* mutation rates as an adaptive strategy; such a maneuver, in all probability, would weaken, if not extinguish, the species.

The discovery of the close relation between DNA repair and chromonemal recombination in bacteria has had a considerable impact on current biological theory. Margulis and Sagan (1986) have suggested that the emergence of mechanisms for the repair of germicidal (254 nm) ultraviolet-light damage in bacteria pre-adapted them for sexuality because at least some of the enzymes involved in post-replication repair also are involved in the recombination of DNA molecules. Others have argued that the adaptive significance of meiotic chromosomal recombination in higher organisms lies primarily in its ability to provide a mechanism for repairing mutations in DNA, rather than in the production of genetic variation in the form of new combinations of alleles (Maynard Smith 1978; Bernstein et al. 1988). Thus, the two fundamental features of eukaryotic sex, recombination and outcrossing, may be adaptive responses to DNA damage and mutagenesis: recombination as a mode of DNA repair and outcrossing as a mechanism for "masking" deleterious recessive mutations in heterozygotes.

The adaptive significance of mismatch repair in maintaining replication fidelity in bacteria is well established. Recently, however, Rayssiguier et al. (1989) have shown that the requirement for DNA sequence homology in general recombination is greatly relaxed in bacterial mutants deficient in "long-patch" mismatch repair (LPMR). They found that in such mutants intergeneric recombination occurs efficiently between *E. coli* and *Salmonella typhimurium*, organisms which are about 20% divergent in DNA sequence and otherwise cannot conjugate. LPMR is anti-recombinagenic, whereas very short-patch mismatch repair is hyper-recombinagenic. Thus, in bacteria, mismatch-stimulated anti-recombination by LPMR may act as a general "proofreading" system that promotes the fidelity of homologous chromonemal recombination and thereby ensures the sterility of intergeneric crosses.

Defining Mutation

The word *mutation* is used in genetics in both broad and narrow senses. Molecular biologists sometimes restrict its meaning to "point" mutations in DNA sequence, that is, base-pair substitutions and frame shifts arising from small deletions and insertions involving only one or a few nucleotides. Cytogeneticists use the word more broadly to encompass chromosomal aberrations and rearrangements as well as point mutations. Sister-chromatid exchanges and various forms of "illegitimate" mitotic recombination, in somatic or germ cells, can also be included in this broader definition of *mutation*. However, there are three main categories of genetic change that arise at three hierarchical levels in the genetic machinery of cells: changes in DNA sequence involving only a "few" nucleotides; changes in karyotype and gross chromosomal rearrangements involving many nucleotides; and changes in the number of different genomes in cells. The distinction between "few" and "many" nucleotides in this context is arbitrary and generally left undefined (Drake 1970).

In discussions of evolutionary theory, it is often convenient to use the word *mutation* in the broadest possible sense to denote *any* genetic change, not caused by segregation or "legitimate" genetic recombination, which, when transmitted to progeny cells or organisms, gives rise to heritable variation. The three categories of genetic change might then be distinguished as *micromutation*, *macromutation*, and *megamutation*. Micromutation is a synonym for "point mutation."

Megamutations are restricted to the acquisition of foreign genomes in hereditary symbioses. Macromutation includes all other types of genetic change at the chromosomal level, including genomic reorganizations produced by mobile genetic elements (Syvanen 1984).

Micromutations likely occur most commonly, and megamutations most rarely, per unit of geological time. Many micromutations are selectively neutral, or nearly neutral; some of these become fixed in populations by random genetic drift (Kimura 1989). Much of this variation can be regarded as "entropic genetic noise" (Reanney 1987). Micromutations seem to be involved largely, but not exclusively, in micro-evolutionary processes which selectively refine existing adaptations. On the other hand, megamutational symbioses appear to have provided the basis for major metabolic innovations in the early evolution of eukaryotic cells (Margulis 1981), in speciation (Breeuwer and Werren 1990), and in the origin of many higher taxa (Bermudes and Margulis 1987).

Macromutations include a wide range of genomic alterations whose immediate selective significance for the organism, if any, often is not obvious (Syvanen 1984; Shapiro 1985). Included in this category are chromosomal changes produced by mobile genetic elements and viruses, gene amplification and deletion, aneuploidy, centric fusions, karyotypic fissions, polyteny, and polyploidization. Macromutations can make possible the evolution of new proteins through gene duplication. In some cases they may be associated with speciation as well as with the control of certain developmental programs, for example, in the somatic reorganization of immunoglobulin genes and mating type switching in yeast.

Mechanisms of Genetic Change

The processes of mutagenesis in germ-line and somatic cells are similar but by no means identical. Unfortunately, it is difficult and laborious to study germ-line mutagenesis. Hence, most current knowledge of mutagenic mechanisms, as summarized here, is derived from studies on microorganisms and somatic cells in tissue culture. Different cellular and molecular mechanisms must have been involved in initiating and stabilizing the symbiotic associations of megamutagenesis in early cellular evolution (Margulis 1981).

Micro- and macromutagenesis encompass a wide variety of mechanisms. At the level of primary DNA sequences and secondary structures, many complex and still poorly understood phenomena have been observed. These include examples of site specificity (DNA sequences preferentially mutated by a given mutagen) and of mutagenic specificity (mutational events preferentially induced by a mutagen), as well as the likely occurrence of transient secondary structural intermediates in misalignment mutagenesis (Drake et al. 1983; Miller 1983; Horsfall et al. 1990). Furthermore, Hanawalt and co-workers have discovered several examples of "genomic heterogeneity" in error-free DNA repair activity (Bohr and Wassermann 1988; Hanawalt 1989). For example, excision repair of UV-induced pyrimidine dimers is much more efficient in the active dihydrofolate reductase (DHFR) gene of CHO cells in culture than in surrounding inactive DNA sequences or in the genome overall. The Hanawalt group has observed further that the selective, rapid repair of the DHFR gene occurs preferentially in the transcribed DNA strand; similar results also were obtained in the induced lactose operon of *E. coli* (Mellon and Hanawalt 1989). The adaptive value of preferential repair in the transcribed strands of active genes seems obvious; the evolutionary significance, if any, of site and mutagenic specificities is by no means clear. However, these observations do suggest that genetic variation is under much finer genetic control than hitherto imagined. Mutagenesis (and recombination) may be specifically constrained in regions of the genome where it would disrupt functions that are essential for cell viability.

Despite the variable probability of mutagenesis and repair in different DNA sequences, it remains an article of faith among most biologists that mutagenesis is not a Lamarckian process and that mutations occur randomly with respect to their effects on phenotype (Davis 1989). Many mutations arise as a consequence of misreplication and misrepair events, and it would appear that the majority of these are selectively neutral or nearly neutral. On the other hand, "bulky lesions" in DNA, such as pyrimidine dimers produced by exposure of cells to ultraviolet light, are capable of blocking DNA replication and causing cell death. In bacteria, dimers left unrepaired by error-free excision processes can serve to induce the error-prone "SOS" response (Walker 1985). This regulated response, and the various kinds of genomic restructuring events and rearrangements associated with the propagation of mobile genetic elements (in both prokaryotes and eukaryotes), are very complex, and in many ways highly specific biochemically (Lambert et al. 1988). They are not "metabolic errors" in replication and repair. Sometimes promoting cell survival in unfavorable circumstances even at the expense of concomitant mutagenesis (McClintock 1984; Syvanen 1984), they are "programmed" homeostatic responses of the cell to various kinds of stress, especially those which are likely to occur repeatedly.

The relative historic importance in evolution of random micromutations, as opposed to "programmed" mutational responses, is still subject to speculation (Reanney 1987). However, the biochemical and regulatory complexity of the mechanisms of macromutagenesis might be taken to suggest that in early cellular evolution misreplication and/or misrepair, together with the megamutations of intracellular symbiont acquisition, were the major sources of evolutionary innovation. As increasingly complex developmental programs emerged, the genetic machinery became more sophisticated, and macromutations may then have become more important material sources of biological diversity.

Mutations and Evolution

In the absence of any more credible explanation, the founders of the "synthetic" theory of evolution generally assumed that macroevolutionary changes were "nothing but" an extrapolation and a magnification of microevolutionary processes. This classical picture of "neo-Darwinian gradualism" has been seriously undermined by a discontinuous model of "punctuated equilibrium" based largely on a straightforward reading of the fossil record (Gould and Eldredge 1977). This latter-day saltationist element in evolutionary theory can be seen metaphorically in current ideas on the origin of eukaryotes through serial endosymbiosis (Margulis 1970, 1981, and this volume). If one is sufficiently poetic, the metaphor can be extended further to speculations on the origin of the genetic translation machinery which may be regarded as a "symbiotic" coupling of nucleic-acid and protein chemistry (Eigen and Schuster 1979). Dyson's (1985) hypothesis of the "double origin" of life, which assumes the independent formation of replicating and metabolizing entities which ultimately fuse, also is consistent with this extrapolation of the concept of symbiosis to the macromolecular level.

The serial endosymbiosis theory (SET) for the origin of eukaryotic cells is sustained by many plausible inferences drawn from a remarkably wide range of morphological, physiological, and biochemical observations on organisms from all five taxonomic kingdoms. Furthermore, comparisons of both DNA and 16S ribosomal RNA sequences from appropriate species indicate that mitochondria and photosynthetic plastids probably arose from eubacterial symbionts in the lineage of cells containing the progenitor of the eukaryotic nuclear genome. Indeed, the long-heterodox *myth des symbiotes* has been incorporated into recent textbook orthodoxy.

Other claims of the SET are still subject to dispute. One is the ingenious proposal that spirochete-like symbionts donated the genetic information required for the formation of undulipodia (eukaryotic cilia, flagella, and related fibrous structures which develop from kinetosomes) to form early eukaryotes (Margulis 1980; Hinkle, this volume). However, the recent discovery of basal body/centriolar DNA in *Chlamydomonas* greatly strengthens this conjecture (Hall et al. 1989). Another is that heterotrophic prokaryotic anaerobes were ancestral to the earliest eukaryotes. This idea is difficult to reconcile with recent comparisons of the complete 16S ribosomal RNA sequences of several hundred organisms (Olsen, 1988) which imply that the eukaryotes, the eubacteria, and the archaeobacteria all evolved from a common ancestor (Woese 1981). The high degree of congruence of the basic domains in the folded stem-loop structures of these RNAs also are consistent with a common origin. Thus, it has been argued that eukaryotes may not have evolved from prokaryotes, but rather all three lineages may have emerged independently from a hypothetical macromolecular system of precellular evolution called the "progenote" (Woese 1983). Doolittle (1978) has suggested further that prokaryotes may be considered more "advanced" than eukaryotes, that is, a rapidly evolving, metabolically efficient lineage with immense regulatory potential for pioneering new niches in exotic environments. (For a more recent discussion of the progenote hypothesis in the context of the "RNA first" theory of the origin of life, and the intron/exon structure of eukaryotic genomes, see Darnell and Doolittle 1986.)

The scenarios of Woese (1983) and of Schuster and Sigmund (1987) for the transition from precellular to cellular replication are similar in some essential respects. They both postulate "symbiotic" associations of systems of polyribonucleotides and polypeptides to account for the origin of translation and the genetic code in primitive replicating systems. Another common theme is a requirement for the early development of efficient mechanisms for the elimination, correction, prevention, and/or bypass of errors produced during the replication and expression of genetic information. However, the origin of an autopoietic metabolic system to support replication, translation, and repair is not explained explicitly in these theories (Margulis and Sagan 1986; Dyson 1985).

The SET identifies the major macroevolutionary steps which gave rise to eukaryotic cells. However, little is known about the cellular and molecular mechanisms involved. One possibility is that the primitive (wall-less) host cell took up bacterial symbionts by phagocytosis but failed to digest them. If this were the case, it would be consistent with the principle that mutations can arise as a result of "metabolic errors." However, our understanding of such mechanisms is so limited that the megamutations postulated in the SET may well be regarded as "frozen accidents" of evolutionary history.[1] Thus, the rhythms of precellular and early cellular evolution appear as a series of macroevolutionary saltations separated by long periods of slow microevolutionary refinement. As more complex chromosomal transactions became possible, genomes became responsive in more sophisticated ways to environmental stresses, with the result that the tempo of evolution, based largely on speciation, speeded up dramatically.

Concluding Remarks

Thirty-five years ago, it was widely believed that the genetic material was intrinsically very stable and stood "isolated" from the routine metabolism of the cell. Any suggested process which entailed the breakdown and resynthesis of segments of chromosomal DNA ran counter to an established orthodoxy. Indeed, this presumed metabolic languor was taken in those far-off days as evidence for the genetic role of DNA. The discovery of excision repair played a seminal role in the formulation of our current picture of the molecular mechanisms of genetic stability and change, a picture based more on biochemical dynamics than on molecular statics (Hanawalt and Haynes 1967; Haynes 1985).

Cells and organisms possess a variety of remarkably precise and specific homeostatic control mechanisms which promote their viability, reproducibility, individuality, and species integrity. Homeostasis is the definitive functional characteristic of life itself. Many of these mechanisms are associated with repairing damaged DNA and maintaining the accuracy of its replication. Others enable the genome to make "programmed" responses to various life-threatening environmental stresses (Shapiro 1985).

In the earliest stages of precellular and cellular evolution, the efficiencies of these mechanisms must have been low. The amount of available genetic variation upon which selection could act would be correspondingly high. Any improvement in the efficiency of these "buffering" systems presumably would confer some selective advantage, even at the cost of increasing biochemical complexity. It is reasonable to assume that in the earliest cells (whatever they were) most mutations arose as a result of "metabolic error" or inefficiencies in the primitive mechanisms of homeostasis. Later, as genomes and their metabolic support systems increased in complexity and sophistication, the programmed responses of macromutagenesis, together with megamutational symbioses, likely became the major sources of evolutionary innovation in the origin of species and higher taxa.

Elucidation of the mechanisms of DNA repair and replicational fidelity has had a far-reaching impact on biology and medicine. New insights into the etiology of genetic disease, aging, cancer, and the origin and evolution of life have arisen directly from work in this area. The discovery of DNA repair, and of the intricate mechanisms which maintain high-fidelity DNA replication, brings into sharp focus the basic unsolved problem of evolutionary biology: the origin of homeostatic regulatory systems at the molecular level (Pattee 1973; Dyson 1985). At present, no one can account for the "spontaneous" origin of such mechanisms in terms of standard physics and chemistry. We can only speculate on how primitive replication systems, with very limited coding capacities, could expand sufficiently to encode the additional information necessary to allow the evolution of long genetic messages. Indeed, the very useful but still anthropocentric metaphors of "information" and "repair" at the molecular level have not yet been linked to the statistical and quantum-mechanical properties of atoms and molecules as we know them. At this deep level, physics and biology have not yet merged. Until these problems are resolved, we can claim only a superficial understanding of the physicochemical origin of the quasi-stable yet environmentally responsive genomes of contemporary organisms.

Acknowledgments

This paper was written in 1989 during the tenure of a Fellowship at the Wissenschaftskolleg zu Berlin.

Note

1. The unknown origin of biochemical chirality—the universal requirement for D-sugars and L-amino acids in biopolymers (Bonner 1972; Cairns-Smith 1982)—may be the most important frozen accident of prebiotic chemistry, inasmuch as it "just happened" to make possible the development of an efficient biochemistry in autopoietic systems.

References

Ames, B. N. 1983. Dietary carcinogens and anticarcinogens. *Science* 221: 1256–1264.

Bermudes, D., and Margulis, L. 1987. Symbiont acquisition as neoseme: origin of species and higher taxa. *Symbiosis* 4: 185–198.

Bernstein, H., Hopf, F. A., and Michod, R. E. 1988. The molecular basis of the evolution of sex. *Advances in Genetics* 24: 323–370.

Bohr, V. A., and Wassermann, K. 1988. DNA repair at the level of the gene. *TIBS* 13: 429–433.

Bonner, W. A. 1972. Origins of molecular chirality. In: Ponnamperuma, C., ed., *Exobiology*. North-Holland.

Breeuwer, A. J. and Werren, J. H. 1990. Microorganisms associated with chromosome destruction and reproductive isolation between two insect species. *Nature* 346: 558–560.

Cairns-Smith, A. G. 1982. *Genetic Takeover*. Cambridge University Press.

Dancoff, S. M., and Quastler, H. 1953. The information content and error rate of living things. In: Quastler, H., ed., *Information Theory in Biology*. University of Illinois Press.

Darnell, J. E., and Doolittle, W. F. 1986. Speculations on the early course of evolution. *Proc. Natl. Acad. Sci. USA* 83: 1271–1275.

Das, S. K., Kunkel, T. A., and Loeb, L. A. 1985. Effects of altered nucleotide concentrations on the fidelity of DNA replication. In: de Serres, F. J., ed., *Genetic Consequences of Nucleotide Pool Imbalance*. Plenum.

Davis, B. D. 1989. Transcriptional bias: A non-lamarckian mechanism for substrate induced mutations. *Proc. Natl. Acad. Sci. USA* 86: 5005–5009.

Demple, B. 1987. Adaptive responses to genotoxic damage: Bacterial strategies to prevent mutation and cell death. *BioEssays* 6: 157–160.

Doolittle, W. F. 1978. Genes in pieces: Were they ever together? *Nature* 272: 581–582.

Drake, J. W. 1970. *The Molecular Basis of Mutation*. Holden-Day.

Drake, J. W., Glickman, B. W., and Ripley, L. S. 1983. Updating the theory of mutation. *American Scientist* 71: 621–630.

Dyson, F. 1985. *Origins of Life*. Cambridge University Press.

Echols, H. 1982. Mutation rate: Some biological and biochemical considerations. *Biochimie* 64: 571–575.

Eigen, M., and Schuster, P. 1979. *The Hypercycle: A Principle of Natural Self-Organization*. Springer-Verlag.

Friedberg, E. C., and Hanawalt, P. C., eds. 1988. *Mechanisms and Consequences of DNA Damage Processing*. Alan R. Liss.

Gould, S. J., and Eldredge, E. 1977. Punctuated equilibria. The tempo and mode of evolution reconsidered. *Paleobiology* 3: 115–151.

Hall, J. L., Ramanis, Z., and Luck, D. J. L. 1984. Basal body/centriolar DNA: Molecular genetic studies in *Chlamydomonas*. *Cell* 59: 121–132.

Hanawalt, P. C. 1989. Concepts and models for DNA repair: From *Escherichia coli* to mammalian cells. *Environ. Molec. Mutagenesis* 14, Suppl. 16: 90–98.

Hanawalt, P. C., and Haynes, R. H. 1967. The repair of DNA. *Scientific American* 216(2): 36–43.

Haynes, R. H. 1985. Molecular mechanism in genetic stability and change: The role of deoxyribonucleotide pool balance. In: de Serres, F. J., ed., *Genetic Consequences of Nucleotide Pool Imbalance*. Plenum.

Haynes, R. H., and Kunz, B. A. 1981. DNA repair and mutagenesis in yeast. In: Strathern, J. N., Jones, E. W., and Broach, J. R., eds., *The Molecular Biology of the Yeast Saccharomyces*, vol. I. Cold Spring Harbor Laboratory.

Haynes, R. H., and Kunz, B. A. 1988. Metaphysics of regulated deoxyribonucleotide biosynthesis. *Mutation Res.* 200: 5–10.

Horsfall, M. J., Gordon, A. J. E., Burns, P. A., Zielenska, M., van der Vliet, M. E. and Glickman, B. W. 1990. Mutational specificity of alkylating agents and the influence of DNA repair. *Environ. Molec. Mutagenesis* 15: 107–122.

Kimura, M. 1989. The neutral theory of molecular evolution and the world view of the neutralists. *Genome* 31: 24–31.

Kirkwood, T. B. L., Rosenberger, R. F., and Galas, D. J., eds. 1986. *Accuracy in Molecular Processes*. Chapman and Hall.

Kunkel, T. A. 1988. Recent studies of the fidelity of DNA synthesis. *Biochim. Biophys. Acta* 951: 1–15.

Kunz, B. A. 1982. Genetic effects of deoxyribonucleotide pool imbalance. *Environ. Mutagenesis* 4: 695–725.

Kunz, B. A., and Haynes, R. H. 1982. Repair and the genetic effects of thymidylate stress in yeast. *Mutation Res.* 93: 353–375.

Küppers, B.-O. 1985. *Molecular Theory of Evolution*. Springer-Verlag.

Lambert, M. E., McDonald, J. F., and Weinstein, I. B., eds. 1988. *Eukaryotic Transposable Elements as Mutagenic Agents*. Banbury Report 30, Cold Spring Harbor Laboratory.

MacPhee, D. G., Haynes, R. H., Kunz, B. A., and Anderson, D., eds. 1988. Genetic aspects of deoxyribonucleotide metabolism. *Mutation Res.* 200: 1–256.

Margulis, L. 1970. *Origin of Eukaryotic Cells.* Yale University Press.

Margulis, L. 1980. Flagella, cilia and undulipodia. *BioSystems* 12: 105–108.

Margulis, L. 1981. *Symbiosis in Cell Evolution.* Freeman.

Margulis, L., and Sagan, D. 1986. *Origins of Sex.* Yale University Press.

Maynard Smith, J. 1978. *The Evolution of Sex.* Cambridge University Press.

McClintock, B. 1984. The significance of responses of the genome to challenge. *Science* 226: 792–801.

Mellon, I., and Hanawalt, P. C. 1989. Induction of the *Escherichla coli* lactose operon selectively increases repair of its transcribed DNA strand. *Nature* 342: 95–98.

Meuth, M. 1984. The genetic consequences of nucleotide pool imbalance in mammalian cells. *Mutation Res.* 126: 107–112.

Miller, J. H. 1983. Mutational specificity in bacteria. *Annual Review of Genetics* 17: 215–238.

Modrich, P. 1987. DNA mismatch correction. *Annual Review of Biochemistry* 56: 435–466.

Olsen, G. J. 1988. Earliest phylogenetic branchings: Comparing rRNA-based evolutionary trees inferred with various techniques. *Cold Spring Harbor Symp. Quant. Biol.* 52: 825–837.

Pattee, H. H. 1973. Physical Problems of the Origin of Natural Controls. In: Locker, A., ed., *Biogenesis, Evolution, Homeostasis*. Springer-Verlag.

Radman, M. 1980. Molecular mechanisms of mutagenesis: A summary review. In: *Conference of Structural Pathology of DNA and the Biology of Ageing*. Harald Boldt Verlag, Boppard.

Rayssiguier, C., Thaler, D. S. and Radman, M. 1989. The barrier to recombination between *Escherichia coli* and *Salmonella typhimurium* is disrupted in mismatch repair mutants. *Nature* 342: 396–401.

Reanney, D. C. 1987. Genetic error and genome design. *Cold Spring Harbor Symp. Quant. Biol.* 52: 751–757.

Sargentini, N. J., and Smith, K. C. 1985. Spontaneous mutagenesis: The roles of DNA repair, replication and recombination. *Mutation Res.* 154: 1–27.

Saul, R. L., and Ames, B. N. 1986. Background levels of DNA damage in the population. In: Simic, M. G., Grossman, L., and Upton, A. C., eds., *Mechanisms of DNA Damage and Repair: Implications for Carcinogenesis and Risk Assessment*. Plenum.

Schrödinger, E. 1944. *What Is Life*? Cambridge University Press.

Schuster, P., and Sigmund, K. 1987. Self-organization of macromolecules. In: Yates, F. E., ed., *Self-Organizing Systems: The Emergence of Order*. Plenum Press.

Shapiro, J. A. 1985. Mechanisms of DNA reorganization in bacteria. *International Review of Cytology* 93: 25–56.

Syvanen, M. 1984. The evolutionary implications of mobile genetic elements. *Annual Review of Genetics* 18: 271–293.

Walker, G. C. 1985. Inducible DNA repair systems. *Annual Review of Biochemistry* 54: 425–457.

Witkin, E. M. 1969. Ultraviolet induced mutation and DNA repair. *Annual Review of Genetics* 3: 525–552.

Woese, C. R. 1981. Archaebacteria. *Scientific American* 244(6): 98–125.

Woese, C. R. 1983. The primary lines of descent and the universal ancestor. In: Bendall, D. S., ed., *Evolution from Molecules to Man*. Cambridge University Press.

5 The Symbiotic Phenotype: Origins and Evolution

Richard Law

It has long been recognized that the existence of complex phenotypic traits in living organisms is a critical issue for Darwin's theory of evolution by natural selection. Darwin (1859) understood this well, but he reasoned that complex traits were entirely consistent with his theory, as long as every small step in the evolutionary process conferred a selective advantage and was inherited.

The adequacy of Darwin's explanation has been a recurring subject of debate, because it is not immediately clear that each intermediate step leading to a new phenotypic trait would possess the required selective advantage. However, most evolutionary biologists have come to see other mechanisms, such as macromutations (e.g. Goldschmidt 1940) and changes in ploidy (Levin 1983), as rare events relative to the "normal" process, in which small phenotypic changes are gradually accumulated. If it is the case that changes are required before a selective advantage can accrue to a new phenotype, these are envisaged to come about as by-products of selection operating on genes with other effects on the phenotype (pleiotropy), through intensification of selection pressures, or, most important, through changes in the functions of structures which already exist (Mayr 1960, 1963; Frazetta 1975).

The study of symbiosis has shown that novel phenotypic traits can come about in another radically different way: through the intimate association of organisms of different species (Margulis 1976). There are some striking cases of this process in the living world, and it deserves serious attention. For instance, the massive exoskeletons of reef-building corals occur only when appropriate coelenterates are living in association with certain dinoflagellate algae (Goreau 1963; Woodhead and Weber 1973). A lichen thallus, with its organized structure and physiology, emerges only when the photobiont and the mycobiont are placed together (Hawksworth 1988). Perhaps most important of all, the properties of aerobic respiration and photosynthesis of eukaryotic cells are most readily understood as the outcomes of ancient endosymbioses with bacteria and cyanobacteria (Margulis 1981). There is a sense in which this process resembles the integration of separate pieces of genetic information into single units, which took place at an early stage in the evolution of life (Szathmáry 1989).

These two contrasting *endogenous* and *symbiotic* sources of new phenotypes are in no sense mutually exclusive. Within a single population, some phenotypic traits may be changing owing to endogenous processes; at the same time, other new phenotypes may come about as a result of symbiosis. Moreover, it is most likely that novel phenotypes arising from symbiosis themselves undergo endogenous change, as each species evolves through natural selection generated by its interaction with the other(s). The issue at stake is not whether one process operates to the exclusion of the other, but what is the *relative* importance of the processes as a source of new phenotypes. For a number of reasons discussed in this essay, I think that innovation is likely to occur less often through symbiosis than through endogenous processes. But this in no way diminishes the great interest of cases in which new phenotypes do come about by symbiosis, unpredictable though such events are.

The Symbiotic Phenotype

I will take a symbiotic phenotypic trait to be one which exists as the direct outcome of the joint expression of genes in both (or all) partners, the partners being different species. The intention of this definition is to focus on phenotypic traits which exist only by virtue of the association of the partners. The definition encompasses traits such as the leghemoglobin protein of the root nodules of legumes, coded in part by the *Rhizobium* genome and in part by the genome of the leguminous host. But it excludes cases where one individual's phenotype is influenced by another's simply because the one is a component of the environment of the other, as in the reduced size of an individual plant when in competition with an individual of another species. It is as well to recognize, though, that we are looking at a continuum in which the separate genetic origins of phenotypic traits of the partners become increasingly difficult to distinguish as the partners lose their separate identities (Smith 1979).

As a matter of definition, I will suppose that the symbiotic phenotypic trait is advantageous to one of the partners. The effect of the trait on the reproductive success of the other(s) is, however, more variable. For instance, the algae whose chloroplasts are retained in a viable state after being eaten by *Elysia viridis* provide the mollusc with the products of photosynthesis (Trench et al. 1973a,b) but clearly do not themselves benefit from the association. On the other hand, the nitrogen-fixing nodule of the legume/*Rhizobium* association, a phenotypic trait which comes about through the joint expression of genes in both partners (Kondorosi and Kondorosi 1986), appears to benefit both the host and the endosymbiont.

Raw Materials

The raw materials for symbiotic phenotypic traits are chance associations between species in the living world. At the most general level, the biosphere can be envisaged as a melting pot for natural experiments, which bring species together in a myriad of different combinations and which try out endless combinations of genotypes of these associating species. But more than just spatial proximity is required, because individuals of different species can be close together and yet be independent of each other. The raw materials consist of those combinations of species, and of genotypes within the species, which have some effect on each other. All other things being equal, symbiotic phenotypes are more likely to have their origins in highly interactive communities, where species are tightly coupled by strong links, than in communities where species interact weakly or not at all.

At present it is difficult to gain even the roughest estimate of how much raw material for symbiotic phenotypes is being generated by fortuitous associations between species. However, the limited information which is emerging from community ecology suggests that it is premature to regard the highly interactive community as a general feature of the living world (Hall et al. 1989). Experimental manipulation of the population density of one species often has little or no effect on the density of other species in the vicinity (Sih et al. 1985; Raffaelli and Milne 1987; Warren 1988; Law and Watkinson 1989). Some kinds of communities are characterized by stronger interactions than others (Sih et al. 1985; Warren 1988). Occasional strong interactions do occur when most others are weak (Paine 1980), but such interactions are evidently far from ubiquitous.

The information from community ecology tells us only about the average effect of one species on another, not about the genetic variation in this effect from one individual to another. Such variation is known to exist (Thompson 1988). For instance, some strains of *Rhizobium* are much more promiscuous than others in the species of Leguminosae which they nodulate (Lange 1961; Crow et al. 1981), and their host range can be altered by single gene mutations (Djordjevic et al. 1985; Horvath et al. 1986). The interaction may be dependent on bringing together a specific combination of genes in the partners, as plant geneticists have found when breeding for pathogen resistance in crops (the gene-for-gene concept; see Flor 1956 and Day 1974). The probability of such an event is a matter of conjecture, but is likely to be low in diverse natural communities where most species are rare and the appropriate genotypes necessarily even rarer. It will presumably be greatest for common species in communities with strong interactions, where the interactions are largely independent of genotype.

In contrast to the symbiotic phenotype, the source of new phenotypes from endogenous processes is genetic variation within populations. Such variation in metric characters is widespread in the living world. The variation stems in the first instance from mutation; data from a variety of sources indicate that the variance in a character caused by new mutations in a single generation is of the order of 0.001 of the environmental variance of the trait (Lande 1976; Hill 1982). This is sufficient to maintain genetic variance in the presence of stabilizing selection (Lande 1976; Bulmer 1980), and it provides a replenishing store of variation on which directional selection can operate under the appropriate environmental conditions (e.g. Yoo 1980).

Continuity

If, at some time, an incipient symbiotic phenotype has come about, the next hurdle for the association is its inheritance. In general, there is no assurance that the integrity of the association will be maintained during reproduction of the partners. This must make the inheritance of the symbiotic phenotype less likely than the inheritance of one of endogenous origin; in the latter case, continuity is ensured by the genetic system.

Nevertheless, there are several factors which increase the likelihood that the offspring of the partners will retain the association.

First, one partner may be inside another; if the host is unicellular, then it passes on its partner to daughter cells when it divides. If the host is an animal, differentiation between somatic and gonadic tissue makes direct transmission of the inhabitant less likely during sexual reproduction, unless the inhabitant is targeted directly to the gonads. *A priori*, one would therefore expect symbioses to be more characteristic of unicellular than of multicellular organisms. It may not be coincidental that the best-documented case of the evolution of a symbiosis, that of *Amoeba proteus* with unidentified bacteria, is a system of this kind (Jeon 1983; Jeon, this volume).

In some invertebrate/algal symbioses and in many lichens, continuity is achieved through asexual propagules containing both partners (Taylor 1973; Hawksworth 1988). In other invertebrate/algal symbioses, such as that involving the hydroid *Myrionemia amboinense*, the eggs carry algal cells with them (Trench 1981). But some symbioses involving the dinoflagellate *Symbiodinium microadriaticum* have to be reassembled after reproduction of the host; since this alga rarely occurs in the free-living state, it is less clear how reassembly is achieved (Taylor 1973; La Barbera 1975; Jameson 1976; Smith 1978). Mechanisms which enable one partner to locate another do exist, and are particularly well developed in associations between flowering plants and their pollinators (Faegri and van der Pijl 1979).

Evidently the integrity of established symbioses through time is maintained by a diverse set of mechanisms. This diversity itself points to the lack of a general solution to the problem of inheritance of symbiotic phenotypes comparable to the inheritance of a phenotypic trait of endogenous origin.

Paths of Coevolution

The evolution of a phenotype through symbiosis is likely to involve some degree of endogenous change in the symbionts, rather than springing ready-made from a fortuitous association of individuals. Given that there exists genetic variation within at least one of the interacting populations which has effects on the fitness of the symbiotic phenotype, natural selection will bring about genetic change within one or more of the species as they interact. Although such evolution may lead to fixation of a new phenotype advantageous to at least one of the partners in symbiosis, other outcomes are also feasible.

To illustrate some of the complexities which can arise in coevolutionary dynamics, consider an interaction between a host species and its endosymbiont. Suppose that there are two phenotypes within each species. In the host species, these are the *cultivator* (which takes up the endosymbiont and nurtures it within its tissues) and the *hunter* (which simply eats the endosymbiont). The endosymbiont phenotypes are the *cooperator* (which promotes uptake by the host) and the *fighter* (which resists the process). The frequency of the cultivator and of the cooperator within their respective populations (p_1 and p_2, respectively) will be the state variables whose dynamics we investigate. We will suppose that the novel symbiotic phenotype under investigation consists of cultivator/cooperator, and that this comes about when coevolution leads to fixation of these phenotypes (i.e., $p_1 = 1$, $p_2 = 1$).

The fitness of an individual depends on its own phenotype and that of the individual of the other species with which it interacts. The effect on fitness can be defined in terms of payoffs, as it is in games within species (Maynard Smith 1982):

		Host		**Endosymbiont**	
		Cultivator	Hunter	Cooperator	Fighter
Payoff to:	Cultivator	0	0	a_1	b_1
	Hunter	0	0	c_1	d_1
	Cooperator	a_2	c_2	0	0
	Fighter	b_2	d_2	0	0

The elements of this matrix are the payoffs for all combinations of phenotypes. Suppose that each individual encounters no more than one individual of the other species, and that it does so with the constant probability e_1 and e_2 (for host and endosymbiont, respectively). By holding this probability constant, we are in effect resetting the population densities to a constant value at the start of each generation, which reduces the dimension of the dynamical system from 4 to 2. Encounters with the other species are assumed to occur at random, so the probability of an encounter with a particular phenotype depends linearly on the frequency of this phenotype. The fitness of an individual, then, is made up of some base value achieved in the absence of any interaction (w_1 and w_2 for host and endosymbiont, respectively), plus increments due to the payoffs from interaction, weighted by the probability of the interaction, as follows:

Cultivator fitness: $w_{11} = w_1 + e_1[a_1p_2 + b_1(1 - p_2)]$
Hunter fitness: $w_{12} = w_1 + e_1[c_1p_2 + d_1(1 - p_2)]$
Cooperator fitness: $w_{21} = w_2 + e_2[a_2p_1 + c_1(1 - p_1)]$
Fighter fitness: $w_{22} = w_2 + e_2[b_2p_1 + d_1(1 - p_1)]$

The parameters are constrained so that $w_{ij} \geqslant 0$ for all i,j. Assuming that generations of the two species are synchronous, that phenotypes contribute to the next generation in proportion to their fitness, that reassociation of the phenotypes must occur afresh in each generation, and that the phenotypes breed true, the following recurrence relations fully describe the evolutionary dynamics of the symbiosis:

$$p_1' = p_1w_{11}/\bar{w}_1,$$
$$p_2' = p_2w_{21}/\bar{w}_2,$$

where $\bar{w}_1$ and $\bar{w}_2$ are mean fitnesses:

$$\bar{w}_1 = w_{11}p_1 + w_{12}(1 - p_1),$$
$$\bar{w}_2 = w_{21}p_2 + w_{22}(1 - p_2).$$

The prime denotes the phenotype frequency at the start of the next generation.

Evolution of the novel phenotype cultivator/cooperator from a hunter/fighter starting point ($p_1 \simeq 0$, $p_2 \simeq 0$) depends very much on the elements of the payoff matrix. Some numerical simulations will illustrate this. Figures 1–4 show some solutions to a model of co-evolution between a host and its endosymbiont as given in the text, using contrasting values in the payoff matrix. The graphs are phase planes, with trajectories for the frequency of the cultivator (p_1) and the cooperator (p_2) through time. Broken lines are isoclines on which the frequency of the cultivator (horizontal lines) and the cooperator (vertical lines) do not change from one generation to the next. Circles mark the initial conditions of the simulations, and the star marks the point to which each tends, if such a point exists. Probabilities of encounter (e_1 and e_2) are set at 0.5 for both species in figures 1 and 2, and at 0.25 for both species in figures 3 and 4. In the simulations, the payoffs to the cultivator and the cooperator of their association (a_1 and a_2) are held at the value 3, so that there is a substantial benefit to both of them arising from their interaction. The base fitnesses (w_1 and w_2) are set to unity, and the encounter probabilities are either 0.5 or 0.25, as stated in the figures.

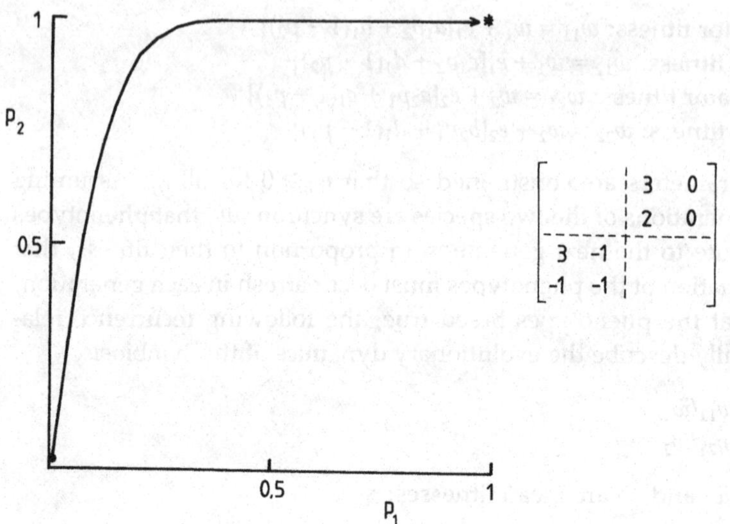

Figure 1

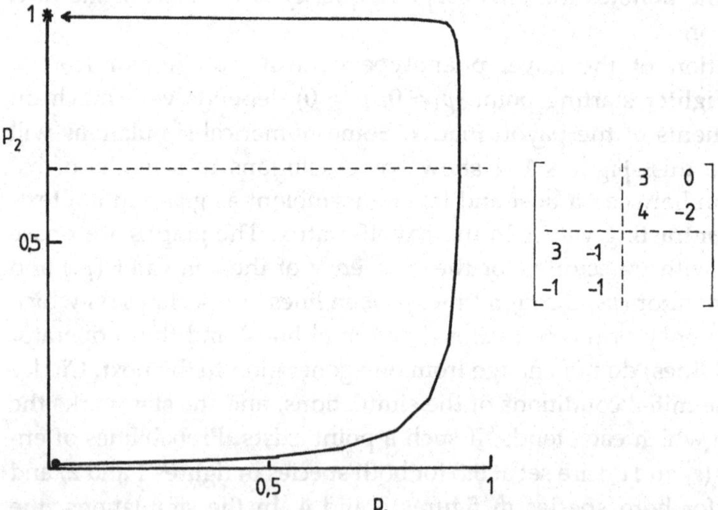

Figure 2

Simulation 1 (figure 1)

Here the fighter always escapes from the cultivator, so that the cultivator gains nothing ($b_1 = 0$) and the fighter experiences some cost in escaping ($b_2 = -1$). The cooperator is sometimes eaten by the hunter, so the hunter gains a benefit ($c_1 = 2$) and the cooperator suffers a cost ($c_2 = -1$). The benefit to the hunter of catching the fighter is balanced by the cost of so doing ($d_1 = 0$), and again there is a cost to the fighter in escaping ($d_2 = -1$). With these payoffs, the cultivator is more fit than the hunter whenever $p_2 > 0$; likewise, the cooperator is more fit than the fighter whenever $p_1 > 0$. Consequently, cultivator and cooperator mutants will spread through their respective populations to fixation, thereby fixing the novel symbiotic phenotype in the population.

Simulation 2 (figure 2)

However, the fitness of cultivator does not necessarily remain greater than that of hunter, as the cooperator phenotype increases in frequency in the other population. For instance, the benefit to the hunter of eating the cooperator could be greater than the corresponding benefit to the cultivator if there are heavy energetic costs of maintaining the endosymbiont ($c_1 = 4$, say). At the same time, the fighter could inflict damage on the hunter so that the latter loses fitness ($d_1 = -2$). Keeping all other payoffs the same as before, both cultivator and cooperator mutants will start to spread, but the hunter will take over from the cultivator when the cooperator reaches a high frequency ($p_2 > 0.67$). In this case we do not end up with the novel symbiotic phenotype.

Simulation 3 (figure 3)

Suppose now that the payoffs to the host remain as in figure 2, but that the payoffs to the endosymbiont are changed. The fighter and the cultivator do not interact at all, so $b_2 = 0$. The cooperator is more likely to be caught by the hunter than is the fighter, and consequently it suffers a greater loss in fitness ($c_2 = -2$). The dynamics are now oscillatory, and they spiral out to the boundary. No corner of the phase space is resistant to invasion, so no phenotype which is resistant to invasion by all others can predominate.

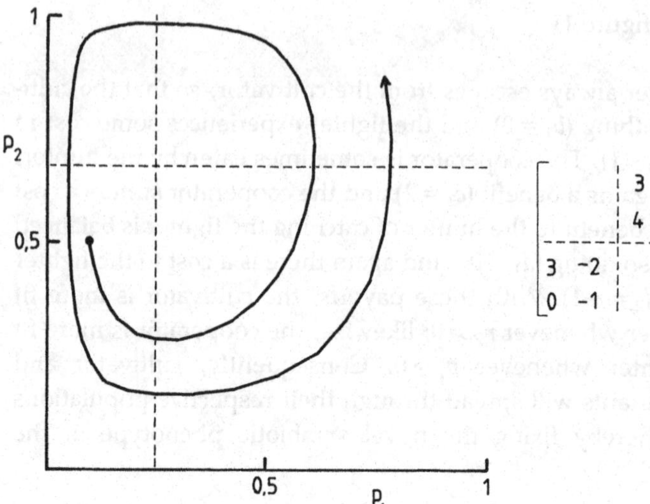

Figure 3

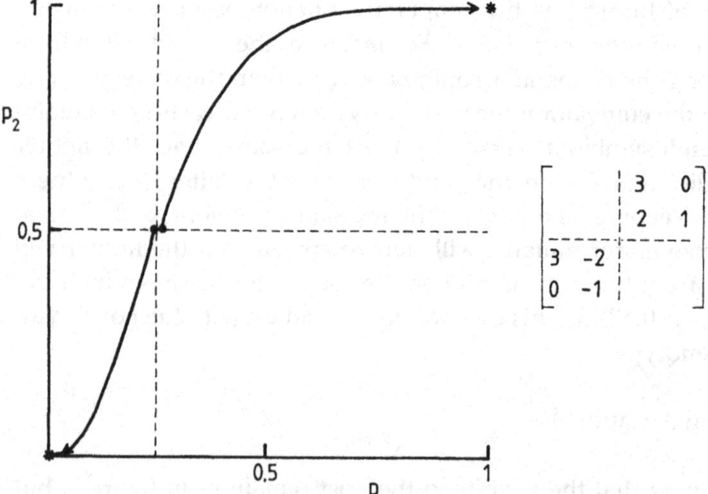

Figure 4

Simulation 4 (figure 4)

In this simulation we suppose that the payoffs to the endosymbiont remain as in figure 3. However, the benefit to the hunter of consuming the cooperator is now less than that to the cultivator of nurturing it ($c_1 = 2$). In addition, the hunter gains some benefit from the fighter instead of losing fitness ($d_1 = 1$). Although cultivator/cooperator is a stable state of the system, so is hunter/fighter. Thus, cultivator and cooperator mutants will not simultaneously spread through their respective populations, and the novel phenotype will not become fixed.

Evidently it is far from certain that a mutually advantageous phenotype can come about through coevolution of symbionts. The system investigated is, of course, a gross simplification of events in the real world. For instance, if one of the symbionts lives mostly inside the other, which in turn is embedded in an antagonistic community, there is an asymmetry in the biotic interactions of the partners and hence in the selection pressures generated by their biotic environments (Law and Lewis 1983; Law 1985). But introducing complications to the model will not, in general, make it better behaved (with the possible exception of a mechanism to retain the integrity of cultivator/cooperator from one generation to the next).

In defense of the symbiotic phenotype, we should recognize that fixation of a strictly endogenous phenotype may itself be prevented by the underlying genetic architecture of the trait. However, the constraints discussed above, which arise when two or more populations separately undergo evolution as a result of their interaction, are additional to those arising from the genetic basis of the trait.

Conclusion

There can be little doubt that some new and complex phenotypic traits stem from an association between individuals of different species, followed by subsequent phenotypic change within the species due to the selection pressures the partners generate on one another. This, then, is a further mechanism to add to those already discussed as potential sources of evolutionary innovation. The crucial issue for evolutionary biology is the relative importance of symbiosis as a source of novel phenotypes when compared with the gradual accumulation of small phenotypic changes within species. The latter process can operate irrespective of whether the conditions for symbiotic innovation are satisfied; it is not conditional on abundance, strong interactions, reassembly of the symbiosis in each generation, or the changes taking place in other interacting but separately evolving populations. This leads me to conclude that symbiosis is less common as a source of evolutionary innovation than changes generated within species. But this is in no way to deny the fact that symbiotic phenotypes do come about from time to time, with profound consequences for the course of evolution.

Acknowledgments

I thank L. Margulis for the invitation to attend the conference on Symbiosis as a Source of Evolutionary Innovation, and T. J. Crawford, D. H. Lewis, and M. H. Williamson for critical comments on the manuscript.

References

Bulmer, M. G. 1980. *The Mathematical Theory of Quantitative Genetics*. Clarendon.

Crow, V. L., Jarvis, B. D. W., and Greenwood, R. M. 1981. Deoxyribonucleic acid homologies among acid-producing strains of *Rhizobium*. *International Journal of Systematic Bacteriology* 31: 152–172.

Darwin, C. 1859. *The Origin of Species by Means of Natural Selection*, first edition. Reprinted by Penguin.

Day, P. R. 1974. *Genetics of Host-Parasite Interaction*. Freeman.

Djordjevic, M. A., Schofield, P. R., and Rolfe, B. G. 1985. Tn5 mutagenesis of *trifolii* host-specific nodulation genes result in mutants with altered host range ability. *Molecular and General Genetics* 200: 463–471.

Faegri, K., and van der Pijl, L. 1979. *The Principles of Pollination Ecology*, third edition. Pergamon.

Flor, H. H. 1956. The complementary genic systems in flax and flax rust. *Advances in Genetics* 8: 29–54.

Frazetta, T. H. 1975. *Complex Adaptations in Evolving Populations*. Sinauer.

Goldschmidt, R. 1940. *The Material Basis of Evolution*. Yale University Press.

Goreau, T. F. 1963. Calcium carbonate deposition by coralline algae and corals in relation to their roles as reef-builders. *Annals of the New York Academy of Sciences* 109: 127–167.

Hall, S. J., Rafaelli, D., and Turrell, W. R. 1990. Predator caging experiments in marine systems: A re-examination of their value. *American Naturalist* (in press).

Hawksworth, D. L. 1988. Coevolution of fungi with algae and cyanobacteria in lichen symbioses. In: Pirozynski, K. A., and Hawksworth, D. L., eds., *Coevolution of Fungi with Plants and Animals*. Academic Press.

Hill, W. G. 1982. Predictions of response to artificial selection from new mutations. *Genetical Research Cambridge* 40: 255–278.

Horvath, B., Kondorosi, E., John, M., Schmidt, J., Torok, I., Gyorgypal, Z., Barabas, I., Wieneke, U., Schell J., and Kondorosi, A. 1986. Organization, structure and symbiotic function of *Rhizobium meliloti* nodulation genes determining host specificity for Alfalfa. *Cell* 46: 335–343.

Jameson, S. C. 1976. Early life history of the giant clams *Tridacna crocea* (Lamarck), *Tridacna maxima* (Roding), and *Hippopus hippopus* (Linnaeus). *Pacific Science* 30: 219–233.

Jeon, K. W. 1983. Integration of bacterial endosymbionts in Amoebae. *International Review of Cytology* Suppl. 14: 29–47.

Kondorosi, E., and Kondorosi, A. 1986. Nodule induction on plant roots by *Rhizobium*. *Trends in Biochemical Sciences* 11: 296–299.

La Barbera, M. 1975. Larval and post-larval development of the giant clams *Tridacna maxima* and *Tridacna squamosa* (Bivalvia: Tridacnidae). *Malacologia* 15: 67–79 .

Lande, R. 1976. The maintenance of genetic variability by mutation in a polygenic character with linked loci. *Genetical Research Cambridge* 26: 221–235.

Lange, R. T. 1961. Nodule bacteria associated with indigenous Leguminosae of South-Western Australia. *Journal of General Microbiology* 26: 351–359.

Law, R. 1985. Evolution in a mutualistic environment. In: Boucher, D. H., ed., *The Biology of Mutualism: Ecology and Evolution*. Croom Helm.

Law, R., and Lewis, D. H. 1983. Biotic environments and the maintenance of sex: Some evidence from mutualistic symbioses. *Biological Journal of the Linnean Society* 20: 249–276.

Law, R., and Watkinson, A. R. 1989. Competition. In: Cherrett, J. M., ed., *Ecological Concepts. The Contribution of Ecology to an Understanding of the Natural World*. Blackwell.

Levin, D. A. 1983. Polyploidy and novelty in flowering plants. *American Naturalist* 122: 1–25.

Margulis, L. 1976. A review: Genetic and evolutionary consequences of symbiosis. *Experimental Parasitology* 39: 277–349.

Margulis, L. 1981. *Symbiosis in Cell Evolution*. Freeman.

Maynard Smith, J. 1982. *Evolution and the Theory of Games*. Cambridge University Press.

Mayr, E. 1960. The emergence of evolutionary novelties. In: Tax, S., ed., *Evolution After Darwin*, vol. 1: *The Evolution of Life*. University of Chicago Press.

Mayr, E. 1963. *Animal Species and Evolution*. Harvard University Press.

Paine, R. T. 1980. Food webs: Linkage, interaction strength and community infrastructure. *Journal of Animal Ecology* 49: 667–685.

Raffaelli, D., and Milne, H. 1987. An experimental investigation of the effects of shorebird and flatfish predation on estuarine invertebrates. *Estuarine, Coastal and Shelf Science* 24: 1–13.

Sih, A., Crowley, P., McPeek, M., Petranka J., and Strohmeier, K. 1985. Predation competition and prey communities: A review of field experiments. *Annual Review of Ecology and Systematics* 16: 269–311.

Smith, D. C. 1978. Symbiosis in the microbial world. In: Norris, J. R., and Richmond, M. H., eds., *Essays in Microbiology*. Wiley.

Smith, D. C. 1979. From extracellular to intracellular: The establishment of a symbiosis. *Proceedings of the Royal Society (London) Series B* 204: 115–130.

Szathmáry, E. 1989. The integration of the earliest genetic information. *Trends in Ecology and Evolution* 2: 200–204.

Taylor, D. L. 1973. The cellular interactions of algal-invertebrate symbiosis. *Advances in Marine Biology* 11: 1–56.

Thompson, J. N. 1988. Variation in interspecific interactions. *Annual Review of Ecology and Systematics* 19: 65–87.

Trench, R. K. 1981. Cellular and molecular interactions in symbioses between dinoflagellates and marine invertebrates. *Pure and Applied Chemistry* 53: 819–835.

Trench, R. K., Boyle, J. E., and Smith, D. C. 1973a. The association between chloroplasts of *Codium fragile* and the mollusc *Elysia viridis*. I. Characteristics of isolated *Codium* chloroplasts. *Proceedings of the Royal Society (London) Series B* 184: 51–61.

Trench, R. K., Boyle, J. E., and Smith, D. C. 1973b. The association between chloroplasts of *Codium fragile* and the mollusc *Elysia viridis*. II. Chloroplast ultrastructure and photosynthetic carbon fixation in *E. viridis*. *Proceedings of the Royal Society (London) Series B* 184: 63–81.

Warren, P. H. 1988. The Structure and Dynamics of a Freshwater Benthic Food Web. D.Phil thesis, University of York.

Woodhead, P. M. J., and Weber, J. N. 1973. The evolution of reef-building corals and the significance of their association with zooxanthellae. In: Ingerson, E., ed., *Proceedings of the Symposium on Hydrogeochemistry and Biogeochemistry*, vol II. Clarke.

Yoo, B. H. 1980. Long-term selection for a quantitative character in large replicate populations of *Drosophila melanogaster*. *Genetical Research Cambridge* 35: 1–17.

6 Symbiosis Inferred from the Fossil Record

David Bermudes and
Richard C. Back

Fossil assemblages are notorious for their many biases. Notably, hard parts are more likely to be preserved than soft ones. The fossil record is largely a history of these hard parts from which past biology must be inferred. There is little direct evidence of symbiosis in the fossil record. Most microbial symbionts do not preserve well, and thus a host and a symbiont are seldom preserved together. Notable exceptions are the fungus-root associations found in Rhynie cherts dating to 400 million years ago (Kidston and Lang 1921; Boullard and Lemoigne 1971). These associations are sometimes interpreted as mutualistic mycorrhizae (Pirozynski and Malloch 1975; Malloch et al. 1980; Pirozynski and Dalpé 1989), although problems remain with interpretations of biological relationships based purely on morphology. Co-preservation insufficiently resolves the question of the influence of a biological relationship, and thus additional information must be sought. Actualistic assumptions have been applied to ecology (Mayr 1963) and genetics (Dobzhansky 1951) for the interpretation of the fossil record and, we believe, should be further extended to incorporate symbiosis. The field of symbiosis has long been a contributor to the understanding of biology (De Bary 1879; Mereschkovskii 1910; Wallin 1927), yet symbiosis remains largely peripheral to contemporary evolutionary theory. Several papers have already discussed the influence of symbiosis on evolutionary history (Taylor 1979, 1983; Cowen 1983; Margulis and Bermudes 1985; Bermudes and Margulis 1987). The purpose of this paper is to identify different ways in which symbiosis may have influenced the fossil record.

New Taxa, Evolutionary Discontinuities, and Rates of Evolutionary Change

Symbionts and hosts engage in numerous types of interactions, including metabolic transfer (Smith 1980; Cook 1983), genetic transfer (Chilton 1983), and the adoption of whole symbionts as organelles (i.e., the origin of plastids and mitochondria through hereditary symbiosis; see Margulis 1981). These interactions often result in emergent biological properties with specific selective advantages. The establishment of a heritable, mutualistic symbiosis has been equated with the origin of neosemes (Bermudes and Margulis 1987) and, it has been suggested, may lead to new taxa by saltation rather than by gradual change. Upon establishment of a symbiosis, natural selection acts on both organisms and on the sum of their properties as a single biological unit. The establishment of the legume/*Rhizobium* symbiosis may be viewed as equivalent to the origin of nitrogen fixation in plants. The success of the legumes includes their symbionts' ability to obtain nitrogen from the atmosphere. The establishment of this symbiosis therefore represents a major evolutionary change.

Several examples of symbiosis suggest that such changes are likely to occur rapidly. Endosymbionts in generative cells of the host are usually heritable, i.e., they are passed from generation to generation cytoplasmically. Examples of heritable symbioses include *Paramecium bursaria*/*Chlorella* (which imparts photosynthetic capability to the host; see Karakashian 1975) and *Paramecium aurelia*/*Caedibacter* (which imparts the killer characteristic to the host; see Pond et al. 1989). These symbioses may be disassociated and reestablished experimentally and have demonstrated that the phenotype endowed by the symbiont is observed immediately. However, as the host and the symbiont have previously interacted, the rate at which the symbiosis is reestablished is undoubtedly higher than the rate of the origin of this symbiosis.

There is a remarkable series of observations and experiments on the establishment of a symbiosis in a protist with no known history of such interaction. Jeon and co-workers (Jeon 1972, 1975, 1983, 1987; Jeon and Ahn 1978; Jeon and Jeon 1976; Lorch and Jeon 1980) have investigated the bacterial parasitism of *Amoeba proteus* and the rapid transition from necrotrophic parasite to obligate heritable symbiont (Jeon, this volume). Initially, amoebae were killed by an unknown species of obligately parasitic bacteria; however, both taxa are now obligate symbionts, and natural selection acts upon them as a unit. The symbiontic association exhibits a new temperature and antibiotic sensitivity and slower growth. One selective advantage of the symbiosis may lie in the ability of the bacteria to kill nonresistant amoebae. This may provide a competitive advantage comparable to the killer trait that *Paramecium* derives from *Caedibacter*.

Obligate heritable intracellular organelles (e.g., mitochondria) have been considered as part of a single organism. When the partners of the amoeba/bacteria symbiosis are considered as a unit, the rate and size of the change are especially striking. The *Amoeba proteus* endosymbiont, if typical of other bacteria, would contain approximately 3000–5000 genes, many of them novel to the amoeba. Furthermore, each gene is amplified by the great numbers of individual bacteria (up to 100,000) infecting each amoeba. Such large and rapid genetic changes have been likened to Goldschmidt's (1933, 1940) concept of a "Hopeful Monster" by Margulis and Bermudes (1985) and are suggestive of one possible mechanism for the discontinuous appearance of species described by Eldredge and Gould (1972) and Gould and Eldredge (1977). Requiring only 200 generations, the evolution of this interaction between amoebae and bacteria from host and parasite to mutually dependent symbionts occurs rapidly, and does not necessarily involve gradual evolution of bacterial genes. Rather, the bacterium is acquired *in toto* as the result of hereditary symbiosis. Hereditary symbiosis, then, by virtue of the establishment of relationships resulting in the acquisition of entire genomes, is by definition a form of macroevolutionary saltation.

Higher taxa are categories of a hierarchical system in which all member species are unified by a fundamental characteristic. Evolutionary changes that distinguish higher taxa are shared by all members of a taxon through evolutionary decent. Therefore, these traits must be established at the origin of, and must be crucial to, that taxon. Many such characteristics can now be reasonably interpreted as the results of separate endosymbioses (e.g., chloroplasts and rhodoplasts; see Raven 1970 and figure 1). Acquisition of a symbiont can be distinguished from a mere correlate or consequence of a speciation event by examination of the importance of that symbiont or symbiosis to that taxon. In the case of a new species, this may be difficult or impossible. However, in the case where a species generates a number of new species, all of which contain the symbiont, and where the symbiont is a distinguishing characteristic and is crucial to the new

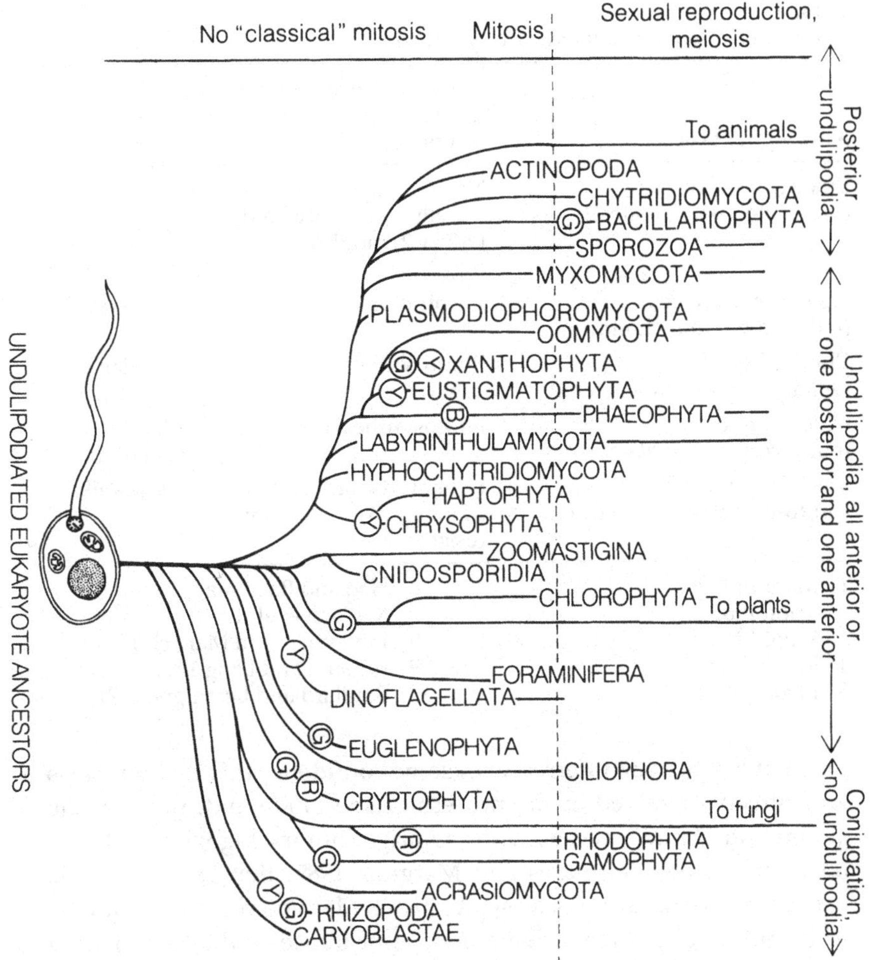

Figure 1
Phyla correlated with the presence of symbiotically derived plastids. The circled letters B,G,R, and Y refer to the acquisition of brown, green, red, and yellow plastids. (From Margulis 1981.)

Table 1
Taxonomic origins correlated with symbiosis.

New higher taxa	Emergent properties (for the host) resulting from symbiosis	In fossil record
Eukaryotes[1,2] (proto-eukaryote + spirochetes)[1]	Nuclear membrane, intracellular motility, and [9(2) + 2] motility organelle[1]	> 1400 Ma[3]
Aerobic eukaryotes[1] (eukaryote + mitochondria)	Respiration	1400 Ma[3]
Rhodophytes[4] (eukaryotes + cyanobacteria)	Photosynthesis	Vendian[5]
Chlorophytes[4] (eukaryotes + *Prochloron*)	Photosynthesis	Middle Cambrian[6]
Vascular plants[7,8] (proto-vascular plant + fungi)	Enhanced water and nutrient uptake, desiccation-resistance	400 Ma[9,10]

1. Margulis 1988
2. Knoll 1983
3. Vidal 1984
4. Raven 1970
5. Enzien 1991
6. Haq and Boersma 1978
7. Malloch et al. 1980
8. Pirozynski and Malloch 1975
9. Kidston and Lang 1921
10. Boullard and Lemoigne 1971

taxon, it can then be inferred with some confidence that the symbiosis was directly involved in the process of speciation and, through the generation of new species, with the origin of new higher taxa. It has been suggested (Bermudes and Margulis 1987) that 28 of 75 phyla (excluding bacteria) possess symbiotically derived organelles unique to or defining the taxon, have their appearance in the fossil record correlated with symbiosis, or consist largely or wholly of symbiotic members. Even if only partly true, this observation would indicate that symbiosis is a major force in evolution, and that it results in the generation of new taxa. Examples of higher taxa correlated with symbiosis are presented in table 1.

Morphology

Morphological changes (especially size) correlated with symbiosis have often been noted (Yonge 1975; Goreau et al. 1979; Lee et al. 1979; Cowen 1983). Here we discuss extant symbioses in Foraminiferida and Hydrozoa as a potential source of changes seen in the fossil record.

The foraminifera, with over 35,000 described species, is one of the most widespread groups of organisms in the ocean. Their wide geological and geographical distribution and their importance in petroleum exploration make them one of the most intensely studied fossil groups (Haq and Boersma 1978). As figure 2 shows, foraminiferans engage in symbioses with diverse photosynthetic microbes (Lee 1980). The symbiotic species of foraminifera tend to be atypically large (up to several centimeters in diameter), late-maturing, and long-lived. Several features of the host's morphology serve to promote the symbiosis. Hosts increase potentially symbiont-containing tissue on the upper side of the body. The calcareous tests of symbiotic forams are divided into many chambers by septa buttressing and supporting the outer wall, which is sufficiently thin to allow light to penetrate. The body is usually flattened, which exposes to light a greater surface area relative to volume. In some genera (*Amphistegina* and *Operculina*), the inner test wall resembles an egg carton, each "cup" holding an individual symbiont (Ross 1974). Symbioses with algae occur in both planktonic and benthic forms. In all cases, the endosymbiont is intracellular within a vacuole. Symbionts are usually localized in the intermediate chambers of the host test, where they may account for 80% of the chamber volume (Lee 1983), and are relatively scarce in the outer chambers, where the food vacuoles are concentrated. In some cases, the symbionts reside on pseudopodia which are displayed outside the test during the day and drawn back in at night (Be et al. 1977).

Foraminifera reproduce by dimorphic alternation of generations. Symbiotic forms evidently suppress sexual reproduction. Asexual reproduction allows direct propagation of the symbiosis. When sexual reproduction occurs, progeny cells must be reinoculated with algae (Rottger 1974). Both members of this symbiosis may live separated from their partner, so the association is by definition facultative. However, in view of the oligotrophic environment in which these organisms occur (Lee 1980), the association may be ecologically obligate.

Nutritional interaction between host and symbiont is believed to occur. Foraminifera are heterotrophs, feeding on a variety of microorganisms by means of pseudopodia and digestive lysosomes, but symbiotic forms survive in the absence of prey if provided with ample light (Hallock 1981; McEnery and Lee 1981). Both aposymbiotic and nonsymbiotic forms perish under such conditions. Although ex-

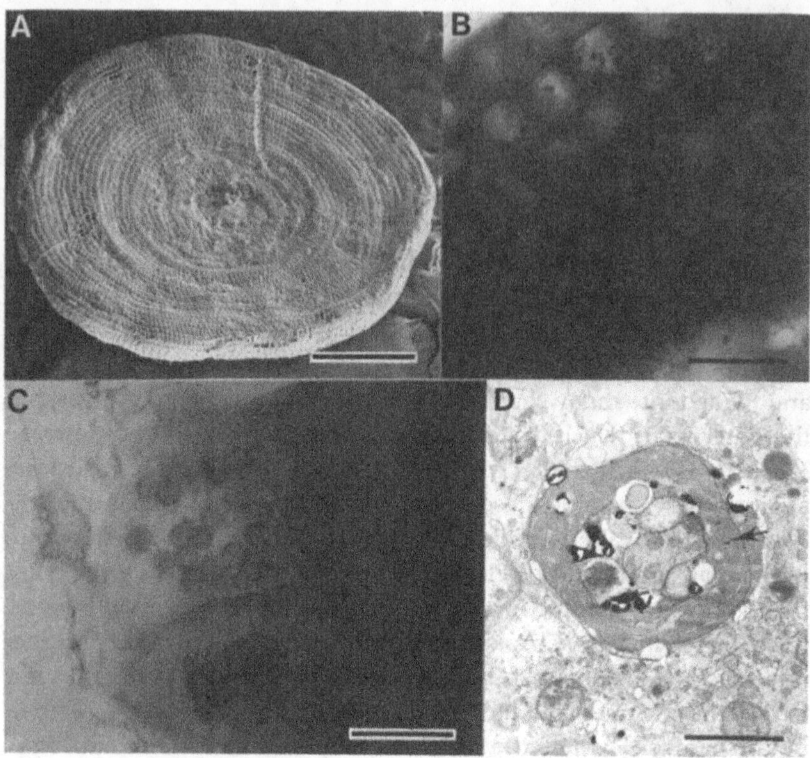

Figure 2
Forams containing photosynthetic symbionts. (Courtesy of J. J. Lee, City College, New York.) (A) Scanning electron micrograph of *Amphisorus hemprichii*. Bar = 1 mm. (B) Light micrograph of *Amphisorus hemprichii* containing dinomastigote (dinoflagellate) symbionts. Bar = 0.2 mm. (C) High-magnification light micrograph of dinomastigote-containing chambers viewed through the test. Bar = 50 μm. (D) Transmission electron micrograph of the dinomastigote endosymbiotic in *Amphisorus hemprichii* showing photosynthetic lamellae (arrow). Bar = 5 μm.

perimental evidence is lacking, it is presumed that inorganic nutrients from host tissue are cycled back to the symbiont. The only external requirements of the mixotroph may be vitamins such as biotin and thiamine. Reproduction is dependent upon an adequate supply of vitamin B12 (Lee et al. 1979), which is presumably derived from ingested food particles.

Increased rates of calcification in symbiotic foraminifera (Lee and Zucker 1969; Duguay and Taylor 1978) account for their large size. As in corals (see below), photosynthesis by the symbiont aids calcification. Increased calcification in foraminifera strengthens the shell, making it more resistant to physical damage. There is also evidence that increased size reduces the rate of mortality (Hallock 1985), indicating increased resistance to predation. This is particularly evident in asexually reproducing individuals. Furthermore, since asexual reproduction is by schizogony, increased parental cell size may increase fecundity.

Scleractinian (reef-building) corals, which are exclusively symbiotic with dinomastigotes (dinoflagellates, zooxanthellae), cover a vast area and provide a specific and important habitat for numerous other organisms. Individual coral polyps are millimeters to centimeters in length; the coral colony can exceed several meters in diameter. The endosymbiotic dinomastigotes reside in the gastrodermal tissue of the host, enclosed in the host's vacuolar membrane. Although the host coral is an active predator and may be grown aposymbiotically, the symbiosis is considered ecologically obligate. Virtually all healthy reef-building corals possess symbiotic dinomastigotes. These dinomastigotes, generally members of the genus *Symbiodinium*, are found in the environment but are not considered to be permanently free-living (Taylor 1973).

The symbiosis between coral and dinomastigotes is nutritional. Photosynthetically fixed carbon, mostly as glycerol, can be recovered from the host's cytoplasm (Muscatine 1967). After a 24-hour incubation in ^{14}C-labeled bicarbonate, 32–45% of the carbon tracer is found in the animal (Muscatine and Cernichiari 1969). Symbiotic corals release less phosphorus and ammonia than do aposymbiotic and nonsymbiotic corals (Cook 1983). These nutrients are apparently transferred to the symbionts (Muscatine 1980).

In symbiotic corals, an important consequence of photosynthesis by *Symbiodinium* is increased calcification. Rates of calcification, measured as ^{45}Ca incorporation, is from 3 to 25 times higher in the light than in the dark or in the presence of the photosynthesis inhibitor DCMU (3-(3,4-dichlorophenyl)-1,1 dimethyl urea) (Goreau 1959, 1961; Pearse and Muscatine 1971). Rates of calcification are also higher than in aposymbiotic or nonsymbiotic corals, and this may help to explain the great abundance of these organisms. Coral bleaching (i.e., expulsion of their symbionts) may have far-reaching ecological consequences, *viz.*, breakdown of the reef-building process (Roberts 1987).

Symbiosis in scleractian corals is believed to have evolved during the late Triassic and may have contributed to their dramatic speciation in the Mesozoic (Stanley 1981). From that time, coral remains show a general ecological trend from non-reef-building in deep waters to reef-building in shallow, photic environments.

Examples of symbioses and their correlations with morphology are presented in table 2.

Ecological and Geographic Distribution

Light is considered to be the primary factor in determining ecological and geographical limits to algal symbioses (Cowen 1983). Distribution of mixotrophs is generally limited to shallow waters or waters of low turbidity (e.g., waters with low inputs of clastic sediments, little resuspension of sediments, and no phytoplankton blooms), which occur particularly in the tropics (where in addition there is little or no seasonality of light supply). These conditions usually correlate with oligotrophic settings, where the selective advantages of transfer of photosynthetic metabolites are the greatest.

Examples of ecological and geographic correlations with symbiosis occur in the foraminifera/alga and coral/*Symbiodinium* associations (discussed above). Exposure to light is critical in promoting these symbioses, and it constrains both the geographical distribution and the anatomy of the host. The large foraminifera containing symbiotic algae evolved independently several times during the Phanerozoic. Symbioses appear to have evolved in the suborder Fusulina during the late Paleozoic (Carboniferous-Permian), and in the suborders Miliolina, Textularina, and Rotalina during the late Mesozoic (Jurassic-Cretaceous) and in the Cenozoic (Hallock 1982).

Occasional transgression of the seas onto the continental masses produce enlarged shelf areas and shallow epicontinental seas. Transgressions, tending to contribute to oligotrophic conditions, reduce in-

Table 2
Symbioses effecting morphology.

Symbiosis	Emergent property	Fossil record
Coral/*Symbiodinium*	Increased colonial size and abundance (see text)[1,2]	Radiation of reef-builders in late Triassic[3]
Foram/algae	Increased size, complex test morphology to accomodate symbiont (see text)[4]	Large forams (suborder Fusulina in late Paleozoic suborders Miliolina, Textularina and Rotalina in late Mesozoic and Cenozoic)[5]
Brachiopods/algae	Increased size, morphological adaptation to accomodate symbiont unusual feeding mechanism[6]	Superfamilies Richthofeniacea and Lyttoniacea during the Permian with these characters[6]
Tridacna/ *Symbiodinium*	Increased size, morphological adaptation for photosynthesis (hinge on bottom)[7]	*Tridacna* with hinge on bottom of shell in late Triassic[7]
Rudist bivalve/algae	Increased size, extreme morphological modification for exposure of symbiont-bearing tissue to light[8]	Dominant reef-builders in late Cretaceous[8]
Mycorrhizae (vasular plants and fungi)	Increased plant size,[9] extreme morphological differentiation of underground organs (roots) to accomodate symbionts[10]	Devonian mycorrhizae[10]
Lichens (fungi + chlorophyte or cyanobacteria)	Curstose, foliose, and fruticose morphology[11]	Mesozoic and Miocene lichen-like structures[12]
Large (over 5 kg) herbivores (e.g., ruminants, hippopotami, or kangaroos[13] and dinosaurs[14] + fermenting bacteria)[13,14]	Forestomach,[15] large size,[14,15] hindgut?[14]	Large mammalian herbivores in the Cenozoic, large herbivorous dinosaurs in the Mesozoic[14]

1. Goreau 1959, 1961
2. Pearse and Muscatine 1971
3. Stanley 1981
4. Lee et al. 1979
5. Hallock 1982
6. Cowen 1970
7. Yonge 1975
8. Cowen 1983
9. Harley and Smith 1983
10. Pirozynski and Dalpé 1989
11. Ahmadjian 1967
12. Smith 1927
13. Moir 1968
14. Farlow 1987
15. McBee 1971

Table 3
Symbioses correlated with ecological and geographic distribution.

Host	Symbiont	Emergent properties	Fossil record
Paramecium bursaria	*Chlorella*	Phototactic behavior (photo-accumulation)[1,2]	None
Vascular plants	Fungi	Enhanced water and nutrient uptake,[3] colonization of land[4,5]	Devonian fungus-root associations in terrestrial environments[8]
Corals	*Symbiodinium*	Ability to inhabit oligotrophic environment[8]	Late Triassic scleractinian shift to photic environments[8]
Sponges	Cyanobacteria	Ability to inhabit oligotrophic environments[9–13]	Precambrian (650 Ma) sponges in oligotrophic environments[13]
Ascomycetes	*Trebouxia*	Enhanced stress resistance, colonization of extreme environments[14,15]	Mesozoic and Miocene lichen-like fossils in terrestrial environments[16]

1. Neiss et al. 1981
2. Reisser and Hader 1984
3. Harley and Smith 1982
4. Pirozynski and Malloch 1975
5. Malloch et al. 1980
6. Kidston and Lang 1921
7. Boullard and Lemoigne 1971
8. Stanley 1981
9. Wilkinson 1987a–c
10. Wilkinson and Vacelet 1979
11. Wilkinson 1983
12. Ziegler and Reitschel 1970
13. Finks 1970
14. Ahmadjian 1967
15. Smith 1975
16. Smith 1927

puts of terrigenous sediments. Transgressions correlate with repeated radiations of symbiotic foraminifera and corals (Hallock 1982, 1985). During periods of regression, retreating seas brought terrigenous clastic sediments and nutrients into the oceanic basin, and the continental shelf habitat was greatly reduced (Newell 1967). Regression was accompanied by global cooling and by increased convection and upwelling in the oceans, which correlates with recurring extinctions of symbiotic foraminifera (Hallock 1985) and corals (Stanley 1981). The fossil record reveals parallel evolutionary trends in larger foraminifera and hermatypic corals, suggesting iterative selection for phototrophic symbioses (Lee et al. 1979).

Examples of symbioses correlated with ecology and geographic distribution are presented in table 3.

Physiology and Chemical Composition

Physiological influences of symbiosis may also be preserved in the fossil record. Morphology provides indirect records of physiological changes resulting from symbiosis. A more direct indication of physiological changes correlated with symbiosis is the fractionation of stable isotopes. Stable isotopes of biologically active elements (e.g., ^{13}C, ^{18}O, ^{15}N, ^{2}H, and ^{34}S) are preferentially segregated by certain metabolic processes. Isotopic fractionation of carbon in photoautotrophs occurs because of discrimination by the carboxylating enzyme involved in CO_2 fixation (O'Leary 1981). This leads to differences in stable-isotope composition between different algal groups and between plants with different photosynthetic metabolisms (e.g., CAM, C-3, and C-4 pathways). Isotopic fractionation in consumers derives from differences in the isotopic composition of food. The ratios of stable isotopes have been interpreted as indicating photosynthesis in primary producers (Berry 1989), bacterial nitrogen fixation in plants (Shearer and Kohl 1986), and particular diets ingested by consumers (Fry and Sheer 1984). Thus, the analysis of stable isotopes may also aid in determining symbiosis in the fossil record.

Increased rates of calcification in symbiotic foraminifera may result in altered $\delta^{13}C$ ratios (Erez 1978). Spero and Deniro (1987) found that carbon fractionation in postgametic foraminifera correlated strongly with light exposure and the presence of algal symbionts. If the host derives a significant portion of its carbon from the symbiont, the host's tissues and skeleton (if built from the same pool) will carry a similar isotopic signature.

In deep-sea cold seeps (figure 3), methane serves as an electron donor for chemosynthetic bacteria symbiotic with bivalve mollusks (Childress et al. 1986). $\delta^{13}C$ values for methane-derived carbonates from these environments are low (Kulm et al. 1986). Findings of fossilized 140-million-year-old cold seeps (Beauchamp et al. 1989) containing serpulid worms and bivalves, the latter with $\delta^{13}C$ values similar to those of bivalves symbiotic with methylotrophs in extant communities, suggest the presence of symbiosis within these fossils.

Examples of physiological and chemical compositional changes associated with symbiosis are presented in table 4.

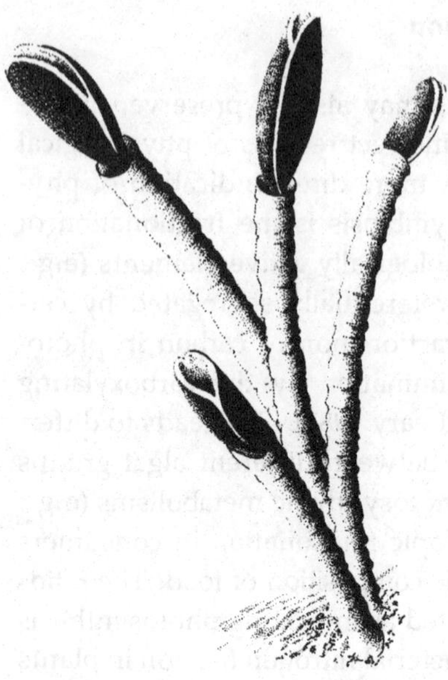

Figure 3
Hydrothermal-vent tube worms containing sulfide-oxidizing bacteria. Some members of fossil vent communities have been found to enriched in ^{12}C; this is consistent with the presence of chemosynthetic bacteria. Drawing by Christie Lyons.

Table 4
Physiological/compositional correlates of symbiosis.

Host	Symbiont	Emergent property	Observations in fossil record
Foraminifera (*Orbulina universa*)	Dinomastigote	$\delta^{13}C$ (enrichment of ^{13}C[1])	?
Legumes	*Rhizobium*	$\delta^{15}N$[2] (lower ^{15}N value in symbiotic plants)	?
Cold-seep bivalves	Chemosynthetic bacteria[3]	$\delta^{13}C$ (enrichment of ^{12}C)[4]	140 Ma similar fossils and fractionation values[5]
Scleractinian corals	*Symbiodinium*	$\delta^{13}C$ (enrichment of ^{12}C)[6]	?

1. Spero and Deniro 1987
2. Shearer and Kohl 1986
3. Childress et al. 1986
4. Kulm et al. 1986
5. Beauchamp et al. 1989
6. Erez 1978

Conclusion

Most reconstructions and interpretations of fossil organism assemblages and paleo-environments must be based on studies of extant organisms. We extrapolate from extant symbioses, inferring what is most likely to have occurred. Indirect evidence pertaining to several important fossil groups is consistent with the hypothesis that symbiosis has contributed substantially to the origin of new taxa, correlated with changes in the morphology, the ecology, the geographic distribution, and the physiology of various organisms. As these influences tend to be interrelated, multiple effects of symbiosis are often observed. In foraminiferans, for example, the effect of symbiosis on physiology apparently led to changes in carbon-isotope composition, in morphology, in ecology, and in geographic distribution. Criteria for the importance of a symbiosis *a posteriori* have generally aided in this assessment, because symbionts are rarely preserved with their host. Given the pervasive phylogenetic, ecological, and physiological influences of symbiosis, we believe that interpretation of evolutionary history requires further incorporation of observations from symbiosis research.

Acknowledgments

This work was supported by National Institute of Environmental Health Sciences National Service Research Award ES 07043 (for D.B.) and by a University of Wisconsin-Milwaukee Graduate School Dissertation Fellowship (for R.C.B.). We wish to thank M. Boraas, M. Enzien, L. Margulis, K. H. Nealson, P. M. Sheehan, P. Strother, F. J. R. Taylor, and R. K. Trench for helpful conversations.

References

Ahmadjian, V. 1967. *The Lichen Symbiosis*. Blaisdell.

Be, A. W., Heinleben, C., Anderson, O. R., Spindler, M., Hacunda, J., and Tuntivate-Choy, S. 1977. Laboratory and field observations of living planktonic foraminifera. *Micropaleontology* 23: 155–179.

Beauchamp, B., Krouse, H. R., Harrisón, J. C., Nassichuk, W. W., and Eliuk, L. S. 1989. Cretaceous cold-seep communities and methane-derived carbonates in the Canadian Arctic. *Science* 244: 53–56.

Bermudes, D., and Margulis, L. 1987. Symbiont acquisition as neoseme: Origin of species and higher taxa. *Symbiosis* 4:185–198.

Berry, J. A. 1989. Studies of mechanisms affecting the fractionation of carbon isotopes in photosynthesis. In: Rundel, P. W., Ehleringer, J. R., and Nasy, K. A., eds., *Isotopes in Ecological Research*. Springer-Verlag.

Boullard, B., and Lemoigne, Y. 1971. Les champignons endophytes du *Rhynia gwynne-vaughanii* K. et L. *Botaniste* 54: 49–89.

Childress, J. J., Fisher, C. R., Brooks, J. M., Kennicutt M. C., II, Bidgare, R., and Anderson, A. E. 1986. A methanotrophic marine molluscan (Bivalvia, Mytilidae) symbiosis: Mussels fueled by gas. *Science* 233: 1306–1308.

Chilton, M.-D. 1983. A vector for introducing new genes into plants. *Scientific American* 248(6): 50–59.

Cook, C. B. 1983. Metabolic interchange in algae-invertebrate symbioses. *International Review of Cytology* Suppl. 14: 177–210.

Cowen, R. 1970. Analogies between the recent bivalve *Tridacna* and the fossil brachiopods *Lyttoneacea* and *Richthofeniacea*. *Palaeogeo*. *Palaeoclimat*. *Palaeoecol*. 8: 329–344.

Cowen, R. 1983. Algal symbiosis and its recognition in the fossil record. In: Tevesz, M. J. S., and McCall, P. L., eds., *Biotic Interactions in Recent and Fossil Benthic Communities*. Plenum.

De Bary, A. 1879. Die Erscheinung der Symbiose. Tagebl. 51, Vers. Deut. Naturforscher und Aerzte zu Cassel, p. 121.

Dobzhansky, T. 1951. *Genetics and the Origin of Species*, third edition, revised. Columbia University Press.

Duguay, L. E., and Taylor, D. L. 1978. Primary production and calcification by the soritid foraminifer *Archais angulatus* (Fichtel and Moll). *Journal of Protozoology* 25: 356–361.

Eldredge, N., and Gould, S. J. 1972. Punctuated equilibria: An alternative to phyletic gradualism. In: Schopf, T. J. M., ed., *Models in Paleobiology*. Freeman, Cooper and Co.

Enzien, M. 1991. Cyanobacteria or Rhodophyta? Interpretation of a Precambrian microfossil. *BioSystems* (in press).

Erez, J. 1978. Vital effect on stable-isotope composition seen in foraminifera and coral skeletons. *Nature* 273: 199–202.

Farlow, J. O. 1987. Speculation about the diet and digestive physiology of herbivorous dinosaurs. *Paleobiology* 13: 60–72.

Finks, R. H. 1970. The evolution and ecologic history of sponges during Palaeozoic times. *Symposium of the Zoological Society of London* 25: 3–22.

Fry, B., and Sheer, E. B. 1984. $\delta^{13}C$ measurements as indicators of carbon flow in marine and freshwater ecosystems. *Contributions in Marine Science* 27: 13–47.

Goldschmidt, R. 1933. Some aspects of evolution. *Science* 78: 539–547.

Goldschmidt, R. 1940. *The Material Basis of Evolution*. Yale University Press.

Goreau, T. F. 1959. The physiology of skeleton formation in corals. I. A method for measuring the rate of calcium deposition by corals under different conditions. *Biological Bulletin* 116: 59–75.

Goreau, T. F. 1961. Problems of growth and calcium deposition in reef corals. *Endeavor* 20: 32–39.

Goreau, T. F., Goreau, N. I., and Goreau, T. J. 1979. Corals and coral reefs. *Scientific American* 241(2): 124–136.

Gould, S. J., and Eldredge, N. 1977. Punctuated equilibria: The tempo and mode of evolution reconsidered. *Paleobiology* 3: 115–151.

Hallock, P. 1981. Light dependence in *Amphistegina*. *J. Foram. Res.* 11: 40–46.

Hallock, P. 1982. Evolution and extinction in larger foraminifera. *Proc. 3rd N. Am. Paleontol. Conv.* 1: 221–225.

Hallock, P. 1985. Why are larger foraminifera large? *Paleobiology* 11: 195–208.

Harley, J. L., and Smith, S. E. 1983. *Mycorrhizal Symbiosis*. Academic Press.

Haq, B. U., and Boersma, A. 1978. *Introduction to Marine Micropaleontology*. Elsevier.

Jeon, K. W. 1972. Development of cellular dependence in infective organisms: Micrurgical studies in amoebas. *Science* 176: 1122–1123.

Jeon, K. W. 1975. Selective effects of enucleation and transfer of heterologous nuclei on cytoplasmic organelles in *Amoeba proteus*. *Journal of Protozoology* 22: 402–405.

Jeon, K. W. 1983. Integration of bacterial endosymbions in amoebae. *International Review of Cytology* Suppl. 14: 29–47.

Jeon, K. W. 1987. Change of cellular "pathogens" into required cell components. *Annals of the New York Academy of Sciences* 503: 359–371.

Jeon, K. W., and Ahn, T. I. 1978. Temperature sensitivity: A cell character determined by obligate endosymbionts in amoebae. *Science* 202: 635–637.

Jeon, K. W., and Jeon, M. S. 1976. Endosymbiosis in amoeba: Recently established endosymbionts have become required cytoplasmic components. *Journal of Cell Physiology* 89: 337–344.

Karakashian, M. W. 1975. Symbiosis in *Paramecium bursaria*. *Symposium of the Society for Experimental Biology* 29: 229–265.

Kidston, R., and Lang, H. W. 1921. On old red sandstone plants showing structure, from the Rhynie chert bed, Aberdeenshire. Part V. The thallophyta occurring in the peat-bed; the succession of the plants throughout a vertical section of the bed, and the conditions of accumulation and preservation of the deposit. *Transactions of the Royal Society of Edinburgh* 52: 855-902.

Knoll, A. H. 1983. Biological interactions and Precambrian eukaryotes. In: Tevesz, M. J. S., and McCall, P. L., eds., *Biotic Interactions in Recent and Fossil Benthic Communities*. Plenum.

Kulm, L. D, Suess, E., Moore, J. C., Carson, B., Lewis, B. T., Ritger, S. D., Kadko, D. C., Thornburg, T. M., Embley, R. W., Rugh, W. D., Massoth, G. J., Langseth, M. G., Cochrane, G. R., and Scamman, R. L. 1986. Oregon subduction zone: Venting, fauna, and carbonates. *Science* 231: 561–566.

Lee, J. J. 1980. Nutrition and physiology of the foraminifera. In: Levandowsky, M., and Hunter, S. H., eds., *Biochemistry and Physiology of Protozoa*, vol. 3, 2nd edition. Academic Press.

Lee, J. J. 1983. Perspective on algal endosymbionts in larger foraminifera. *International Review of Cytology Supplement* 14: 49–78.

Lee, J. J., McEnery, M. E., Kahn, E. G., and Schuster, F. L. 1979. Symbiosis and the evolution of larger foraminifera. *Micropaleontology* 25: 118–140.

Lee, J. J., and Zucker, W. 1969. Algal flagellate symbiosis in the foraminifer *Archaias*. *Journal of Protozoology* 16: 71–81.

Lorch, I. J., and Jeon, K. W. 1980. Resuscitation of amoebae deprived of essential symbiotes: Microsurgical studies. *Journal of Protozoology* 27: 423–426.

Malloch, D. W., Pirozynski, K. A., and Raven, P. H. 1980. Ecological and evolutionary significance of mycorrhizal symbioses in vascular plants (a review). *Proceedings of the National Academy of Sciences* 77: 2113–2118.

Margulis, L. 1981. *Symbiosis in Cell Evolution*. Freeman.

Margulis, L. 1988. Serial endosymbiotic theory (SET): Undulipodia, mitosis and their microtubule systems preceded mitochondria. *Endocytobiosis and Cell Research* 5: 133–162.

Margulis, L. and Bermudes, D. 1985. Symbiosis as a mechanism of evolution: Status of cell symbiosis theory. *Symbiosis* 1: 101–124.

Mayr, E. 1963. *Animal Species and Evolution*. Harvard University Press.

McBee, R. H. 1971. Significance of intestinal microflora in herbivory. *Annual Review of Ecology and Systematics* 2: 165–176.

McEnery, M. E., and Lee, J. J. 1981. Cytological and fine structural studies of three species of symbiont-bearing larger foraminifera from the Red Sea. *Micropaleontology* 27: 71–83.

Mereschkovskii, K. 1910. Theorie der zwei Plasmaarten als Grundlage der Symbiogenesis, einer neuen Lehre von der Entstehung der Organismen. *Biologische Centralblatt* 30: 352–367.

Moir, R. J. 1968. Ruminant digestion and evolution. *Handbook of Physiology,* Section 6: 2673–2694.

Muscatine, L. 1967. Glycerol excretion by the symbiotic algae in corals and *Tridacna* and its control by the host. *Science* 156: 516–519.

Muscatine, L. 1980. Uptake, retention and release of dissolved inorganic nutrients by marine algal-invertebrate associations. In: Cook, C. B., Pappas, P. W., and Rudolf, E. D. eds., *Cellular Interactions in Symbiosis and Parasitism.* Ohio State University Press.

Muscatine, L., and Cernichiari, E. 1969. Assimilation of photosynthetic products of zooxanthellae by a coral reef. *Biological Bulletin* 137: 506–523.

Newell, N. D. 1967. Revolutions in the history of life. *Geol. Soc. Amer. Spec. Paper* 89: 63–91

Neiss, D., Reisser, W., and Weissner, W. 1981. The role of endosymbiotic algae in photoaccumulation of green *Paramecium bursaria. Planta* 152: 268–271.

O'Leary, M. H. 1981. Carbon isotope fractionation in plants. *Phytochemistry* 20: 553–567.

Pearse, V. B., and Muscatine, L. 1971. Role of symbiotic algae (zooxanthellae) in coral calcification. *Biological Bulletin* 141: 350–363.

Pirozynski, K. A., and Dalpé, Y. 1989. Geological history of the Glomaceae with particular reference to mycorrhizal symbiosis. *Symbiosis* 7: 1–36.

Pirozynski, K. A., and Malloch, D. W. 1975. The origin of land plants: A matter of mycotrophism. *BioSystems* 6: 153–164.

Pond, F. R., Gibson, I., Lalucat, J., and Quackenbush, R. L. 1989. R-body-producing bacteria. *Microbiological Reviews* 53: 25–67.

Raven, P. H. 1970. A multiple origin for plastids and mitochondria. *Science* 169: 641–646.

Reisser, W., and Hader, D.-P. 1984. Role of endosymbiotic algae in photokinesis and photophobic responses of ciliates. *Photochemistry and Photobiology* 39: 673–678.

Roberts, L. 1987. Coral bleaching threatens Atlantic reefs. *Science* 238: 1228–1229.

Ross, C. A. 1974. Evolutionary and ecological significance of large calcareous Foraminiferida (Protozoa), Great Barrier Reef. *Proceedings of the 2nd International Coral Reef Symposium* 1: 327–333.

Rottger, R. 1974. Larger foraminifera: Reproduction and early stages in *Heterostegina depressa*. *Marine Biology* 26: 5–12.

Shearer, G., and Kohl, D. H. 1986. N_2-fixation in field settings: Estimations based on natural ^{15}N abundance. *Australian Journal of Plant Physiology* 13: 699–756.

Smith, A. L. 1927. *Lichens*. Cambridge University Press.

Smith, D. C. 1975. Symbiosis and the biology of lichenised fungi. In: Jennings, D. H., and Lee, D. L., eds., *Symbiosis*. Symposia of the Society for Experimental Biology XXIX. Cambridge University Press.

Smith, D. C. 1980. Mechanisms of nutrient movement between lichen symbionts. In: Cook, C. B., Pappas, P. W., and Rudolph, E. D., eds., *Cellular Interactions in Symbiosis and Parasitism*. Ohio State University Press.

Spero, H. J., and Deniro, M. J. 1987. The influence of symbiont photosynthesis on $\delta^{13}C$ and $\delta^{13}C$ values of planktonic foraminiferal shell calcite. *Symbiosis* 4: 213–228.

Stanley, G. D., Jr. 1981. Early history of scleractinian corals and its geological consequences. *Geology* 9: 507–511.

Taylor, D. L. 1973 . The cellular interactions of algal-invertebrate symbiosis. *Advances in Marine Biology* 11: 1–56.

Taylor, F. J. R. 1979. Symbionticism revisited: A discussion of the evolutionary impact of intracellular symbioses. *Proceedings of the Royal Society of London*, Series B 204: 267–286.

Taylor, F. J. R. 1983. Some eco-evolutionary aspects of intracellular symbiosis. *International Review of Cytology* Supplement 14: 1–28.

Vidal, G. 1984. The oldest eukaryotic cells. *Scientific American* 250 (2): 48–57.

Wallin, I. E. 1927. *Symbionticism and the Origin of Species*. Williams and Wilkins.

Wilkinson, C. R. 1983. Net primary productivity in coral reef sponges. *Science* 219: 410–410.

Wilkinson, C. R. 1987a. Significance of microbial symbionts in sponge evolution and ecology. *Symbiosis* 4: 135–146.

Wilkinson, C. R. 1987b. Productivity and abundance of large sponge populations on Flinders Reef flats, Coral Sea. *Coral Reefs* 5: 183–188.

Wilkinson, C. R. 1987c. Interocean differences in size and nutrition of coral reef sponge populations. *Science* 236: 1654–1657.

Wilkinson, C. R., and Vacelet, J. 1979. Transplantation of marine sponges to different conditions of light and current. *Journal of Experimental Marine Biology and Ecology* 37: 191–104.

Yonge, C. M. 1975. Giant clams. *Scientific American* 232(4): 96–105.

Ziegler, B., and Rietschel, S. 1970. Phylogenetic relationships of fossil calci-sponges. *Symposium of the Zoological Society of London* 25: 23–40.

II Microbial Symbioses

7 Bacterial Evolution without Speciation

Sorin Sonea

All of biology has been plagued by the lack of a modern concept of bacteria (Wilson and Miles 1975). Too many views have been heavily tinted by a eukaryotic bias, particularly in the case of evolution. A more appropriate and coherent approach to the bacterial world is now in progress (Campbell 1988; Reanney 1976; Reanney et al. 1982; Sonea 1971, 1987; Sonea and Panisset 1976, 1983). The important role of temporary, adaptable symbioses as a bacterial way of life has recently been described (Sonea 1988a,b). During the 2 billion years of evolution preceding the origin of eukaryotes, these symbioses left a decisive mark on bacteria, and subsequently on the entire biosphere. This paper describes the probable impact of bacterial symbioses on evolution, from the beginning of life to the conquest of the continents by terrestrial forms of life. Comparing the effect of symbioses on bacteria (as temporary, recurring phenomena) with that on eukaryotes (as permanent, irreversible phenomena) may offer a clearer view of the entire origin and evolution of organisms and species.

How Symbioses Originated and Evolved in Bacteria

The divisions of the first cells on Earth probably did not result in two identical copies, and therefore in many offspring some enzymes could not be synthesized. Figure 1 shows that a possible way to correct this problem was by having the cell's genes grouped, a few hundred at a time, in self-replicating DNA molecules (replicons); each replicon could have been represented by multiple copies within the cell. The probability of a daughter cell's not inheriting at least one specimen from each type of replicon would thus have been diminished. If this did happen, however, it might have been corrected by two mechanisms. A surplus copy of a replicon could cross into a cell

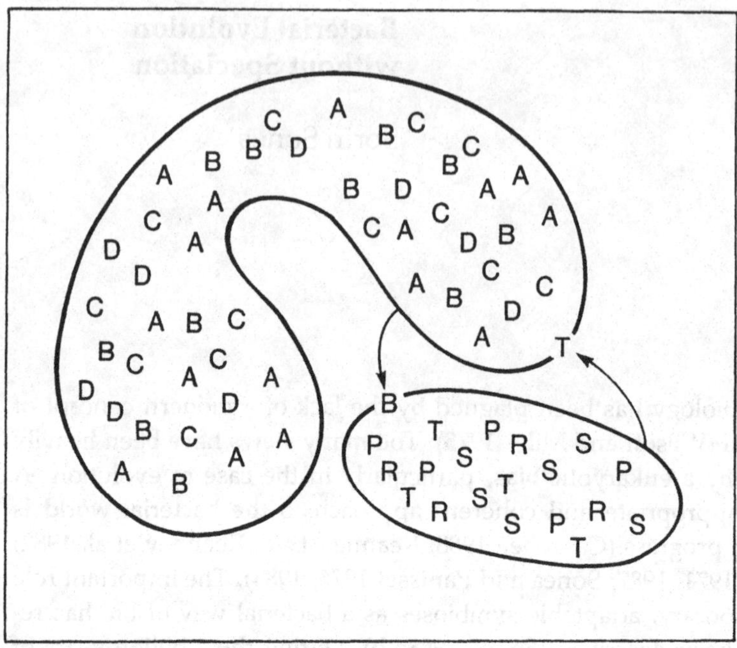

Figure 1
Probable gene transmissions in early bacteria. A cell containing a few copies of each of its middle-size or small replicons (A, B, C, D, etc.) divides them unequally in the two daughter cells. One of the latter is receiving (by transfection) one copy of replicon T from another type of cell, to which it transmits one of its B replicons.

in which it was needed; a similar mechanism (transfection) allows gene exchanges between different strains in nature. Complementary metabolism may have been another way of compensating for a missing enzyme; the product of such an enzyme which was synthesized in a neighboring cell could cross into cells unable to produce it. This is a common mechanism in mixed microbial communities (Bull and Slater 1982; Cohen and Rosenberg 1989). Bacteria seem to have constantly used both gene exchanges and associations of metabolically complementary strains as mechanisms of reciprocal support.

During the first 100 million years of life on Earth, many new genes were synthesized by random mutations. These genes were probably exchanged between strains, and therefore cellular variation in early bacteria increased by geometrical progression with the help of horizontal gene transfer (Anderson 1966; Sonea 1987; Sonea and Panisset 1983). Among these bacteria, the replicon containing the genes responsible for division seems to have gradually accumulated all the essential genes for bioenergetic activity, ending up as the only stable replicon found today in a bacterium: the large replicon, also called the "chromosome." It is probable that nonessential genes were progressively concentrated in smaller replicons and therefore became more adapted to gene exchange. Their offspring, prophages and plasmids, have constantly offered temporary help to bacterial cells (Bergh et al. 1989; Gunsalus et al. 1975; Reanney 1976; Richmond 1970; Sonea 1971, 1987; Sonea and Panisset 1976). The prophages appeared very early (Ackermann and DuBow 1987), along with their self-transmissible forms, the temperate phages, which are still erroneously considered by many to be viruses. Unlike viruses, temperate phages actively participate in possibly useful gene exchanges among bacterial cells (Sonea 1987). Their role in helping pathogenic bacteria, by transduction, to resist antibacterial drugs has been well established in the last 30 years (Lacey and Kruczenik 1986; Lion and Skurray 1987; Richmond and John 1964; Skurray et al. 1988). There are strong indications that prophages and their temperate phages play an important role in the soil (Reanney et al. 1982), and also in gene exchanges in the oceans (Bergh et al. 1989). Prophages may be found in any bacterial category, including Archaeobacteria (Ackermann and DuBow 1987; Zillig et al. 1988). Practically every bacterial strain in nature may be visited successively by tens of types of prophages (Sonea 1987). A small proportion of plasmids can transmit themselves to different strains; the others may be passively transferred by transfection, transduction, or conjugation (Broda 1979).

All these small replicons carry "converting" genes (Richmond 1970), which the "visited" cells may find useful. Small replicons can be eliminated (curing) when their presence is no longer beneficial; they are therefore "disposable" (Karska-Wysocki and Sonea 1973). In any niche, selective pressures favor the appropriate, adaptable combination of a cell with a chosen assortment of small replicons. In nature, any bacterium contains one or more small replicons on a temporary basis. Thus, the bacterial world could benefit increasingly from a flow of genes among its subunits. Phyletically distant from the cell's stable genes, the visiting small replicons may be considered as temporary endosymbionts. Since exchange among strains may occur frequently and happens inside the global genetic bacterial pool, it might also be considered a peculiar form of reversible sexuality. Bacteria solved many different problems by finding the appropriate endosymbioses for different circumstances. Obtaining resistance to antibacterial drugs (Datta 1975; Imanaka et al. 1981; Mitsuhashi et al. 1963; Watanabe 1963), heavy metals (Novick and Roth 1968; Richmond and John 1964; Schütt 1989; Smith 1967), and toxic industrial waste (Broda 1979) are the best-known examples. Moreover, the capacity of pathogenic bacteria to invade animals or plants is often associated with the existence of such temporary endosymbioses involving prophages or plasmids (Ackermann and DuBow 1987; Bishai and Murphy 1988; Broda 1979; Dobardzic and Sonea 1971).

Bacterial problems are usually solved by temporary symbioses in a computer-like way. The common genetic pool is the information bank, and the ever-present selective pressures serve as "engines" and selection systems. The right answers are found for different problems when genes arriving in different cells of a strain, or in a newly arrived strain of a community, are able to add suitable information. The bacterium or entire community which receives these genes will function better and multiply more than others. The capacity of cells and small replicons to multiply allows for any needed amplification (Sonea 1987, 1988a,b; Sonea and Panisset 1976, 1983). Biological circuits have also been suggested (Reanney et al. 1982).

At the surface of any bacterium lie copies of several kinds of receptors for conjugation pili and the "tails" of temperate phages. These receptors specifically attract self-transmissible small replicons and help them to penetrate a receiving cell. A bacterium is thus a two-way broadcasting station for hereditary information, and all bacterial strains are "solicited" by visiting genes to form endosymbioses. As already mentioned, entire cells may also "solicit" bacterial communities, and, if useful, they may end up as members of an improved temporary ectosymbiosis. Therefore, bacterial cells and communities are opportunistic chimeras, symbioses of basic elements.

The bacterial world has evolved in a clonal way from the first ancestral cell, and its genes are available to all bacteria. Its evolution presents a remarkable continuity, and may be traced from the clumsy clone at the start of life on Earth to a complex entity with a large variety of bioenergetic processes in its subunits (cells and small replicons), which are thus able to produce the most appropriate temporary symbioses.

The dispersed global bacterial organism seems to be the only bacterial entity that is genetically and functionally stable, although its constituents may be often reshuffled. This dispersed super-organism has a rich and apparently chaotic structure; however, it performs complex functions. Its long evolution, partially based on a common global genome, is also its ontogenesis, roughly equivalent to embryogenesis in plants or animals. Moreover, the global bacterial organism, as any animal or plant, forms a single clone (Sonea and Panisset 1983). Living for 2 billion years in the confined environment of the accessible parts of our planet, this global organism increased the solidarity of its subunits by changing this meganiche into a "milieu intérieur," a bacteria hive, by extensively using temporary symbioses in different local environments and coordinating them by communication. It is a perfect example for the Gaia hypothesis (Lovelock and Margulis 1974) of how life modifies the environment.

The evolution of the bacterial world contrasts with that of the familiar eukaryotic one. Essential for the survival of the earliest cells, temporary bacterial symbioses became established as the prokaryotes' strategy for problem-solving. The number of bacteria and their variety increased to the point where selection pressures were all-pervading. Cells started to become smaller, and soon reached the smallest possible volume, having a minimum of genes, and being able to divide in the shortest possible time. Cell walls became generalized for the same reason. Compelled to keep a minimum of intracellular DNA, the number of genes on the stable, large replicon (the chromosome) became insufficient for life in nature.

Today, this replicon usually possesses fewer genes than are necessary for a bacterial strain to exist alone in any natural environment. Plasmids and/or prophages are needed to complete the bacterium in the most appropriate way. They do not accumulate in large numbers, but usually replace each other (Karska-Wysocki and Sonea 1973). Thus, a bacterium is not a unicellular organism; it is an incomplete cell (a share cell), belonging to different chimeras (symbioses) according to circumstances. As it usually lives in metabolically mixed communities of strains which exhibit division of labor (ectosymbioses) (Bull and Slater 1982; Cohen and Rosenberg 1989; Sonea and Panisset 1976; Troussellier and Legendre 1981), a bacterium is the functional equivalent of a specialized animal or plant cell. It is remarkable that each one of these types of cells participates as a subunit of a clonal multicellular organism. The tendency of bacterial cells to associate in opportunistically combined mixtures (ectosymbioses) was probably always present, and its success increased in time with the greater metabolic variety of cells. Among the very few exceptions to the generalized life in communities are bacteria that take part in exclusive endosymbioses with eukaryotes (Margulis 1981) and pathogenic strains. All strains have gradually lost autonomy during the last 2.5–3 billion years, until now they are probably as interdependent as the cells of an animal. The global bacterial organism seems to have reached its own type of adulthood. Its cells no longer synthesize new genes, just as they do not keep "neutral" DNA (which would slow down their capacity to keep abreast of the other strains).

Bacteria do not form species; instead of genetic isolation, which exists in plants and is even more complete in animals, any bacterium may obtain genes from other types of strains. Gene exchange is a generalized phenomenon involving mostly plasmids and prophages. This extensive intercellular communication using genetic information is complemented by exchanges of strains between communities and probably correlates the homeostatic properties of the pervading mixed communities for the biosphere's benefit. All bacteria, prophages, and plasmids are basic subunits of the global bacterial organism, which is endowed with a general solidarity (Sonea and Panisset 1978). This global entity is the only bacterial one which should be considered an organism; its different types of strains function as if they were its "differentiated" cells (Sonea 1971). From the early unicellular organisms, which helped each other through temporary symbioses 3 billion years ago, the entire bacterial world has changed into a single, pluricellular organism composed of a large variety of specialized cells and small replicons, which are its incomplete subunits. This contrasts with the centrifugal eukaryotic evolution which is directed towards more diversified and genetically isolated organisms.

Impact of Bacterial Symbioses on Origin and Evolution of Eukaryotes

Since their origin, the eukaryotes have been immersed in the all-pervading, dispersed, global bacterial organism. Regardless of where the eukaryotic ancestor cell is placed phyletically, it was probably a predator cell; the main mechanism for its evolution seems to have been the rapid successive additions to it of outside genes present in different bacteria, which, probably for the first time on our planet, resulted in a permanent endosymbiosis (Margulis 1970). One overlooked aspect of this phenomenon is that these acquired symbionts

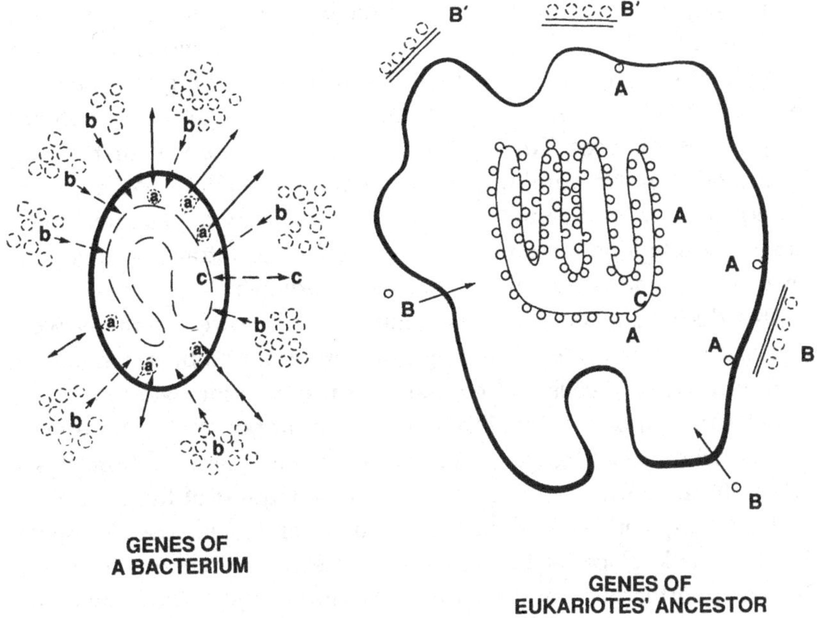

Figure 2
Different types of gene stability in a bacterium and in the evolving eukaryotic ancestor. Small replicons: (**a**) temporary intracellular (two of them are inserted in the more stable large replicon (**c**) and four are in the cytoplasm); (**b**) from outside the cell but able to visit it; (**A**) intracellular (79 of them, mostly prophages, are inserted in the large replicon (**C**), and only three are free in the cytoplasm (although membrane-bound, as are all bacterial replicons); (**B**) small replicons that will be soon the last ones to be admitted for permanent symbioses; (**B'**) outside small replicons that will not be admitted in the cell, as the latter will be permanently "closed" to gene exhanges.

had a long uninterrupted past of participating in temporary symbioses. Thus they were preselected, and able to bring along easily matching groups of genes, contributing to the eukaryotes' rapid success.

It is highly probable that the emerging nucleus also started as part of a permanent endosymbiosis involving the large replicon of the eukaryotes' ancestor cell and the former "visiting" plasmids and prophages (Sonea 1972). The latter have a tendency to integrate frequently in large replicons by additive recombinations, and are therefore much more resistant to "curing" than plasmids; thus, they were better prepared to become permanent symbionts (figure 2). Once the nucleus was completed, the early eukaryotic cell had no further use for other "visiting" small replicons, and, subsequently, lost its ability to obtain genes from the global potential bacterial genome (Sonea and Panisset 1976, 1983). This newly independent eukaryotic cell ended up by blossoming into a full organism whose offspring could evolve in many directions without being restricted by the rules of temporary adaptable symbioses. For the first time there were unicellular organisms, and they were followed by obligatory speciation. With the exception of new endosymbioses, synthesis of new genes after many random mutations remains the main source of intracellular genetic enrichment, and it continues today in eukaryotes.

This decisive change in life on Earth was followed in many well-established eukaryotes by new, permanent endosymbioses involving different bacterial cells, which like their predecessors were fully prepared for temporary symbioses. The most important of these symbionts were cyanobacteria, ancestors of the plants' chloroplasts, which became the main elements for the conquest of the continents by the living world. Many other permanent endosymbioses involving bacteria took place as isolated but interesting phenomena in protozoa, invertebrates, and so forth (Margulis 1981). Along with the evolution toward multicellular eukaryotes, there were also many unicellular eukaryotes (algae and so on), which joined the bacteria in mixed communities. They lived together in the old bacterial way as temporary ectosymbioses (Bull and Slater 1982; Cohen and Rosenberg 1989).

Conclusion

The entire evolution of life becomes more coherent when both bacteria and eukaryotes are considered. Widespread temporary and adaptable symbioses, not cells devoid of small replicons, are the viable basic bacterial elements in nature. Bacterial cells and small replicons (plasmids and prophages) have therefore evolved as standardized subunits or construction blocks of the needed reshufflings for such symbioses at the cellular or mixed-community level. They form a global gene pool, a source of a nearly infinite variety for temporary symbioses. The resulting bacterial world has shown a unifying tendency in its evolutionary pattern, and has developed as a complex, dispersed, single global organism. The origin of eukaryotes was based on permanent associations of a variety of bacterial subunits which were previously parts of the global bacterial organism, already conditioned to work together. When these permanent cumulative endosymbioses were eventually completed, they produced different cells, revolutionary in structure and functions. Uni- and multicellular organisms formed a rich variety of species. Thus, symbiosis has played a major role in evolution. Since it is temporary in bacteria and permanent in eukaryotes, the fact that it resulted in contrasting evolutive effects is highly significant. In bacteria the continuous development of a complex clone occurred, which thus became a single, dispersed, pluricellular global organism, devoid of unicellular organisms and species. In eukaryotes a centrifugal evolution took place, resulting in billions of organisms and millions of species.

References

Ackermann, H.-W., and DuBow, M. S. 1987. *Viruses of Prokaryotes*. CRC Press.

Anderson, E. S. 1966. Possible importance of transfer factors in bacterial evolution. *Nature* 209: 637–638.

Bergh, O., Borshein, K. Y., Bratbak, G., and Heldal, M. 1989. High abundance of viruses found in acquatic environments. *Nature* 340: 467–468.

Bishai, W. R., and Murphy, J. R. 1988. Bacteriophage gene products that cause human disease. In: Calendar, R., ed., *The Bacteriophages*. Plenum.

Broda, P. 1979. *Plasmids*. Freeman.

Bull, A. T., and Slater, J. H., eds. 1982. *Microbial Interactions and Communities*. Academic Press.

Campbell, A. 1988. Phage evolution and speciation. In: Calendar, R., ed., *The Bacteriophages*. Plenum.

Cohen, Y., and Rosenberg, E., eds. 1989. *Microbial Mats*. ASM.

Datta, N. 1975. Epidemiology and classification of plasmids. In: Schlessinger, D., ed., *Microbiology 1974*. ASM.

Dobardzic, R., and Sonea, S. 1971. Production d'hemolysines et conversions lysogène chez *Staphylococcus aureus*. *Ann. Inst. Pasteur* 120: 42–49.

Gunsalus, I. C., Hermann, M., Toscano, W. D., Katz, D., and Garg, C. K. 1975. Plasmids and metabolic diversity. In: Schlessinger, D., ed., *Microbiology 1974*. ASM.

Imanaka, T., Fuji, M., and Arba, S. 1981. Isolation and characterization of antibiotic resistance plasmids from thermophilic bacilli and construction of deletion plasmids. *J. Bact*. 146: 1091–1097.

Karska-Wysocky, B., and Sonea, S. 1973. Sensibilité à l'action léthale des rayons UV d'une souche de *Staphylococcus aureus* en relation avec le nombre de prophages. *Rev. Canad. Biol.* 32: 151–156.

Lacey , R. W., and Kruczenyk, S. C. 1986. Epidemiology of antibiotic resistance to *Staphylococcus aureus*. *J. Antimicr. Chemother.* 18, suppl. C: 207–217.

Lion, B. R., and Skurray, R. S. 1987. Antimicrobial resistance of *Staphylococcus aureus*: Genetic basis. *Microbiol. Rev.* 51: 88–134.

Lovelock, J. E., and Margulis, L. 1974. Atmospheric homeostasis by and for the biospere: The Gaia hypothesis. *Tellus* 26: 2–10.

Margulis, L. 1970. *Origin of Eukaryotic Cells*. Yale University Press.

Margulis, L. 1981. *Symbiosis in Cell Evolution*. Freeman.

Mitsuhashi, S., Morimura, M., Kono, K., and Oshima, H. 1963. Elimination of drug resistance of *Staphylococcus aureus* by treatment with acriflavine. *J. Bact.* 86: 162–164.

Novick, R. P., and Roth, C. 1968. Plasmid linked resistance to inorganic salts in *Staphylococcus aureus*. *J. Bact.* 95: 1335–1342.

Reanney, D. C. 1976. Extrachromosomal elements as possible agents of adaptation and development. *Bact. Rev.* 40: 552–590.

Reanney, D. C., Roberts, W. P., and Kelly, W. J. 1982. Genetic interactions among microbial communities. In: Bull, A. T., and Slater, J. H., eds., *Microbial Interactions and Communities*. Academic Press.

Richmond, M. H. 1970. Plasmids and bacterial evolution. *Symp. Soc. Gen. Micr.* 24: 59–85.

Richmond, M. H., and John, M. 1964. Co-transduction by staphylococcal phage of the genes responsible for penicillinase synthesis and resistance to mercury salts. *Nature* 202: 1360–1361.

Schütt, C. 1989. Plasmids in the bacterial assemblage of a distrophic lake: Evidence for plasmid encoded nickel resistance. *Microb. Ecol.* 17: 49–62.

Skurray, R. A., Rouch, D. A., Lyon, B. R., Gillespie, M. E., Byrne, J. M., Messerotti, L. J., and May, J. W. 1988. Multi-resistant *Staphylococcus aureus*: Genetics and evolution of epidemic Australian strains. *J. Antimicr. Chemother.* 21, suppl. C: 19–39.

Smith, D. H. 1967. R-factors mediate resistance to mercury, nickel and cobalt. *Science* 156: 1114–1116.

Sonea, S. 1971. A tentative unifying view of bacteria. *Rev. Can. Biol.* 30: 239–244.

Sonea, S. 1972. Bacterial plasmids instrumental in the origin of eukaryotes? *Rev. Can. Biol.* 31: 61–63.

Sonea, S. 1987. Bacterial viruses, prophages and plasmids, reconsidered. *Ann. N.Y. Acad. Sci.* 503: 251–260.

Sonea, S. 1988a. A bacterial way of life. *Nature* 331: 216.

Sonea, S. 1988b. Dynamic symbioses: The bacterial way of life (in French). *Medicine/Sciences* 4: 378–381.

Sonea, S., and Panisset, M. 1976. Manifesto for a new bacteriology (in French). *Rev. Canad. Biol.* 35: 103–167.

Sonea, S., and Panisset, M. 1978. L'évolution des infections bactériennes et la solidarité génétique des bactéries. *Méd. et Hyg.* 36: 2074–2081.

Sonea, S., and Panisset, M. 1983. *A New Bacteriology*. Jones and Bartlett.

Troussellier, M., and Legendre, P. 1981. A functional evenness index for microbial ecology. *Microbial Ecology* 7: 282–296.

Watanabe, T. 1963. Infective heredity of multiple drug resistance in bacteria. *Bact. Rev.* 27: 87–115.

Wilson, G., and Miles, A. 1975. *Topley and Wilson's Principles of Bacteriology, Virology and Immunity*. Edward Arnold.

Zillig, W., Reiter, W. D., Palm, P., Gropp, F., Neumann, H., and Rettenberger, M. 1988. Viruses of archaebacteria. In: Calendar, R., ed., *The Bacteriophages*. Plenum.

8 Predation as Prerequisite to Organelle Origin: *Daptobacter* as Example

Ricardo Guerrero

Predation, as a general phenomenon, is defined as the attack of one type of organism, the predator, on another type, the prey, followed by the death of the prey; that is, the simultaneous destruction and use of the prey body as the trophic source of the predator. Predator-prey interaction is a mode of nutrition that implies a victor and a heterospecific victim. Because of the small size of the bacteria that display this mode of nutrition, and the consequent difficulty of observing them, most biologists are unaware of the ancient history of predatory-prey relationships prior to the existence of animals and even of protoctists (Guerrero et al. 1986).

In the course of ecological studies of the bacterial communities of sulfurous karstic lakes, we observed and characterized two kinds of predatory bacteria thriving on highly dense populations of purple sulfur bacteria of the family Chromatiaceae: *Vampirococcus*, a strictly anaerobic, Gram-negative ovoid which lacks a flagellum (figure 1), and *Daptobacter*, a facultatively anaerobic, Gram-negative straight rod with a single polar flagellum (figure 2).

Vampirococcus and *Daptobacter* were first observed in October 1978 in Lake Estanya (Huesca, Spain), and later in other Spanish lakes and in aquatic systems elsewhere. They were found attached to several species of *Chromatium* when the natural population of this purple sulfur bacterium reached highly concentrated numbers (up to 10^6 and 10^7 cells per cm^3). These blooms were due to the simultaneous presence of adequate amounts of light and sulfide, and permitted the observation of relatively high numbers of the predator (up to 10^3 and 10^4 cells per cm^3), both directly (by light and electron microscopy of natural samples) and indirectly (by the detection of lytic plaques on thin layers of *Chromatium*-dominated mixed cultures) (Guerrero et al. 1987).

Transmission electron micrographs revealed that whole cells of *Daptobacter* are capable of penetrating directly into the cytoplasm of its prey, *Chromatium*. Hence, all the components of one motile bacterium may enter a another bacterium while remaining intact. As phagocytosis is unknown in prokaryotes, the intercellular penetration of bacterial cells by *Daptobacter* may be analogous to early events in the symbiotic acquisition of cell organelles. If prey bacteria avoided digestion by predators, stable symbioses between the two types of organisms might have developed. Such attenuation of associations from necrotrophy to symbiotrophy occurred in the transition of bacteria from lethal invader to required endosymbiont in amoebae (Jeon, this volume). Once established, such bacterial associations might have many emergent properties, including phagocytosis.

Predation, and resistance to it, probably evolved in the Archean eon in anaerobic bacterial communities dominated by phototrophic anoxygenic bacteria. Direct evidence of stratified microbial communities exists in the fossil record of western Australia and southern Africa (Schopf 1983). The best interpretation of these layered rocks is that they are remnants of stratified microbial mat communities of which at least some components were anaerobic phototrophs (Stolz 1990). It is safe to conclude that the conditions for the growth and development of anaerobic prokaryotic predation had already evolved 3 billion years ago (Guerrero et al. 1986). The discovery of an intracellular bacterium inside a microbial mat phototroph (probably *Thiocapsa*) from the laminated community in Laguna Figueroa, Baja California, was made by John F. Stolz and David Chase (see figure 7c of Guerrero and Mas 1989, reprinted here as figure 3). This intertidal microbial community is known to have pre-Phanerozoic analogues (Margulis et al. 1980). The discovery of a *Daptobacter*-like Gram-negative organism inside a *Chromatium*-like phototroph further enhances the likelihood that such predatory relationships among prokaryotes have been in continuous existence throughout the geological record.

The threat of desiccation due to the drying of puddles, ponds, lakes, and seashores provided the major selection pressures for the origin of the most impressive bacterial morphological diversity: endospores (of *Clostridium* and *Bacillus*) and cysts (of myxobacteria). The high temperatures typical of sulfurous springs and active volcanoes provided other severe selection pressures for the development of resistant propagules, spores, ensheathed cells, and the like. Penetrating ultraviolet solar radiation no doubt selected for UV-resistant bac-

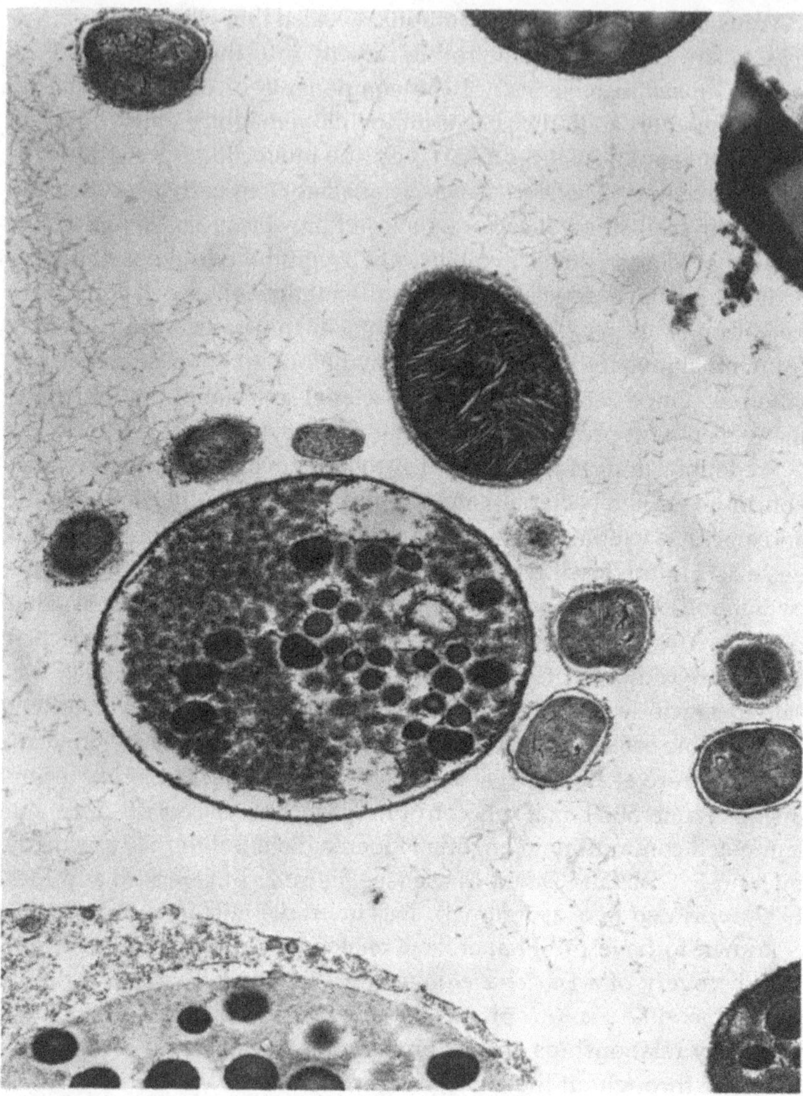

Figure 1
Predatory prokaryotes from the densely inhabited phototrophic community, dominated by *Chromatium*, at the surface of Lake Cisó, Girona, Spain. Two *Vampirococcus* cells, one in division, are attached to the *Chromatium*, from which, it is presumed, they derive nutriment. They apparently do not associate with other phototrophs, such as the one shown in thin section here. (Courtesy of D. Chase.)

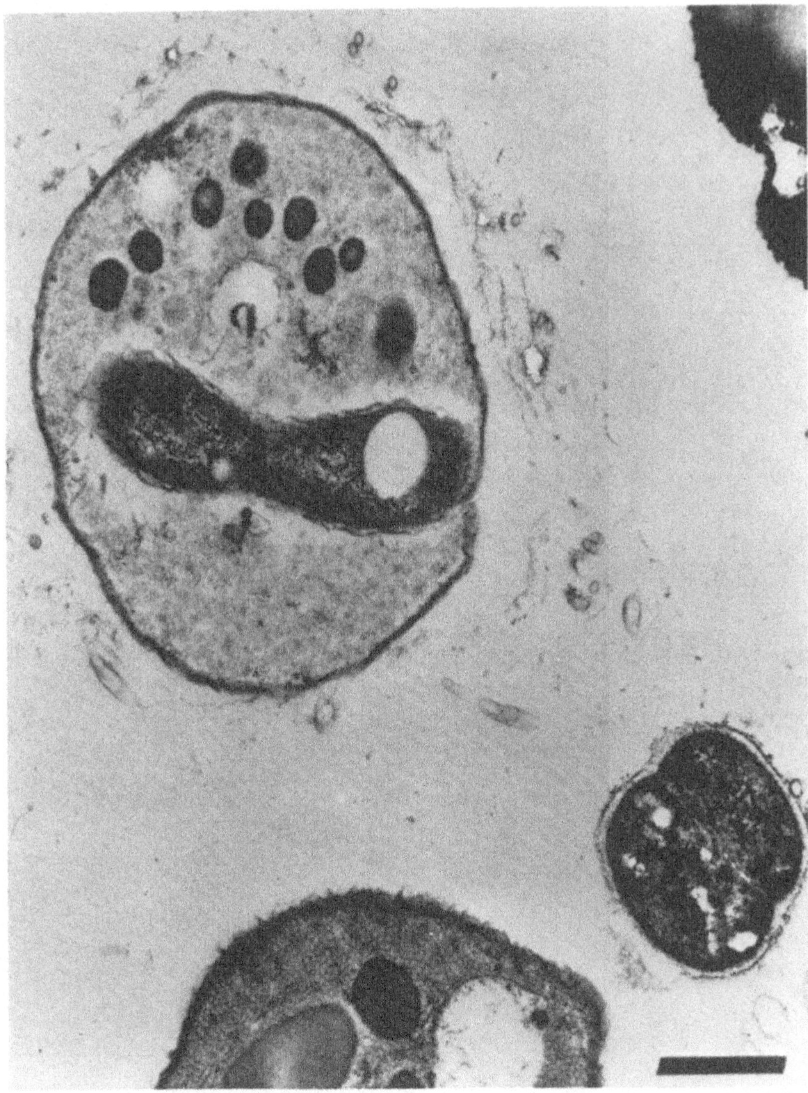

Figure 2
Daptobacter is cytoplasmic in location. It divides in the degraded cytoplasm of its *Chromatium* prey. Although the thylakoid-rich cytoplasm degenerates as *Daptobacter* (a facultative heterotroph) grows, the storage granules remain intact. Bar = 1 μm. (Courtesy of I. Esteve and D. Chase.)

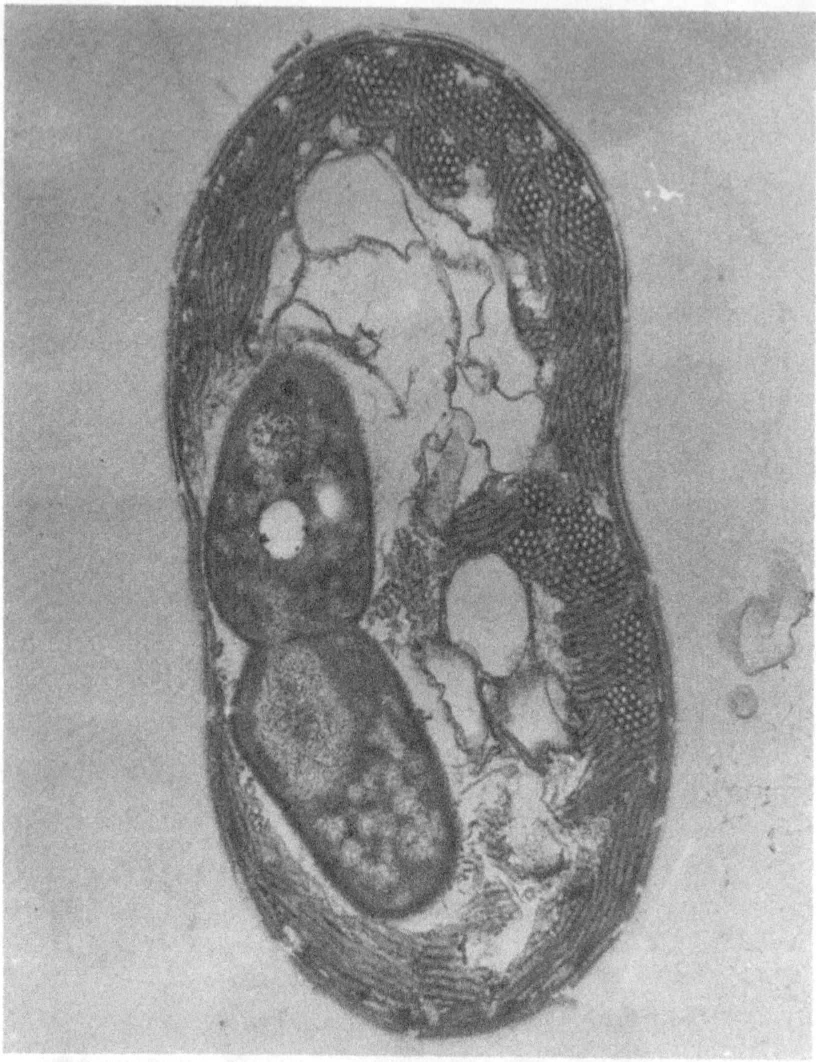

Figure 3
Chromatium harboring unidentified bacterium in division. These organisms were fixed by David Chase in the field; they are from the microbial scum on the surface of the channels in North Pond, in Baja California. From the ultrastructure it appears unlikely that the internalized bacterium is *Daptobacter*; thus, it may represent still another type of predatory bacterium. Bar = 1 μm. (Courtesy of J. F. Stolz.)

teriophages and radio-resistant lineages of prokaryotes. These severe sources of selection represent abiotic pressures that affected all members of entire communities of bacteria simultaneously as elevated temperatures, unattenuated radiation, and local desiccation restricted microbial population growth.

Biotic selection pressures, which are fundamentally different from the abiotic ones in that they require organismic interaction, can also be recognized. Among these, predation, pathogenesis, crowding, and shading clearly have prokaryotic origins; they are observed today in highly organized microbial communities (Margulis et al. 1986; Guerrero and Mas 1989). Predation concerns us here because it is a prerequisite to the thesis that symbiosis is a source of evolutionary novelty: predation is likely to have been the immediate precursor of the origin of the eukaryotic cell. There are many features of bacterial predation (Guerrero et al. 1987): the chemical detection of the prey; the chemotactic attraction of the predator by the prey; the attachment of the predator to the surface of the prey; the penetration of the prey by the predator; the production of proteases, lipases, and other degradative enzymes; the decomposition of the nucleic acid and proteins of the prey; and the reutilization of the prey's constituents for the growth and reproduction of the predator under the genetic instruction of the latter. With the discovery, beginning in the 1960s, of at least three distinct genera of predatory bacteria—*Bdellovibrio* (Burnham and Conti 1984; figures 4 and 5 here), *Vampirococcus* (figures 1 and 5), and *Daptobacter* (Guerrero et al. 1986; figures 2 and 5 here)—it has become clear that biotic relations generally attributed to complex eukaryotes (e.g., mammals or cephalopod mollusks), such as predation and attempted resistance to it, are bacterial virtuosities already established in the record of pre-Phanerozoic life.

What is the relation of bacterial predation to organelle origin? Can a scenario be developed that plausibly describes the transition of prey from victim to endosymbiont? I suggest that at least two classes of eukaryotic organelles, respiratory and motility, have a directly detectable legacy from such prokaryotic predation. Extant intracellular structures, such as mitochondria, hydrogenosomes, kinetosomes, and axonemes of undulipodia (and other "xenosomes," in the terminology of Corliss 1987), are derived from ancient biotic relations among bacteria in which the stringent selection pressures of death by predation were resisted by the potential prey. The survival of attacked and debilitated prey in coexistence with the would-be victor sig-

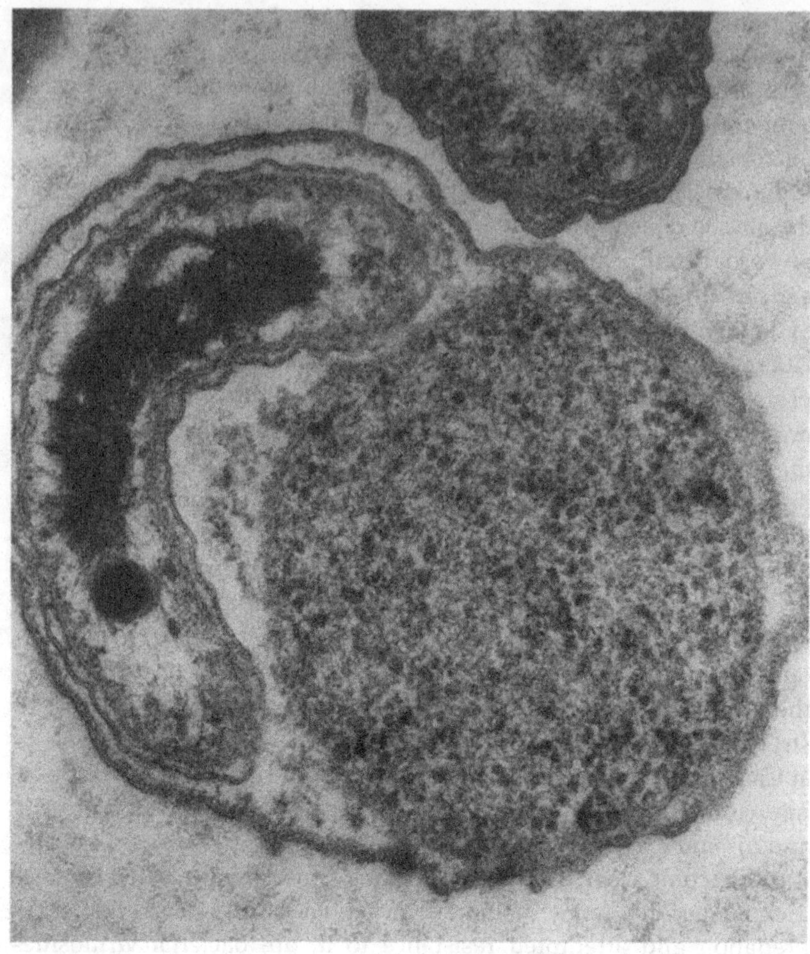

Figure 4
Bdellovibrio is periplasmic in location. It divides in the space between the inner plasma membrane and the outer membrane of the Gram-negative cell wall of its heterotrophic host. Prior to the formation of equal-sized cells, it grows, like a long sausage, to fill the periplasm of its prey.

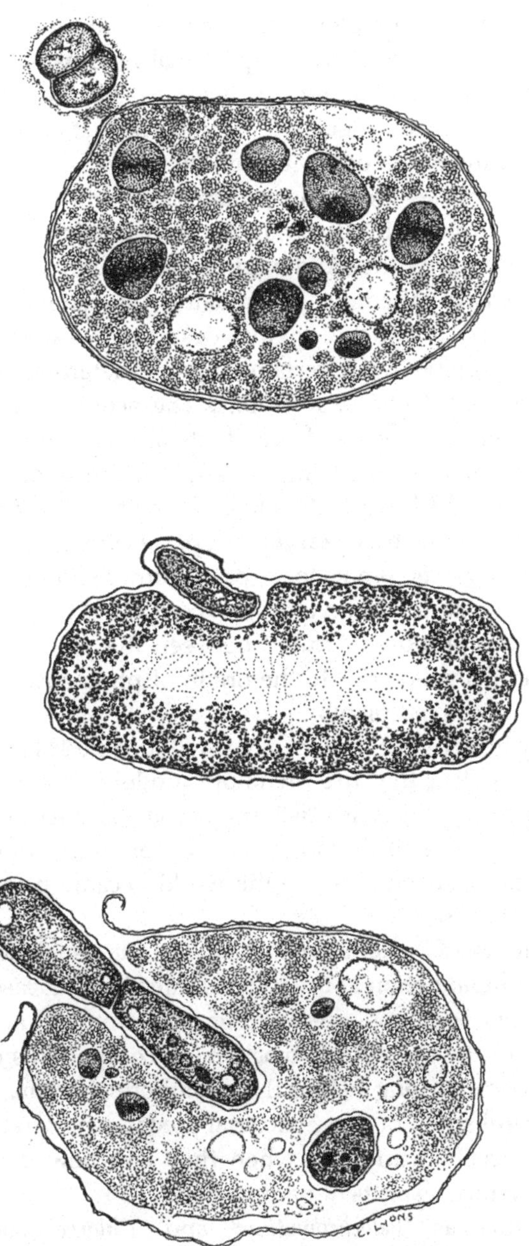

Figure 5
These schematic drawings indicate the epibiotic, periplasmic, and cytoplasmic locations of *Vampirococcus*, *Bdellovibrio*, and *Daptobacter*, respectively. (Drawings by Christie Lyons, reprinted courtesy of Harcourt Brace Jovanovich.)

naled the origin of the cellular complexity that is the distinguishing mark of eukaryosis. The two bacterial types—potential vanquished as surrounded "endocytobiont" and potential victor as surrounding "host"—persisted in an intimate, regulated coexistence that was an absolute prerequisite for the origin of any eukaryotic ancestor.

The origin of all eukaryotes from the prokaryotic "truce" was a phenomenon of heterotrophic bacteria. If mitochondria and microtubule-based motility organelles did originate from *Daptobacter*- and *Spirochaeta*-like abortive attempts at predatory invasions, as hypothesized by Margulis (1988), the original eukaryotes—already capable of intracellular motility—were certainly hungry heterotrophs long before any became algal photoautotrophs. The acquisition of phototrophic modes of metabolism by descendants of these eukaryotes was a later phenomenon, involving subsequent ingestion of green (*Prochloron*; Lewin and Cheng 1989) and blue-green (cyanobacteria) phototrophic prokaryotes by eukaryotes capable of ingestion. Since ingestion by phagocytosis, pinocytosis, or any other sort of endocytosis is unknown in extant prokaryotes, it is concluded to be an emergent property of the proposed "truce" between the predatory bacterium and its potential prey prior to the origin of photoautotrophy in the ancestors of algae and plants.

Reports of predatory behavior of bacteria are listed in table 1. From perusal of the table it is clear that the predatory mode of bacterial behavior is nearly universal. Since it exists in soil, lakes, microbial mats, and seawater today, it is likely to have been even more widespread in the pre-Phanerozoic eon, before the world became populated by eukaryotic predators.

The heterospecific attack of bacteria on animals and plants labeled "pathogenesis" is the phenomenon at the basis of infectious disease. The elements of the phenomenon—recognition, penetration, and growth at the expense of the constituent biochemical components of the prey—existed in bacteria long before the evolution of animals, plants, and fungi. Careful study of these bacterial predator-prey relations in nature is especially warranted when they are viewed as model systems for the critical aspects of pathogenesis.

Bacterial consortia such as "*Pelochromatium*" and "*Chlorochromatium*" (Pfennig 1989) attest to the possibility of stable and effective permanent bacterial association, yet all these documented associations are extracellular. Predatory prokaryotes totally destroy their prey: the only intracellular associations of bacteria known to date are

Table 1
Predation in prokaryotes.

Prey	Predator	References
Phototrophic		
Chromatium	*Vampirococcus*[1]	Guerrero et al. 1986
purple sulfur bacteria	*Daptobacter*	Guerrero et al. 1986
Chlorella[2]	*Vampirovibrio*[3]	Gromov and Mamkaeva 1972
Chemoautotrophic or mixotrophic		
Thiothrix[4]	unidentified rod	Larkin et al. 1990
Heterotrophic		
Escherichia, Spirillum, Pseudomonas, other Gram-negative bacteria	*Bdellovibrio*	Stolp 1981; Burnham and Conti 1984
Micrococcus	*Ensifer*	Casida 1982

1. Anaerobic. Only observed feeding from the outside of its prey.
2. Chlorella is a green alga (eukaryote).
3. This bacterial predator, *V. chlorellavorus*, was believed originally to belong to the genus *Bdellovibrio*.
4. The nutritional mode of the species observed by Larkin et al. has not been determined.

necrotrophic. Why is stable intracellular symbiosis unknown in prokaryotes? Several possible responses to this question exist:

• Stable intracellular symbionts in bacteria exist but have not yet been detected.

• Stable intracellular bacterial symbiosis leads so quickly to membrane hypertrophy and origin of nuclear membranes that it goes undetected, but in fact it is the explanation for organisms such as *Entamoeba* and *Pelomyxa* (protists which lack mitochondria).

• Stable intracellular bacterial symbiosis is theoretically unstable, and therefore extended associations are not possible.

Current methods do not permit distinction of these possibilities. Yet, although infrequent in bacteria, permanent stable coevolved microbial associations are defining characteristics of members of the protoctista. Following the logic of Sonea (this volume), one can conclude that temporary shifting microbial associations dominated the Archean and Proterozoic eons (from 3.5 billion until 650 million years ago) and are defining characteristics of members of the Archaeobacterial and Eubacterial realms.

Bacterial evolution and ecology are fundamentally different in certain respects from the evolution and ecology of eukaryotes. Species, in eukaryotes, are identifiable, classifiable, and stable. They are marked by appearance and subsequent disappearance in the fossil record. Since all eukaryotes represent coevolved associations of microorganisms, the fact of eukaryotic species verifies the stability of the component microbial associations. In contrast, prokaryote speciation itself is in doubt (Sonea and Panisset 1983; Sonea, this volume), and the rampant prevalence of phage, plasmids, transposons, and other small bacterial genomes indicate that bacterial ecology and evolution are marked by the expression of temporary alliances of bacterial genomes. Attempted predation followed by truce (i.e., resistance to predation and subsequent survival of bacterial predator-prey associates) is at least one of the selective processes that lead to new levels of individuality (Margulis and Guerrero 1991). The sort of individuality that is characteristic of eukaryotes in general and protists in particular is probably a direct legacy of prokaryotic predatory modes of nutrition.

References

Burnham, J. C., and Conti, S. F. 1984. *Bdellovibrio*. In: Krieg, N. R., and Hold, J. G., eds., *Bergey's Manual of Systematic Bacteriology*, vol. 1. Williams & Wilkins.

Casida, L. E. 1982. *Ensifer adhaerens* gen. nov., sp. nov.: A bacterial predator of bacteria in soil. *Internat. J. Syst. Bacteriol*. 32: 339–345.

Corliss, J. O. 1987. Protistan phylogeny and eukaryogenesis. *International Review of Cytology* 100: 319–370.

Gromov, B. V., and Mamkaeva, K. A. 1972. Electron microscope examination of *Bdellovibrio chlorellavorus* parasitic on cells of the green alga *Chlorella vulgaris*. *Tsitologiya* 14: 256–260.

Guerrero, R., and Mas, J. 1989. Multilayered microbial communities in aquatic ecosystems: Growth and loss factors. In Cohen, Y., and Rosenberg, E., eds., *Microbial Mats: Physiological Ecology of Benthic Communities*. American Society for Microbiology.

Guerrero, R., Pedros-Alió, C., Esteve, I., Mas, J., Chase, D., and Margulis, L. 1986. Predatory prokaryotes: Predation and primary consumption evolved in bacteria. *Proceedings of the National Academy of Sciences* 83: 2138–2142.

Guerrero, R., Esteve, I., Pedrós-Alió, C., and Gaju, N. 1987. Predatory bacteria in prokaryotic communities. The earliest trophic relationships. *Ann. N.Y. Acad. Sci.* 503 :238–250.

Larkin, J. M., Henck, M. C., and Burton, S. D. 1990. Occurrence of a *Thiothrix* sp. attached to mayflay larvae and presence of parasitic bacteria in the *Thiothrix* sp. *Appl. Env. Microbiol.* 56: 357–361.

Lewin, R. A., and Cheng, L., eds. 1989. *Prochloron: A Microbial Enigma*. Chapman and Hall.

Margulis, L. 1988. Serial endosymbiotic theory (SET): Undulipodia, mitosis and their microtubular systems preceded mitochondria. *Endocytobiosis and Cell Research* 5 :133–162.

Margulis, L., and Guerrero, R. 1991. Two plus three equal one: Individuals emerge from bacterial communities. In: Thompson, W. I., ed., *Gaia: A Way of Knowing*, vol. 2: *Emergence: The Science of Becoming*. Lindisfarne.

Margulis, L., Barghoorn, E. S., Ashendorf, D., Banerjee, S., Chase, D., Francis, S., Giovannoni, S., and Stolz, J. 1980. The microbial community in the layered sediments at Laguna Figueroa, Baja California, México: Does it have precambrian analogues? *Precambrian Research* 11: 93–123.

Margulis, L., Chase, D., and Guerrero, R. 1986. Microbial communities. *BioScience* 36: 160–170.

Pfennig, N. 1989. Phototrophic green sulfur bacteria living in consortia with other microorganisms. In: Staley, J. T., Bryant, M. P., Pfennig, N., and Holt, J. D., eds., *Bergey's Manual of Systematic Bacteriology*, vol. 3. Williams and Wilkins.

Schopf, J. W., ed. 1983. *Earth's Earliest Biosphere: Its Origin and Evolution*. Princeton University Press.

Sonea, S., and Panisset, M. 1983. *A New Bacteriology*. Jones and Bartlett.

Stolp, H. 1981. The genus *Bdellovibrio*. In: Starr, M. P., Stolp, H., Trüper, H. G., Balows, A., and Schlegel, H. G., eds., *The Prokaryotes: A Handbook on Habitats, Isolation, and Identification of Bacteria*. Springer-Verlag.

Stolz, J. F. 1990. Distribution of phototrophic microbes in the flat laminated microbial mat at Laguna Figueroa, Baja California, Mexico. *BioSystems* 23(4): 354–359.

9 Amoeba and x-Bacteria: Symbiont Acquisition and Possible Species Change

Kwang W. Jeon

In a recently established amoeba-bacteria symbiosis (Jeon 1983a), new bacterial symbionts became integrated in the host amoebae such that the hosts became dependent on the symbionts within a few years. The presence of endosymbionts has caused changes in several phenotypic characters of the host cells. When the bacteria first infected the D strain of *Amoeba proteus*, in 1966 (Jeon and Lorch 1967), they were virulent and killed newly infected amoebae. However, some of the infected amoebae survived and became dependent on their newly acquired endosymbionts within a few years (Jeon 1972). At present, symbiont-bearing xD amoebae contain about 42,000 symbionts per cell (about 10% of an amoeba's volume) on the average, and grow well (Jeon 1983a, 1986, 1987). The symbionts, called x-bacteria, differ from other bacterial inclusions of amoebae reported earlier by several workers (Roth and Daniels 1961; Rabinovitch and Plaut 1962; Daniels 1964) in their morphology, in the membranes enclosing them, in their relative number, and in their relationship to the host nuclear genome (Jeon 1975; Jeon and Lorch 1979). When x-bacteria isolated from xD amoebae are introduced into symbiont-free amoebae of the same or different strains, they quickly establish a stable symbiosis and influence the physiology and phenotypic characters of their new host cells (Jeon and Ahn 1978; Lorch et al. 1985). This amoeba-bacteria symbiosis differs from other known cases in that the history of the establishment of the endosymbiosis is known, the symbionts are required for the host's survival, and cellular character changes (including the host's dependence) can be reproduced within a short period of time under laboratory conditions. Thus, the system offers an unusual opportunity for the study of host-symbiont interactions and the establishment of new associations.

This article summarizes what has been learned about the amoeba-bacteria symbiosis and demonstrates that endosymbiosis provides a major mechanism for cellular phenotypic changes and the ensuing "speciation."

Hosts and Endosymbionts

Amoebae

The original host was the D strain of *Amoeba proteus*, which had been first collected in Scotland (Lorch and Danielli 1953) and then continuously cultured in the laboratory. The D amoebae had been known to contain some symbiont-like particles of unknown origin and properties (Hawkins and Wolstenholme 1967), but the present hosts, xD amoebae, became associated with different symbionts, x-bacteria, in 1966 (Jeon and Lorch 1967). Other freshwater free-living amoebae can also act as hosts and establish new symbiosis with x-bacteria (Jeon and Jeon 1982).

Symbionts

The following summarizes what is known about the bacterial symbionts.

They are typical Gram-negative rods (0.5×2 μm in size) enclosed in host-derived vesicles called *symbiosomes* (Roth et al. 1988). (Figure 1; cf. Morgan et al. 1967 and Chang and Trager 1974.)

x-bacteria harbor two kinds of plasmid DNAs of about 59 kb and 21 kb, respectively (Han and Jeon 1980).

Both during the initial infection and after the establishment of the symbiosis, some or all x-bacteria avoid digestion by amoebae, first by being resistant to digestive enzymes of amoebae (Han and Jeon 1980) and then apparently by preventing lysosomal fusion with symbiosomes (Jeon and Jeon 1976; Ahn and Jeon 1979, 1982), as in some other known cases (Hart and Young 1979; Weidner and Sibley 1985). Isolated x-bacteria are also resistant to lysing agents such as enzymes and mild detergents (e.g., Triton-X 100 and Brij) (Han and Jeon 1980).

x-bacteria isolated from xD amoebae may be transferred into other cells either by microinjection (Jeon and Jeon 1976) or by induced phagocytosis (Ahn and Jeon 1979; Kim and Jeon 1986). Isolated bacteria do not multiply *in vitro*, but they stay viable for several days at 4°C.

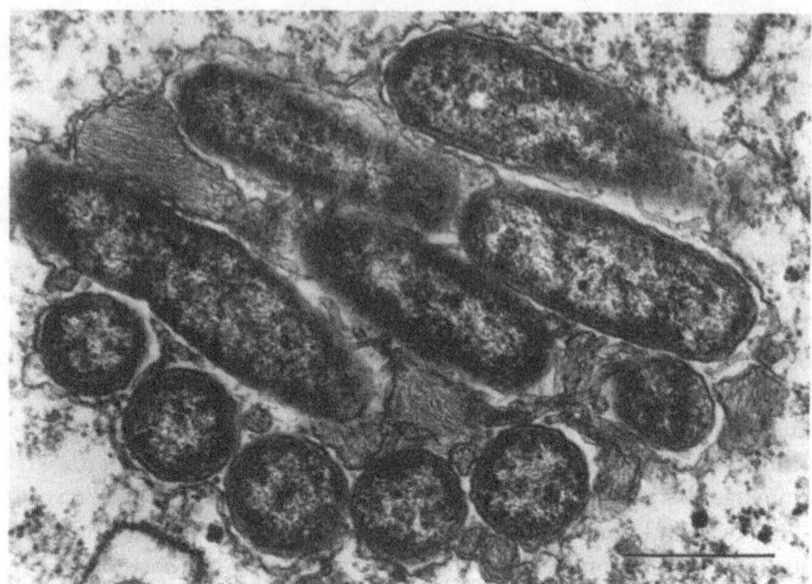

Figure 1
Electron micrograph of a symbiosome of xD amoeba containing bacterial endosymbionts. The inter-bacterial space is filled with inclusions of fibrous matter. The bar equals 0.5 μm. From Jeon 1987; reproduced with permission.

Symbiotic bacteria from xD amoebae treated *in vivo* with substances that interfere with DNA replication (such as ethidium bromide or acridine orange) for 7 to 14 days at sublethal concentrations fail to infect D amoebae (Park 1983). Also, isolated x-bacteria treated with ethidium bromide or acridine orange for 3 hours or longer *in vitro* fail to infect D amoebae. Thus, plasmid DNAs appear to be needed for bacterial infectivity.

x-bacteria are able to infect only amoebae. Thus, they establish stable symbiosis with *Chaos carolinensis* (Jeon and Jeon 1982), amoebae belonging to a different genus, but they do not grow inside other cells, such as ciliates or *Xenopus* eggs (Jeon 1983a).

x-bacteria synthesize and release a large amount of 87-kD protein (trimer of 29-kD polypeptides) into the host cytoplasm (Kim and Jeon 1986).

Effect of Symbionts on the Host

The presence of x-bacteria cause the following physiological changes in host amoebae.

During the initial phase (for up to 12 months) of experimental infection with bacteria isolated from xD amoebae, newly infected amoebae grow faster than either established xD amoebae or symbiont-free D amoebae (Jeon and Ahn 1978).

Symbiont-bearing xD amoebae are more sensitive to starvation than are normal amoebae (Jeon 1983a).

xD amoebae are more sensitive to overfeeding. Many of them become irreversibly damaged after being fed with excess *Tetrahymena* (food organisms) for only two consecutive days (Jeon 1983a).

xD amoebae cannot grow at temperatures above 26°C, which is one degree higher than their optimum growth temperature, and all die within 2 weeks when kept at this temperature (Jeon and Ahn 1978; Lorch and Jeon 1980).

xD amoebae are more sensitive to crowding than are normal amoebae. D amoebae may keep growing at a cell density of over 3000/cm^2 for several days, but xD amoebae cannot remain healthy if the density is over 1000/cm^2. At this or higher density, xD amoebae detach from the substratum, become round, and cytolyze within a few days (Jeon 1983a).

The nuclei of xD amoebae are not compatible with the cytoplasm of D amoebae, the original strain (Jeon and Jeon 1976; Lorch and Jeon 1982). When D amoebae are infected with X-bacteria by microinjection or by induced phagocytosis, their nuclei become incompatible with the cytoplasm of D amoebae within a few weeks (Lorch and Jeon 1981, 1982). Thus, x-bacteria cause nucleocytoplasmic incompatibility among amoebae.

The nuclei of xD amoebae are not only incompatible with the cytoplasm of D amoebae, but also exert a strong lethal effect on D amoebae (Lorch and Jeon 1981). When the nucleus of an xD amoeba is transplanted into a D amoeba containing its own nucleus, the latter loses viability. When D amoebae are newly infected with x-bacteria, their nuclei acquire the ability to exert a lethal effect within a few cell cycles (Lorch and Jeon 1982). Results of gel-filtration and microinjection studies showed that the nuclear lethal effect is caused by a protein of about 200 kD, newly synthesized by xD amoebae as a result of the symbiosis (Lorch et al. 1985).

The host amoebae are dependent on symbionts for survival. The host's dependence on symbionts has been demonstrated in several different ways, including nuclear transplantation and selective removal of symbionts from the host. When nuclei of xD amoebae are transplanted into the cytoplasm of symbiont-free amoebae, the resulting cells (xD_nD_c) are not viable, whereas the reciprocal heterotransplants (D_nxD_c) are viable (Jeon and Jeon 1976). The nonviable xD_nD_c cells can be rescued by transferring x-bacteria into them, for example by microinjection (Jeon and Jeon 1976). If x-bacteria are selectively removed from xD amoebae, the latter lose viability, but the aposymbiotic amoebae may be rescued by reintroducing x-bacteria (Lorch and Jeon 1980).

We have examined amoebae under the electron microscope to monitor possible structural changes that might be occurring as a result of symbiosis. Except for the presence of symbiosomes, no structural changes have been detected.

Proteins in the Host-Symbiont Interactions

Symbiont-bearing amoebae contain at least three specific new proteins: one with 29-kD components (Ahn and Jeon 1983), a second with a molecular weight over 200 kD, which is synthesized by the amoebae and has a lethal effect on D amoebae (Lorch et al., 1985), and a third, probably a glycoprotein, with a molecular weight of about 150 kD, which is present only on symbiosome membranes (SMs) and which appears to play a role in the prevention of lysosomal fusion with symbiosomes (Choi and Jeon 1989). The SMs are originally derived from phagolysosomal membranes but do not fuse with lysosomes.

The xD-Specific Protein

A 29-kD xD protein is synthesized by the symbiotic bacteria and transported to the host cytoplasm (Kim and Jeon 1986). We obtained a monoclonal antibody (mAb) against the xD protein and used it to determine the distribution of xD protein (Kim and Jeon 1987a) and to isolate and purify the protein from xD amoeba cytosol. The native xD protein had an M_r of about 87,000, and was apparently a trimer of 29-kD polypeptides. We then cloned a 1.8-kb x-bacterial DNA fragment coding for the 29-kD polypeptide and determined the nucleotide sequence of the gene (Park and Jeon 1988, 1990). The gene has an open reading frame of 759 beginning with an ATG initiation codon. It has a consensus sequence TATAGA at −12 position from the initiation codon. The deduced amino-acid composition shows that about 42% of it is made up of hydrophobic amino acids. The protein is transported from the symbiotic bacteria to the host cytoplasm across bacterial and vesicle membranes, but the mechanism for its transport across the two membrane systems is not known. The protein synthesized in *E. coli* transformed with the cloned 29-kD gene is also exported to the medium (Park and Jeon 1988).

Strain-Specific Lethal Factor

xD amoebae also contain a protein with M_r above 200 kD, which exerts a lethal effect when injected into D amoebae (Lorch et al. 1985). This protein, which apparently corresponds to the strain-specific lethal factors which had been found earlier to be present in different strains of amoebae (Jeon and Lorch 1969), is synthesized by xD amoebae as a result of the symbiosis. A newly infected D amoeba contained an active lethal factor against other D amoebae as early as 2 weeks after the induced infection. Unfortunately, there is no available quantitative assay for this protein, and the kinetics of its synthesis cannot be easily studied.

Amoeba Actin Selectively Accumulated by Symbionts

Amoeba actin (43 kD) is selectively accumulated in symbiosomes and attaches to the surface of x-bacteria, as is shown by indirect-immunofluorescence and immunogold-staining methods (Kim and Jeon 1987b); however, it is not known how the protein is transported into symbiosomes or accumulated by x-bacteria, or what it does. Using a monoclonal antibody against amoeba actin as a probe, we compared the properties of amoeba actin with those of actins from other sources, including vertebrate muscles, *Acanthamoeba*, and *Naegleria* (Ahn and Jeon 1988). The results showed that the properties of amoeba actin differ from those of other actins, the peptide fragments of amoeba actin showing several unique bands in SDS gels.

Discussion

Intracellular symbiosis has been studied in a wide variety of cell systems, ranging from free-living cells to cells in tissues of plants and animals (Richmond and Smith 1979; Schwemmler and Schenk 1980; Cook et al. 1980; Fredrick 1981; Schenk and Schwemmler 1983; Jeon 1983b; Ahmadjian 1986; Lee and Fredrick 1987; Smith and Douglas 1987). However, much remains unclear about the mechanisms for each of the steps involved in the establishment of stable endosymbiosis, viz., initial recognition, entry of symbionts into the cells, symbionts' avoidance of digestion or ejection by the host, and sustained multiplication of both partners (Bannister 1979; Smith 1980; Kippert 1987). In intracellular symbiosis, host cells usually provide a suitable "shelter" and supply important material needs for endosymbionts, enabling the latter to grow successfully. However, living cells are hostile environments for other cells (Moulder 1979, 1985), and hence potential intracellular symbionts must have traits that enable them to overcome these adverse environmental conditions and to colonize the host cell. In some cases, endosymbionts have no defined functions and are innocuous (Beale and Jurand 1969; Preer 1971). In other cases, however, the symbionts harm their hosts (parasitism) and the symbiotic association ends with the death of the host partners. Examples in which the host cells derive benefits from their endosymbionts (e.g., photosynthate, products of nitrogen fixation, vitamins) are known (Karakashian 1963; Muscatine and Lenhoff 1965; Chang et al. 1975; Verma and Stanley 1987; Ishikawa 1987). A prolonged symbiotic association may result in a functional integration between symbionts and the host cell and hence may increase the complexity of the host-cell structure, as is proposed to have happened in the origin of mitochondria and plastids of eukaryotic cells according to the theory of serial endosymbiosis (Margulis 1970, 1981; Smith 1980; Taylor 1979, 1987; Whatley 1981; Sitte 1983; Cavalier-Smith 1987; Delihas and Fox 1987).

The study of symbiont integration and of changes in cellular character caused by endosymbiosis is significant for the following reasons. First, symbiosis is an important biological phenomenon in terms of genetic novelty because it draws genomes from the entire biosphere (Jeon and Danielli 1971; Sheinin 1983; Cavalier-Smith 1987), and thus genetic changes arising from endosymbiosis are far greater in magnitude than those which may result intrinsically from mutation, hybridization, or changes in ploidy (Taylor 1979, 1987). Genetic studies show that gene transfer occurs between symbionts and hosts (Prakash and Atherly 1986) as well as between cell organelles and cell nuclei (Gellissen and Michaelis 1987) and among organellar genomes (Stern 1987) or plasmids (Schofield et al. 1987). Second, the mechanisms whereby endosymbionts avoid digestion by their hosts are of primary importance in understanding defense mechanisms against diseases caused by infective agents in animals (Quinn 1984). Third, the membrane differentiation seen during the formation of specialized symbiont-containing vesicles (Ahn and Jeon 1982, 1985) is important in understanding selective lysosomal fusion (Pfeifer 1987).

On the basis of the structural and physiological changes brought about by endosymbionts in xD amoebae as described above, one could consider the symbiont-bearing xD strain a new "species" of *Amoeba*. However, until evidence for genetic differences between D and xD amoebae is obtained, it would be more prudent to treat xD amoebae as belonging to a variant strain.

The following are some of the remaining questions that must be answered to understand host-symbiont relationships in amoebae fully.

Why are x-bacteria dependent on amoebae as their habitat? (All attempts to grow them outside amoebae have failed.) Has their genome lost some of its coding functions by transfer to the host genome, as in the case of mitochondria and chloroplasts?

Why are host amoebae dependent on their symbionts for survival? Is the symbiont-synthesized 29-kD xD protein required by the host xD amoebae for their survival? If the xD protein is a required component, what is the mode of its function? What are the characteristics and roles of the xD protein? Is it also involved in altering any of the characters of host amoebae?

What are the origin and the characteristics of the SM-specific antigen? Is the antigen a "fusion-preventing" factor? If so, how does it work? Where is it synthesized, how and when is it inserted into the SM, and what role does it play in the prevention of lysosomal fusion with symbiosomes?

Are the plasmid DNAs of x-bacteria involved in rendering x-bacteria resistant to digestion by lysosomal enzymes? What are their roles in conferring infectivity upon x-bacteria? Do they code for specific proteins?

Are the precursor bacteria from which x-bacteria arose still present in the environment—for example, in amoeba culture dishes? If so, what are the basic differences between them and x-bacteria?

What is the mechanism for the induction of the synthesis of a new lethal factor, a protein larger than 200 kD? Is a permanent genomic change of amoebae brought about as a result of symbiosis?

Acknowledgments

I thank L. E. Roth for reading the manuscript. The research work was supported by a grant from the National Science Foundation.

References

Ahmadjian, V. 1986. *Symbiosis: An Introduction to Biological Associations*. University Press of New England.

Ahn, T. I., and Jeon, K. W. 1979. Growth and electron microscopic studies on an experimentally established bacterial endosymbiosis in amoebae. *Journal of Cellular Physiology* 98: 49–58.

Ahn, T. I., and Jeon, K. W. 1982. Structural and biochemical characteristics of the plasmalemma and vacuole membranes in amoebae. *Experimental Cell Research* 137: 253–268.

Ahn, T. I., and Jeon, K. W. 1983. Strain-specific proteins of symbiont-containing *Amoeba proteus* detected by two-dimensional electrophoresis. *Journal of Protozoology* 30: 713–715.

Ahn, T. I., and Jeon, K. W. 1985. Recycling of membrane proteins during endo- and exo-cytosis in amoebae. *Experimental Cell Research* 160: 54–60.

Ahn, T. I., and Jeon, K. W. 1988. Electrophoretic profiles and peptide maps of *Amoeba proteus* actin, as studied using a monoclonal antibody as the probe. In *Proceedings of the IV International Congress on Cell Biology*.

Bannister, L. H. 1979. The interactions of intracellular protista and their host cells, with special reference to heterotrophic organisms. *Proceedings of the Royal Society of London* B204: 141–163.

Beale, G. H., and Jurand, A. 1969. The classes of endosymbionts of *Paramecium aurelia*. *Journal of Cell Science* 5: 65–91

Cavalier-Smith, T. 1987. The simultaneous symbiotic origin of mitochondria, chloroplasts, and microbodies. *Annals of the New York Academy of Sciences* 503: 55–71.

Chang, K. P., Chang, C. S., and Sassa, S. 1975. Heme biosynthesis in bacterium-protozoan symbioses: Enzyme defects in host hemoflagellates and complemental role in their intracellular symbiotes. *Proceedings of the National Academy of Sciences* 72: 2979–2983.

Chang, K. P., and Trager, W. 1974. Nutritional significance of symbiotic bacteria in two species of hemoflagellates. *Science* 183: 532–533.

Choi, E. Y., and Jeon, K. W. 1989. The presence of a spectrin-like protein on symbiosome membranes of symbiont-bearing *Amoeba proteus* as studied with monoclonal antibodies. *Endocytobiosis and Cell Research* 6: 99–108.

Cook, C. B., Pappas, P. W., and Rudolph, E. D., eds. 1980. *Cellular Interactions in Symbiotic and Parasitic Relationships.* Ohio State University Press.

Daniels, E. W. 1964. Electron microscopy of centrifuged *Amoeba proteus. Journal of Protozoology* 11: 281–290.

Delihas, N., and Fox, G. E. 1987. Origins of the plant chloroplasts and mitochondria based on comparisons of 5S ribosomal RNAs. *Annals of the New York Academy of Sciences* 503: 92–102.

Fredrick, J. F., ed. 1981. *Origins and Evolution of Eukaryotic Intracellular Organelles.* New York Academy of Sciences.

Gellissen, G., and Michaelis, G. 1987. Gene transfer: Mitochondria to nucleus. *Annals of the New York Academy of Sciences* 503: 391–401.

Han, J. H., and Jeon, K. W. 1980. Isolation and partial characterization of two plasmid DNAs from endosymbiotic bacteria in *Amoeba proteus. Journal of Bacteriology* 141: 1466–1469.

Hart, P. D., and Young, M. R. 1979. The effect of inhibitors and enhancers of phagosome-lysosome fusion in cultured macrophages on the phagosome membranes of ingested yeasts. *Experimental Cell Research* 118: 365–375.

Hawkins, S. E., and Wolstenholme, D. R. 1967. Cytoplasmic DNA-containing bodies and the inheritance of certain cytoplasmically determined characters in amoeba. *Nature* 214: 928–929.

Ishikawa, H. 1987. Nucleotide composition and kinetic complexity of the genomic DNA of an intracellular symbiont in the pea aphid *Acyrthosiphon pisum. Journal of Molecular Evolution* 24: 205–211.

Jeon, K. W. 1972. Development of cellular dependence on infective organisms: Micrurgical studies in amoebas. *Science* 176: 1122–1123.

Jeon, K. W. 1975. Selective effects of enucleation and transfer of heterologous nuclei on cytoplasmic organelles in *Amoeba proteus. Journal of Protozoology* 22: 402–405.

Jeon, K. W. 1983a. Integration of bacterial endosymbionts in amoebae. *International Review of Cytology Suppl.* 14: 29–47.

Jeon, K. W., ed. 1983b. *Intracellular Symbiosis*. Academic Press.

Jeon, K. W. 1986. Bacterial endosymbionts as extrachromosomal elements in amoebas. In: Wickner, R. B., Hinnebusch, A., Labowitz, A., Gunsalus, I. C., and Hollaender, A., eds., *Extrachromosomal Elements in Lower Eukaryotes*. Plenum.

Jeon, K. W. 1987. Change of cellular "pathogens" into required cell components. *Annals of the New York Academy of Sciences* 503: 359–371.

Jeon, K. W., and Ahn, T. I. 1978. Temperature sensitivity: A cell character determined by obligate endosymbionts in amoebas. *Science* 202: 635–637.

Jeon, K. W., and Danielli, J. F. 1971. Micrurgical studies with large free-living amebas. *International Review of Cytology* 30: 49–89.

Jeon, K. W., and Jeon, M. S. 1976. Endosymbiosis in amoebae: Recently established endosymbionts have become required cytoplasmic components. *Journal of Cellular Physiology* 89: 337–347.

Jeon, K. W., and Jeon, M. S. 1982. Experimental cross-infection of *Chaos carolinensis* with endosymbiotic bacteria from *Amoeba proteus*. *Journal of Protozoology* 29: 493A.

Jeon, K. W., and Lorch, I. J. 1967. Unusual intra-cellular bacterial infection in large, free-living amoebae. *Experimental Cell Research* 48: 236–240.

Jeon, K. W., and Lorch, I. J. 1969. Lethal effect of heterologous nuclei in amoeba heterokaryons. *Experimental Cell Research* 56: 233–238.

Jeon, K. W., and Lorch, I. J. 1979. Compatibility among cell components in the large, free-living amoebae. *International Review of Cytology Suppl.* 9: 45–62.

Karakashian, S. J. 1963. Growth of *Paramecium bursaria* as influenced by the presence of algal symbionts. *Physiological Zoology* 36: 52–68.

Kim, H. B., and Jeon, K. W. 1986. Protein synthesis by bacterial endosymbionts in amoebae. *Endocytobiosis and Cell Research* 3: 299–309.

Kim, H. B., and Jeon, K. W. 1987a. Actin-like host protein accumulated within symbiont-containing vesicles of amoebae as studied using a monoclonal antibody. *Endocytobiosis and Cell Research* 4: 151–166.

Kim, H. B., and Jeon, K. W. 1987b. A monoclonal antibody against a symbiont-synthesized protein in the cytosol of symbiont-dependent amoebae. *Journal of Protozoology* 34: 393–397.

Kippert, F. 1987. Endocytobiotic coordination, intracellular calcium signaling, and the origin of endogenous rhythms. *Annals of the New York Academy of Sciences* 503: 476–495.

Lee, J. J., and Fredrick, J. F., eds. 1987. *Endocytobiology III*. New York Academy of Sciences.

Lorch, I. J., and Danielli, J. F. 1953. Nuclear transplantation in amoebae. I. Some species characters of *Amoeba proteus* and *Amoeba discoides*. *Quarterly Journal of Microsccopical Science* 94: 445–460.

Lorch, I. J., and Jeon, K. W. 1980. Resuscitation of amoebae deprived of essential symbiotes: Micrurgical studies. *Journal of Protozoology* 27: 423–426.

Lorch, I. J., and Jeon, K. W. 1981. Rapid induction of cellular strain specificity by newly acquired cytoplasmic components in amoebas. *Science* 211: 949–951.

Lorch, I. J., and Jeon, K. W. 1982. Nuclear lethal effect and nucleocytoplasmic incompatibility induced by endosymbionts in *Amoeba proteus*. *Journal of Protozoology* 29: 468–470.

Lorch, I. J., Kim, H. B., and Jeon, K. W. 1985. Symbiont-induced strain-specific lethal effect in amoebae. *Journal of Protozoology* 32: 745–746.

Margulis, L. 1970. *Origin of Eukaryotic Cells*. Yale University Press.

Margulis, L. 1981. *Symbiosis in Cell Evolution*. Freeman.

Morgan, C., Rosenkranz, H. S., Carr, H. S., and Rose, H. M. 1967. *Journal of Bacteriology* 93: 1987–2002.

Moulder, J. W. 1979. The cell as an extreme environment. *Proceedings of the Royal Society of London* B204: 199–210.

Moulder, J. W. 1985. Comparative biology of intracellular parasitism. *Microbiological Reviews* 49: 298–337.

Muscatine, L., and Lenhoff, H. M. 1965. Symbiosis of hydra and algae. II. Effects of limited food and starvation on growth of symbiotic and aposymbiotic hydra. *Biological Bulletin* 129: 316–328.

Park, M. S., and Jeon, K. W. 1988. A symbiont gene coding for a protein required for the host amoeba: Cloning and expression in phage-transformed *E. coli*. *Endocytobiosis and Cell Research* 5: 215–224.

Park, M.S., and Jeon, K. W. 1990. Nucleotide sequence of a symbiotic bacterial gene coding for a protein used by the host *Amoeba proteus*. *Endocytobiosis and Cell Research* 7: 37–44.

Part, Y. M. 1983. Role of plasmid DNAs in the protection of endosymbiotic X-bacteria from digestion by amoeba. Master's thesis, University of Tennessee, Knoxville.

Pfeifer, U. 1987. Functional morphology of the lysosomal apparatus. In: Glaumann, H., and Ballard, F. J., eds., *Lysosomes: Their Role in Protein Breakdown*. Academic Press.

Prakash, R. K., and Atherly, A. G. 1986. Plasmids of *Rhizobium* and their role in symbiotic nitrogen fixation. *International Review of Cytology* 104: 1–24.

Preer, J. R. 1971. Extrachromosomal inheritance: Hereditary symbionts, mitochondria, chloroplasts. *Annual Review of Genetics* 5: 361–406.

Quinn, P. 1984. Intercellular membrane fusion. In: Dingle, J. T., Dean, R. T., and Sly, W., eds., *Lysosomes in Biology and Pathology*. North-Holland.

Rabinovitch, M., and Plaut, W. 1962. Cytoplasmic DNA synthesis in *Amoeba proteus*. I. On the particulate nature of the DNA-containing elements. *Journal of Cell Biology* 15: 525–534.

Richmond, M. H., and Smith, D. C., eds. 1979. The cell as a habitat. *Proceedings of the Royal Society of London* B204: 115–286.

Roth, L. E., and Daniels, E. W. 1961. Infective organisms in the cytoplasm of *Amoeba proteus*. *Journal Biophysical Biochemical Cytology* 9: 317–323.

Roth, L. E., Jeon, K., and Stacey, G. 1988. Homology in endosymbiotic systems: The term "Symbiosome". In: Palacios, R., and Verma, D. P. S., eds., *Molecular Genetics of Plant-Microbe Interactions*. APS Press.

Schenk, H. E. A., and Schwemmler, W., eds. 1983. *Endocytobiology II. Intracelluar Space as Oligogenetic Ecosystem*. Walter de Gruyter.

Schofield, P. R., Gibson, A. H., Dudman, W. F., and Watson, J. M. 1987. Evidence for genetic exchange and recombination of *Rhizobium* symbiotic plasmids in a soil population. *Applied and Environmental Microbiology* 53: 2942–2947.

Schwemmler, W., and Schenk, H. E. A., eds. 1980. *Endocytobiology: Endosymbiosis and Cell Biology; a Synthesis of Recent Research*. Walter de Gruyter.

Sheinin, R. 1983. Strategies utilized by papovaviruses to modify the genome of host cells. In: Schenk, H. E. A., and Schwemmler, W., eds., *Endocytobiology II; Intracellular Space as Oligogenetic Ecosystem*. Walter de Gruyter.

Sitte, P. 1983. General organization of the eucyte and its bearings on cytosymbiosis and cell evolution. In: Schenk, H. E. A., and Schwemmler, W., eds., *Endocytobiology II; Intracellular Space as Oligogenetic Ecosystem*. Walter de Gruyter.

Smith, D. C. 1980. Principles of the colonisation of cells by symbionts as illustrated by symbiotic algae. In: Schwemmler, W., and Schenk, H. E. A., eds., *Endocytobiology: Endosymbiosis and Cell Biology*. Walter de Gruyter.

Smith, D. C., and Douglas, A. E. 1987. *The Biology of Symbiosis*. Edward Arnold.

Stern, D. B. 1987. DNA transposition between plant organellar genomes. *Journal of Cell Science Suppl*. 7: 145–154.

Taylor, F. J. R. 1979. Symbioticism revisited: A discussion of the evolutionary impact of intracellular symbioses. *Proceedings of the Royal Society of London* B204: 267–286.

Taylor, F. J. R. 1987. An overview of the status of evolutionary cell symbiosis theories. *Annals of the New York Academy of Sciences* 503: 1–17.

Verma, D. P. S., and Stanley, J. 1987. Molecular interactions in endosymbiosis between legume plants and nitrogen-fixing microbes. *Annals of the New York Academy of Sciences* 503: 284–294.

Weidner, E., and Sibley, L. D. 1985. Phagocytized intracellular microsporidian blocks phagosome acidification and phagosome-lysosome fusion. *Journal of Protozoology* 32: 311–317.

Whatley, J. M. 1981. Chloroplast evolution—ancient and modern. *Annals of the New York Academy of Sciences* 361: 154–165.

Taylor, F. J. R. 1987. An overview of the status of evolutionary cell symbiosis theories. *Annals of the New York Academy of Sciences* 503: 1–17.

Verma, D. P. S., and Stanley, J. 1987. Molecular interactions in endosymbiosis between legume plants and nitrogen-fixing microbes. *Annals of the New York Academy of Sciences* 503: [illegible]–104.

Weidner, E., and Sibley, L. D. 1985. Phagocytized intracellular microsporidian blocks phagosome acidification and phagosome-lysosome fusion. *Journal of Protozoology* 32: 311–317.

Whatley, J. M. 1981. Chloroplast evolution—ancient and modern. *Annals of the New York Academy of Sciences* 361: 154–155.

III

Symbiosis in Cell Evolution

10 Status of the Theory of the Symbiotic Origin of Undulipodia (Cilia)

Gregory Hinkle

One of the most striking characteristics punctuating the division between the prokaryotes and the eukaryotes is the eukaryotic undulipodium and its underlying structure, the kinetosome. There are no known prokaryotes with structures resembling or even intermediate in form with eukaryotic undulipodia. Nor are there eukaryotes known with motility organelles resembling bacterial rotary motors. Two mutually exclusive theories have been proposed to account for the evolution of the eukaryotic organelle. The theory of direct filiation maintains that undulipodia arose through a development of microtubular structures within the lineage leading to eukaryotes. (For details of the argument see Lee and Fredrick 1987.) Most proponents of the direct-filiation theory believe that microtubules first evolved to separate chromosomes during mitosis and meiosis, and were subsequently redeployed in some manner in the form of a motility organelle—the undulipodium or some precursor (Pickett-Heaps 1974). The symbiotic theory proposes an origin of the undulipodium via a symbiosis between a motile bacterium and a less mobile (i.e., nonmitotic, perhaps nucleus-containing) host which lacked microtubules. (See Margulis 1981 for an extended treatment of this subject.) Contrary to the direct-filiation theory, the symbiotic theory posits that the motility organelle, the undulipodium, formed first, and was subsequently redeployed in the mitotic apparatus.

A Brief Statement on the Term *Undulipodia*

Undulipodia refers to all [9(2) + 2] (where the numbers refer to the arrangement of the microtubules in cross-section) organelles, such as cilia, and is preferable to *flagella* or *eukaryotic flagella*, terms which con-

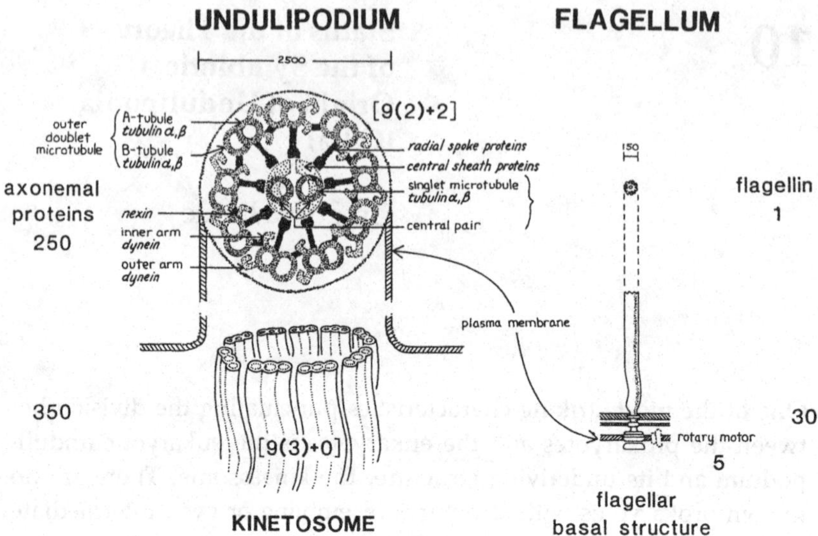

Figure 1
Comparison of undulipodium and flagellum, with numbers of proteins in each structure and arrangement of microtubules shown in brackets.

fuse the eukaryotic structure with bacterial flagella (figure 1). Since there is no evidence that undulipodia and bacterial flagella are evolutionarily homologous structures, the use of the term *flagella* for both organelles is overtly misleading.

Attempts to find the bacterial precursors of undulipodia have focused on spirochete bacteria. Electron micrographs of negatively stained spirochetes regularly have 7–21-nm tubules inserted at the base of the flagellar rotary motors (Hovind-Hougen 1976). In addition, a number of permanent motility symbioses involving spirochetes and otherwise much less mobile protists have been described (e.g., *Mixotricha paradoxa*; see Grimstone and Cleveland 1964 and figure 2). Using a polyclonal anti-tubulin antiserum (raised against guinea pig brain tubulin), Fracek and Stolz (1985) found that whole cells of *Spirochaeta bajacaliforniensis* were immunoreactive. Subsequent analysis using Western blots of whole-cell spirochete extracts demonstrated the presence of two proteins recognized by the anti-tubulin antiserum (Obar 1985; Bermudes et al. 1987).

Eukaryotic tubulin is regularly purified by taking advantage of its unusual property of assembly into tubules at 37°C and dissolution at 4°C. Utilizing a series of cold and warm centrifugations, tubulin can

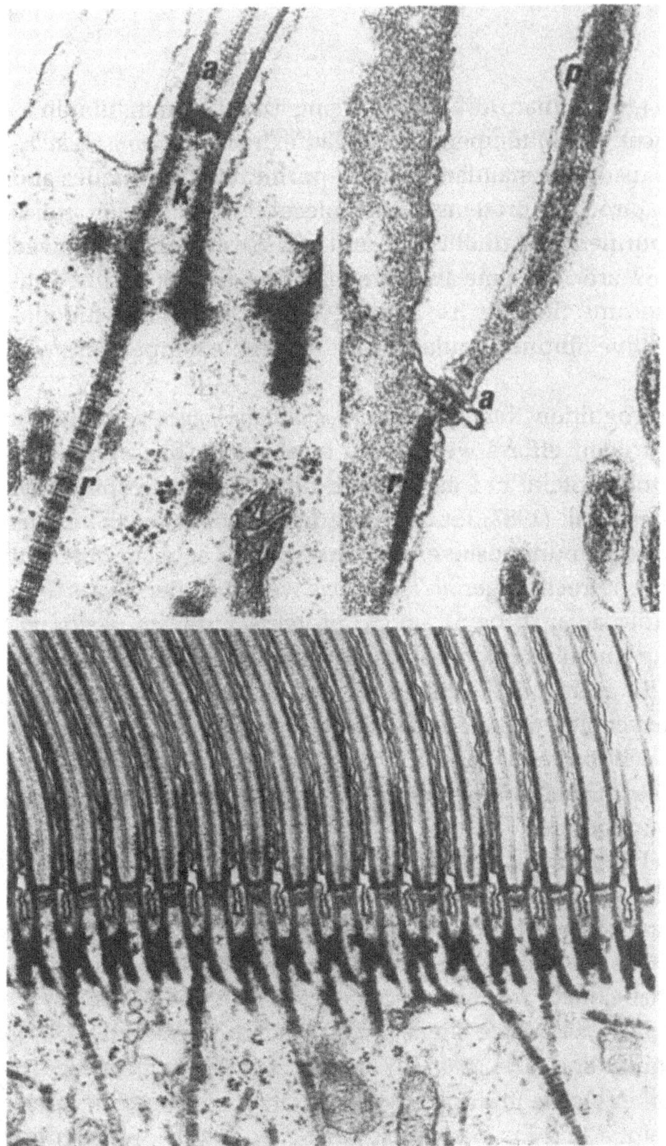

Figure 2
Comparison of spirochete attachment sites with clam gill cilia. Top left: Transmission electron micrograph of a clam gill cilium (× 40,000). r = root fiber, k = kinetosome, a = axostyle. Top right: Attachment of a spirochete to a *Pyrsonympha*-like protist from the hindgut of the California subterranean termite *Reticulitermes hesperus*. r = root-fiber-like structure, p = protoplasmic cylinder of spirochete, a = attachment site. Transmission electron micrograph (× 33,000). Bottom: A row of clam gill cilia. Transmission electron micrograph (× 22,000). Top left and bottom courtesy of Fred Warner, University of Syracuse, and W. H. Freeman, San Francisco; top right courtesy of David Chase and W. H. Freeman, San Francisco.

be purified to greater than 90% purity. Using *bona fide* brain tubulin as a control, Fracek (1984) temperature-cycled extracts of *Spirochaeta bajacaliforniensis* using the standard tubulin-purification techniques and buffers. Two spirochete proteins, with molecular weights of 65 and 45 kD, were copurified and dubbed S1 and S2. Western blots showed that S1 and S2 are the same two proteins recognized by the anti-tubulin antiserum. Soluble S1 and S2 also formed anti-tubulin-antibody-sensitive fibrous bundles *in vitro* when the temperature was raised to 37°C.

After the recognition that *Spirochaeta bajacaliforniensis* contained a tubulin-like protein, efforts were made to elucidate the amino-acid sequence of the protein and its possible function in the spirochete cells. Bermudes et al. (1987) found that other members of the genus *Spirochaeta* were immunopositive with anti-tubulin antisera, but that members of the spirochete genus *Treponema* were not. Bermudes then raised antibodies against the S1 and S2 proteins and used *in situ* immunochemistry in an attempt to localize the proteins in *S. bajacaliforniensis*. Anti-S1 and anti-S2 epitopes were located throughout the length of deoxycholate-treated cells. The anti-S1 and anti-S2 antisera did not react with bacterial flagella or with brain tubulin. Both S1 and S2 appear to be located in the periplasm of the spirochete; their function remains unknown.

Tzertzinis (1989) made a genomic library of *Spirochaeta bajacaliforniensis* in a PEX expression plasmid and screened the library with an anti-tubulin antiserum. He obtained a single immunopositive clone containing a 60-base-pair insert. The sequence of 20 amino acids derived from this cloned portion of spirochete DNA has 35% (7 of 20 residues identical) sequence similarity with amino acids 113–134 of pig alpha tubulin and 30% (6 of 20) similarity with pig beta tubulin (table 1). In other words, the spirochete sequences show greater similarity to pig alpha tubulin than to pig beta tubulin. Using the 60-bp sequence as a probe, we are in the process of obtaining the entire gene sequence.

Tzertzinis (1989) also obtained additional amino-acid-sequence data of the S1 protein from *Spirochaeta bajacaliforniensis* by directly sequencing purified cyanogen-bromide-derived fragments. (The amino terminus is apparently blocked to Edman degradation.) He was able to sequence six peptides with a total of 80 amino-acid residues. Though each of the peptides that were sequenced contained an immunogenic epitope recognized by both anti-tubulin and anti-S1 antisera, none of

Table 1
Alignment of inferred spirochete sequence with tubulin sequences. Amino-acid sequence inferred from a 60-bp sequence of *Spirochaeta bajacaliforniensis* DNA cloned in a PEX expression plasmid and screened with an anti-tubulin antiserum. (From Tzertzinis 1989.)

Sp 289	L	Q	**V**	**D**	E	D	**L**	—	**R**	Q	**R**	**R**	V	D	A	L	Q	L	**G**	**L**	**Q**
Pig α	E	I	I	**D**	L	V	**L**	D	**R**	I	**R**	K	L	A	D	Q	C	T	**G**	**L**	**Q**
Pig β	E	L	**V**	**D**	S	V	**L**	D	V	V	**R**	K	E	S	E	S	C	D	C	**L**	**Q**

The pig α and β tubulin sequences are amino acids 113 and 134, as detailed in Little and Seehaus 1988. Identical residues of spirochete and tubulin sequences are in boldface. One gap (—) is introduced in the spirochete sequence.

the peptides had obvious sequence similarity with any known eukaryotic tubulin sequence. Presumably the recognized epitopes are in regions of the peptides that were not sequenced. Some of the Sl sequences obtained from Edman degradation of cyanogen bromide peptides have similarity with known DNA binding proteins, such as *E. coli* RNA polymerase and bacteriophage SPO1 DNA-binding protein. Claiming that this similarity is any reflection of the function of S1 in the cell at this time would be only speculation.

Further evidence supporting the symbiotic origin of undulipodia comes from an ongoing genetic study of motility mutants in the unicellular alga *Chlamydomonas reinhardtii*. Motility mutants of *C. reinhardtii* (about 20 have been isolated) have unusual patterns of segregation and linkage (Dutcher 1986). All the motility mutants segregate in a 2:2 Mendelian manner and are on the same linkage group (group 19). None of the 200+ other genes already mapped in *C. reinhardtii* are located on linkage group 19. Such a degree of clustering of related phenotypes is very unusual. Also, the 20 mutants map in a circle, which is unusual for a eukaryotic linkage group and is reminiscent of the linkage groups/genophores of mitochondria and plastids. Lastly, the order of the genes on linkage group 19, alone among all the known linkage groups, is sensitive to recombination if the temperature is raised during a brief window of time prior to meiosis and zygospore formation. Again, this is highly unusual and suggests that linkage group 19 is fundamentally different from the other 18 well-characterized linkage groups of *Chlamydomonas*.

Linkage group 19 was recently shown to reside external to the nucleus. Using *in situ* DNA hybridization in *Chlamydomonas reinhardtii*, Hall et al. (1989) presented data demonstrating the presence of DNA (linkage group 19) within the kinetosomes, the [9(3) + 0] structures underlying the axonemes of the undulipodia. Pulse-field electrophoresis showed that the size of the kinetosomal DNA is in the range of 7 megabases—an enormous amount of DNA, equal to that of an entire prokaryotic genome. Whether the kinetosomal DNA is translated to protein has not yet been determined, although 2-D gel electrophoresis and ultrastructural evidence suggest that a number of the motility mutants of *C. reinhardtii* lack kinetosomal proteins (Dutcher 1986). Although nothing is known about the presence of histones, nucleosomes, etc., because the segregation pattern of the mutants was Mendelian (2:2), linkage group 19 was thought to be another nuclear chromosome. (The segregation pattern for the other cytoplasmic linkage groups, the mitochondria and the plastid genomes, is generally 4:0 or 0:4.) Although the mechanics of segregation at the cellular level are still unclear, the autonomous replication of the kinetosomes and presumably the DNA therein provides a plausible mechanism for the 2:2 Mendelian segregation patterns seen for all the motility mutants of C. *reinhardtii*.

The genetic and physical autonomy of kinetosomes, which has long been recognized, remains an enigma (Wheatley 1982) for those cell biologists unaware of or unsympathetic to the theory of the symbiotic origin of undulipodia. The symbiotic origin of two classes of extranuclear, DNA-containing organelles—mitochondria and plastids—has been largely accepted. From the perspective of the symbiotic theory, the recent discovery of kinetosomal DNA is not surprising. Indeed, proponents of the symbiotic theory have long predicted the presence of extranuclear nucleic acids as a means of explaining such non-Mendelian phenomena as cortical inheritance in ciliates (Grell 1973) and motility mutants in *Chlamydomonas* (Dutcher 1986). Although the semiautonomy of plastids and mitochondria had long been recognized, the regular existence of DNA in mitochondria and plastids proved to be the convincing evidence supporting their symbiotic origin. The discovery of kinetosomal DNA and the genetic autonomy of kinetosomes argue for symbiosis as the source of this evolutionary novelty.

The origin of undulipodia remains a mystery. The presence of a tubulin-like protein in free-living spirochetes and the discovery of DNA integrally associated with kinetosomes in C. *reinhardtii* support a symbiotic origin for this organelle. With the tools provided by advances in molecular biology, a definitive answer should not be long pending.

References

Bermudes, D., Fracek, S. P., Laursen, R. A., Margulis, L., Obar, R., and Tzertzinis, G. 1987. Tubulinlike protein from *Spirochaeta bajacaliforniensis*. In: Lee, J. J., and Fredrick, J. F., eds., *Endocytobiology III*. New York Academy of Sciences.

Dutcher, S. K. 1986. Genetic properties of linkage group XIX in *Chlamydomonas reinhardtii*. In: Wickner, R. B., Hinnebusch, A., Lambowitz, A. M., Gunsalus, I. C., and Hollaender, A., eds., *Extrachromosomal Elements in Lower Eukaryotes*. Plenum.

Fracek, S. P., Jr. 1984. Tubulin-like Proteins of *Spirochaeta bajacaliforniensis*, a New Species from a Microbial Mat at Laguna Figueroa, Baja California del Norte, Mexico. Ph.D. dissertation, Boston University.

Fracek, S. P, Jr., and Stolz, J. F. 1985. *Spirochaeta bajacaliforniensis* sp. nov. from a microbial mat community at Laguna Figueroa, Baja California Norte, Mexico. *Arch. Microbiol.* 142: 317–325.

Grell, K. 1973. *Protozoology*. Springer-Verlag.

Grimstone, A. V., and Cleveland, L. R. 1964. The structure of *Mixotricha* and its associated microorganisms. *Transactions of the Royal Society of London, Series B* 159: 668–686.

Hall, J., Hall, J. L., Ramanis, Z., and Luck, D. J. L. 1989. Basal body/centriolar DNA: Molecular genetic studies in *Chlamydomonas. Cell* 59: 121–132.

Hovind-Hougen, K. 1976. Ultrastructure of cells of *Treponema pertenue* obtained from experimentally infected hamsters. *Acta Pathol. Microbiol. Scand. B* 84: 101–108.

Lee, J. J., and Fredrick, J. F. 1987. *Endocytobiology III*. New York Academy of Sciences.

Little, M., and Seehaus, T. 1988. Comparative analysis of tubulin sequences. *Comp. Biochem. Physiol.* 90B: 665–670.

Margulis, L. 1981. *Symbiosis in Cell Evolution*. Freeman.

Obar, R. 1985. Purification of Tubulin-like Protein from a Spirochete. Ph.D. dissertation, Boston University.

Pickett-Heaps, J. D. 1974. Evolution of mitosis and the eukaryotic condition. *Biosystems* 6: 37–48.

Tzertzinis, G. 1989. Immunochemical Characterization and Partial Amino Acid Sequence of Tubulin-like Protein from *Spirochaeta bajacaliforniensis*. Ph. D. dissertation, Boston University.

Wheatley, D. N. 1982. *The Centriole: A Central Enigma of Cell Biology*. Elsevier.

Note added in proof Further peptide-sequence analysis by Tzertzinis and Obar has shown that the spirochete protein S1 belongs to a class of heat-shock proteins with scant amino acid sequence similarity to tubulin but sharing many properties with tubulin, including temperature-dependent polymerizeration and nucleotide binding.

11 *Cyanophora paradoxa* Korschikoff and the Origins of Chloroplasts

Robert K. Trench

The Serial Endosymbiosis Theory (SET) of the origins of chloroplasts (Margulis 1970, 1981) and the various modifications thereof (Taylor 1974; Gray and Doolittle 1982) propose that some chloroplasts originated through polyphyletic associations between various cyanobacteria and the precursors of eukaryotic cells (Raven 1970). Since chloro plasts are genetically semiautonomous, relying on nuclear encoded genes, cytoplasmic translation, and transport of gene products for their biosynthesis (Ellis 1981; Bottomly and Bohnert 1982), a central component of the SET involves the transfer of genes from the incorporated cyanobacteria to the nuclei of their hosts.

The existence of modern examples of symbioses involving cyanobacteria and eukaryotic hosts, for example *Cyanophora paradoxa, Glaucocystis nostochinearum, Paulinella chromatophora,* and *Gloeochaete wittrockiana* (Trench 1982), has often led to the interpretation that these associations represent a replay of the series of events postulated by the SET for the origin of chloroplasts. In fact, the photosynthetic units occurring in the organisms mentioned above, combined with *Cyanidium caldarium*, have been interpreted as representing all possible stages in the postulated process—that is, symbioses between cyanobacteria and eukaryotes (Trench 1982), cyanobacteria "on the way to becoming chloroplasts" (Fredrick 1981; Seckbach et al. 1983), and *de facto* chloroplasts (Herdman and Stanier 1977; Aitken and Stanier, 1979). Where the existence of a peptidoglycan wall associated with the "cyanelle" has been documented, either ultrastructurally or chemically (Kies 1980; Aitken and Stanier 1979), its significance has been postulated to represent "a vestige of their recent evolutionary past" (Aitken and Stanier 1979). Some of the examples mentioned above also demonstrate a vacuolar membrane enclosing the inclusions; others do not.

Most significant advances in the study of these cyanomes have been made on two fronts: in the mapping of the cyanellar genome, and in *in vivo* and *in vitro* studies of the biosynthesis of *C. paradoxa* cyanellar proteins. Most of the gene-mapping studies, exclusively conducted with *C. paradoxa* (Pringsheim strain; until recently, the Kies strain; see Breiteneder et al. 1988), have been elegantly reviewed by Wasmann et al. (1987). In many ways, it is unfortunate that studies of the genome of the host have not progressed at the same pace. Studies of protein biosynthesis have lagged, but recent reports (e.g., Burnap and Trench 1989b,c) corroborate much of the evidence from gene-mapping studies.

Briefly, the circular genome of the cyanelles of *C. paradoxa* has a complexity of 127 kb (136 kb in the Kies strain) containing an inverted repeat and two single copy regions. The cyanellar DNA is present in about 40 copies. The distribution of genes in the cyanellar chromoneme is indicated in figure 1. Some 23 genes have been mapped to the cyanellar chromoneme; major gaps exist in location of the genes for the ribosomal proteins, where only S19 from the small subunit of the ribosome and L2 from the large subunit have been detected (table 1). (See addendum.)

Studies of protein biosynthesis (Bayer and Schenk 1986; Burnap and Trench 1989b) have indicated that, consistent with the reduced genome size, cyanellae import some 80% of their constituent proteins from the host. The results of host-RNA-directed *in vitro* translation studies indicate that linker polypeptide L1 of the phycobilisome, the gamma subunit of coupling factor CF1, and subunit II of PS I are synthesized in the cytoplasm as precursor molecules which are 5–8 kD larger than their mature sizes. Antibodies against the psbA gene product (the D1 protein) precipitated a polypeptide from the translation products of cyanellar-RNA-directed synthesis which is about 1.5 kD larger than the mature protein (Burnap and Trench 1989b,c).

Studies of gene distribution and polypeptide biosynthesis indicate that the cyanelles of *C. paradoxa* are analogous to chloroplasts and can be regarded as semi-autonomous, but should they be regarded as *de facto* chloroplasts? Regarding cyanelles as chloroplasts creates two problems. First, since the cyanelles possess a peptidoglycan wall (and presumably the genes encoding enzymes involved in its synthesis), then, by analogy, free-living cyanobacteria should be thought of as "free-living chloroplasts" (an untenable notion!). Second, major systematic problems arise, as in the proposal to create the Glauco-

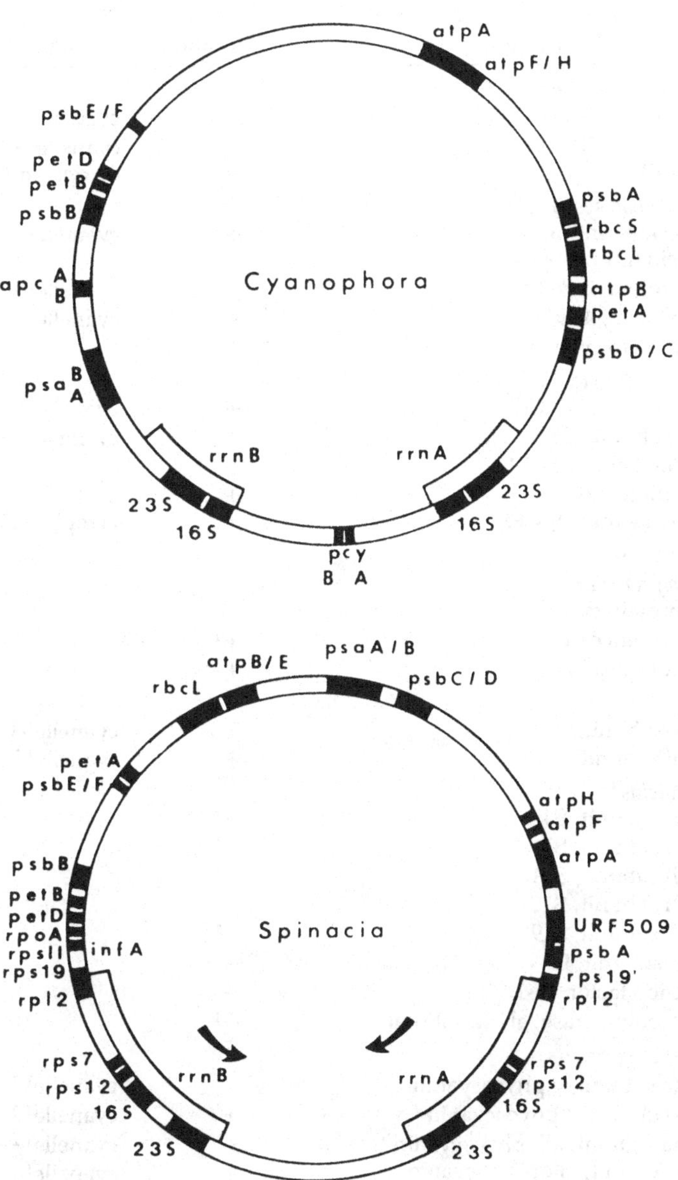

Figure 1
Comparative chromosome structures of cyanelle DNA and spinach chloroplast DNA. The symbols of genes (see table 1) are included at the positions mapped or sequenced. (From Wasmann et al. 1987; reprinted with permission.)

Table 1
Proteins coded in the cyanellar genome and their inferred sites of synthesis.

Gene symbol	Description	Cyanelle location	Site of synthesis
Components of photosystem I			
psaA	apoprotein P700a	+	cyanelle[1,2]
psaB	apoprotein P700b	+	cyanelle[1,2]
	subunit II	−	**cytoplasm[2]**
Components of photosystem II			
psbA	herbicide-binding protein (D1 protein)	+	cyanelle[1,2]
psbB	apoprotein of P680	+	cyanelle[1,2]
psbC	apoprotein, 44 kD	+	cyanelle[1,2]
psbE	apoprotein cyt b-559	+	
ATP-synthetase complex			
atpA	gamma subunit, CF1	+	cyanelle
atpB	beta subunit, CF1	+	cyanelle
atpE	epsilon subunit, CF1	+	
atpF	subunit I, CF1	+	
	gamma subunit, CF1	−	**cytoplasm[2]**
atpH	subunit III, CF_o	+[1]	
Cytochrome-b/f complex			
petA	apoprotein cy-f	+[1]	
petB	apoprotein cy-b	+[1]	
petD	b_6/f complex IV[6]	+[1]	
Rubisco			
rbcL	large subunit	+	cyanelle[1,2]
rbcS	small subunit	+	cyanelle[1,2]
Ribosomal proteins[3]			
rps4	small subunit, S4	−[1]	
rps7	small subunit, S7	−[1]	
rps11	small subunit, S11	−[1]	
rps12	small subunit, S12	−[1]	
rps19	small subunit, S19	+[1]	
rpl2	large subunit, L2	+[1]	
infA	intiation factor, IF-1	−[1]	
rpoA	RNA polymerase, alpha subunit	−[1]	
Phycobilisome components			
pcyA	alpha subunit, c-phycocyanin	+	cyanelle[1,2]
pcyB	beta subunit, c-phycocyanin	+	cyanelle[1,2]
apcA	alpha subunit, allophycocyanin	+	cyanelle[1,2]
apcB	beta subunit, allophycocyanin	+	cyanelle[1,2]
	anchor polypeptide	+(?)	cyanelle
	linker polypeptide, L1		**cytoplasm[2]**
	linker polypeptide, L2		**cytoplasm[2]**
	linker polypeptide, L3		**cytoplasm[2]**
	linker polypeptide, L4		**cytoplasm[2]**

1. Wasmann et al. 1987
2. Burnap and Trench 1989b,c
3. See addendum on page 148.

cystophyta (Kies and Kramer 1986)—in my opinion, an unnatural assemblage. The analysis of 5S and 16S RNAs and the study of phylogenetic relatedness conducted by Wolters and Edelman (1988) place the host or nonphotosynthetic portion of *Cyanophora* among the euglenoids. Similar data should be gathered for the other cyanomes. Burnap and Trench (1989b) consider cyanomes as examples of a photosynthetic eukaryote losing its chloroplasts and acquiring cyanobacteria. The duplication of genes (in the host genome and in the cyanelles) resulted in a loss of cyanellar genes, producing a result analogous to the chloroplast.

The observation that several cyanellar polypeptides are synthesized in the host cytoplasm as precursor molecules larger than their mature size suggests the hypothesis that an intercompartmental protein-translocation mechanism, similar to that found in *bona fide* chloroplast systems, operates during the biosynthesis of the cyanelles. Cytoplasmically synthesized chloroplast proteins are made as precursors which are post-translationally transported across the outer and inner membranes of the chloroplast envelope by energy-dependent processes. After their importation, the precursors are proteolytically cleaved at the N-terminus to remove a 3–7-kD peptide, termed the "transit sequence," that appears to contain structural information required for the targeting of the proteins to their proper locations within the chloroplast (Smeekenes et al. 1986; Koharn and Tobin 1989). It is not known how transport from host cytoplasm into the cyanellae is effected in *C. paradoxa*. Cyanellar proteins synthesized in the host cytoplasm would have to traverse the host vacuolar membrane, the peptidoglycan wall, and the plasmalemma of the cyanelle (Trench 1982) in order to gain access to the cytoplasm of the cyanelle. It should be borne in mind, however, that the D1 protein, synthesized within the cyanelle, is also made as a precursor molecule larger than the mature protein. Perhaps the common feature is that these proteins are all temporarily, or ultimately, membrane-associated, and are synthesized with a "transit sequence" and/or a "target sequence."

But the origins of the genes coding the "transit" sequences remain obscure. If it is assumed that the cyanobacterial precursor of the chloroplast was not involved in the importation of externally synthesized polypeptides, then it is unlikely that such "transit" genes would have been transferred to the host nucleus from the cyanobacteria. The alternative then is that these "transit" genes originated in the host. Perhaps further comparisons of "transit" sequences will provide insight into this problem.

Addendum

Since this chapter was written, three papers that bear on the subject have appeared. In addition to the small subunit ribosomal protein S19 and the large subunit ribosomal protein L2 (table 1), Evrard et al. (1990) and Bryant and Stirewalt (1990) have located in the cyanellar genome of *C. paradoxa* the genes for the large-subunit ribosomal proteins rp133, rp122, rp135, rp120, rp15, and a portion of rp16 and those for the small-subunit ribosomal proteins rps18, rps8, and rps12. Homologues of rp135, rp15, and rp16 are not found in the genome of plant chloroplasts.

Using the inhibitor approach, Bayer and Schenk (1989) found evidence that the biosynthesis of ferredoxin was insensitive to cycloheximide amd α-amanitin, but was inhibited by chloramphenicol, lincomycin and rifampicin, indicating cyanellar synthesis and implying that the gene for ferredoxin is located in the cyanellae. This is also unlike the situation in plant chloroplasts.

These new pieces of evidence are not inconsistent with the hypothesis advanced by Burnap and Trench (1989c).

Additional References

Bayer, M. G., and Schenk, H. E. A. 1989. Ferredoxin of *Cyanophora paradoxa* Korsch. is encoded on cyanellar DNA. *Curr. Genet.* 16: 311–313.

Bryant, D. A., and Stirewalt, V. L. 1990. The cyanelle genome of *Cyanophora paradoxa* encodes ribosomal proteins not encoded by the chloroplast genomes of higher plants. *FEBS Lett.* 259: 273–280.

Evrard, J. L., Kuntz, M., and Weil, J. H. 1990. The nucleotide sequence of five ribosomal protein genes from the cyanelles of *Cyanophora paradoxa*: Implications concerning the phylogenetic relationship between cyanelles and chloroplasts. *J. Mol. Evol.* 30: 16–25.

References

Aitken, A., and Stanier, R. Y. 1979. Characterization of peptidoglycan from the cyanellae of *Cyanophora paradoxa*. *J. Gen. Microbiol.* 112: 219–223.

Bayer, M., and Schenk, H. E. A. 1986. Biosynthesis of proteins in *Cyanophora paradoxa*: Protein import into the endocyanelle analyzed by micro two-dimensional gel electrophoresis. *Endocyt. C. Res.* 3: 197–202.

Bottomly, W., and Bohnert, H. J. 1982. The biosynthesis of chloroplast proteins. In: Parthier, B., and Boulter, D., eds., *Encyclopedia of Plant Physiology*, new series, volume 14B. Springer-Verlag.

Breiteneder, H., Seiser, C., Löffelhardt, W., Michalowski, C., and Bohnert, H. 1988. Physical map and protein gene map of cyanelle DNA from the second known isolate of *Cyanophora paradoxa* (Kies-strain). *Curr. Genet.* 13: 199–206.

Burnap, R. L., and Trench, R. K. 1989a. The biogenesis of the cyanellae of *Cyanophora paradoxa*. I. Polypeptide composition of the cyanellae. *Proc. Roy. Soc. Lond. B* 238: 53–72.

Burnap, R. L., and Trench, R. K. 1989b. The biogenesis of the cyanellae of *Cyanophora paradoxa*. II. Pulse-labelling of cyanellar polypeptides in the presence of transcriptional and translational inhibitors. *Proc. Roy. Soc. Lond. B* 238: 73–87.

Burnap, R. L., and Trench, R. K. 1989c. The biogenesis of the cyanellae of *Cyanophora paradoxa*. III. *In vitro* synthesis of cyanellar polypeptides using separated cytoplasmic and cyanellar RNA. *Proc. Roy. Soc. Lond. B* 238: 89–102.

Ellis, R. J. 1981. Choroplast proteins: Synthesis, transport and assembly. *Annual Rev. Plant Physiol* 33: 111–137.

Ellis, R. J. 1982. Inhibitors for studying chloroplast transcription and translation *in vivo*. In: Edelman, M., Hallick, R. B., and Chua, N.-H., eds., *Methods in Chloroplast Molecular Biology*. Elsevier.

Fredrick, J. F. 1981. The biosynthesis of storage glucans in prokaryotic and eukaryotic algae. *Ann. N.Y. Acad. Sci.* 361: 426–434.

Gray, M. W., and Doolittle, W. F. 1982. Has the endosymbiont hypothesis been proven? *Microbiol. Rev.* 46: 1–42.

Herdman, M., and Stanier, R. Y. 1977. The cyanellae: Chloroplast or endosymbiotic prokaryote? *FEMS Microbiol. Lett.* 1: 7–12.

Kies, L. 1980. Morphology and systematic position of some endocyanomes. In: Schenck, H. E. A., and Schwemmler, W., eds., *Endocytobiology*. Walter de Gruyter.

Kies, L., and Kramer, B. P. 1986. Typification of the Glaucocystophyta. *Taxon* 35: 128–133.

Koharn, B. D., and Tobin, E. M. 1989. A hydrophobic, carboxy-terminal region of a light-harvesting chlorophyll a/b protein is necessary for stable integration into thylakoid membranes. *Plant Cell* 1: 159–166.

Margulis, L. 1970. *Origin of Eukaryotic Cells*. Yale University Press.

Margulis, L. 1981. *Symbiosis in Cell Evolution*. Freeman.

Raven, P. 1970. A multiple origin for plastids and mitochondria. *Science* 169: 641–645.

Seckbach, J., Fredrick, J. F., and Garbary, D. J. 1983. Auto- or exogenous origin of transitional algae: An appraisal. In: Schenk, H. E. A., and Schwemmler, W., eds., *Endocytobiology II*. Walter de Gruyter.

Smeekenes, S., Bauerle, C., Hageman, J., Keegstra, K., and Weisbeek, P. 1986. The role of the transit peptide in the routing of precursors towards different chloroplast compartments. *Cell* 46: 365–375.

Taylor, F. J. R. 1974. Implications and extensions of the serial endosymbiosis theory for the origin of eukaryotes. *Taxon* 23: 229–258.

Trench, R. K. 1982. Physiology, biochemistry and ultrastructure of cyanellae. In: Round, F. E., and Chapman, D. J., eds., *Progress in Phycological Research*, volume I. Elsevier.

Wasmann, C. C., Löefferhardt, W., and Bohnert, H. J. 1987. Cyanelles: Organization and molecular biology. In: Fay, P., and van Baalen, C., eds., *The Cyanobacteria: A Comprehensive Review*. Elsevier.

Wolters, J., and Edelman, V. A. 1988. Cladistic analysis of ribosomal RNAs: The phylogeny of eukaryotes with respect to the endosymbiotic theory. *BioSystems* 21: 209–214.

IV Animals and Bacteria

12 Serial Endosymbiosis Theory and Weevil Evolution: The Role of Symbiosis

Paul Nardon and Anne-Marie Grenier

Does symbiosis play a role in evolution? If so, what is the extent of this role? Here we attempt to address these two questions, using as our example the case of the weevil, *Sitophilus oryzae*.

To assert that symbiosis plays a role in evolution, we must establish that either the establishment or the rupture of a symbiotic association has important consequences for the associated partners. *Sitophilus oryzae* (figure 1) provides a simple example. The host, a weevil, and the symbionts,[1] Gram-negative bacteria, are genetically, morphologically, and physiologically distinct organisms. Despite the observation that the weevil can develop and reproduce in the absence of its symbiont (thus the symbiosis is not obligatory), the bacteria are perfectly integrated into the weevil, which controls both their location and their number (Nardon 1978a,b; Nardon and Wicker 1983). The bacteria behave as permanent organelles of the female germ line (Scheinert 1933; Nardon 1971). Apart from the ovaries, they are found in both sexes only in a larval bacteriome located at the junction of the foregut and the midgut but not connected to the lumen (figure 2). Since the bacterium has not been cultivated outside the weevil, we consider the symbiosis to be obligatory for the growth of these bacteria.

During metamorphosis, the bacteriocytes migrate back along the intestine and are incorporated into the anterior midgut caeca. But they are rapidly eliminated, and in adults older than 3 weeks the bacteria are found only in the ovaries, the apical bacteriomes, the trophocytes, and the oocytes (Mansour 1930; Schneider 1956; Nardon and Wicker 1981). In the egg, the bacteria are scattered among the yolk spheres. They are particularly numerous at the posterior pole (figure 3), where they lie in close contact with the oosome (Nardon 1971), allowing the earliest differentiated germ cells in both males and females to be infected at the very beginning of embryogenesis. For

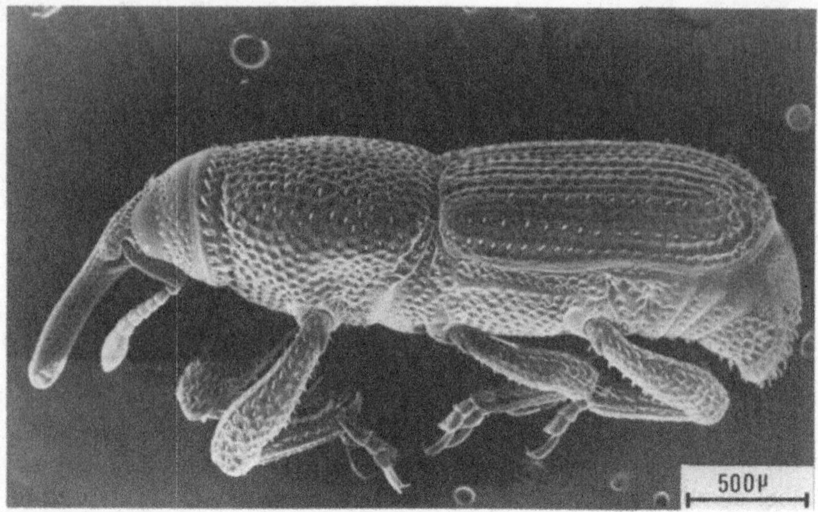

Figure 1
The weevil *Sitophilus oryzae* L.

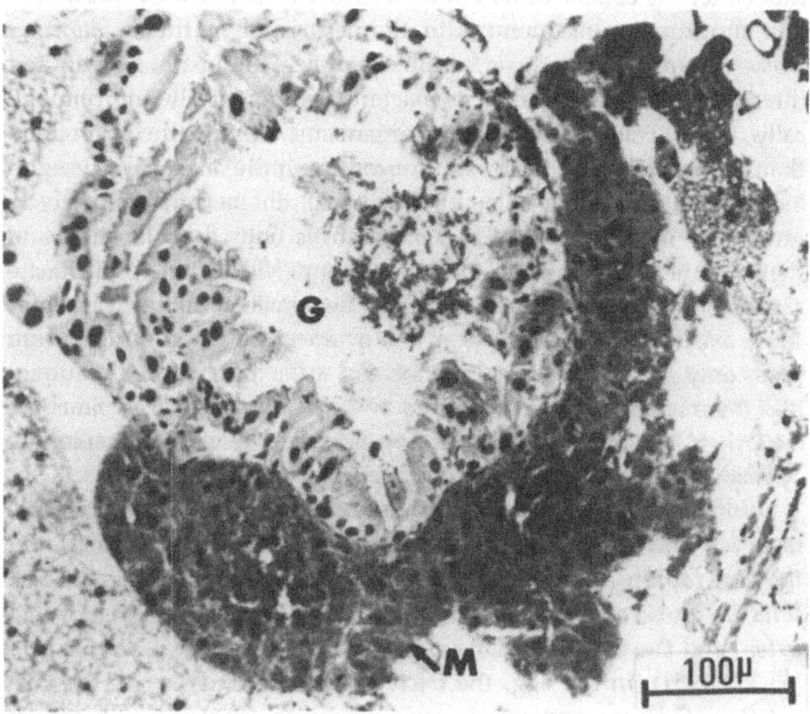

Figure 2
Transverse section of a *S. oryzae* larva stained with toluidine blue. G = midgut, M = larval bacteriome.

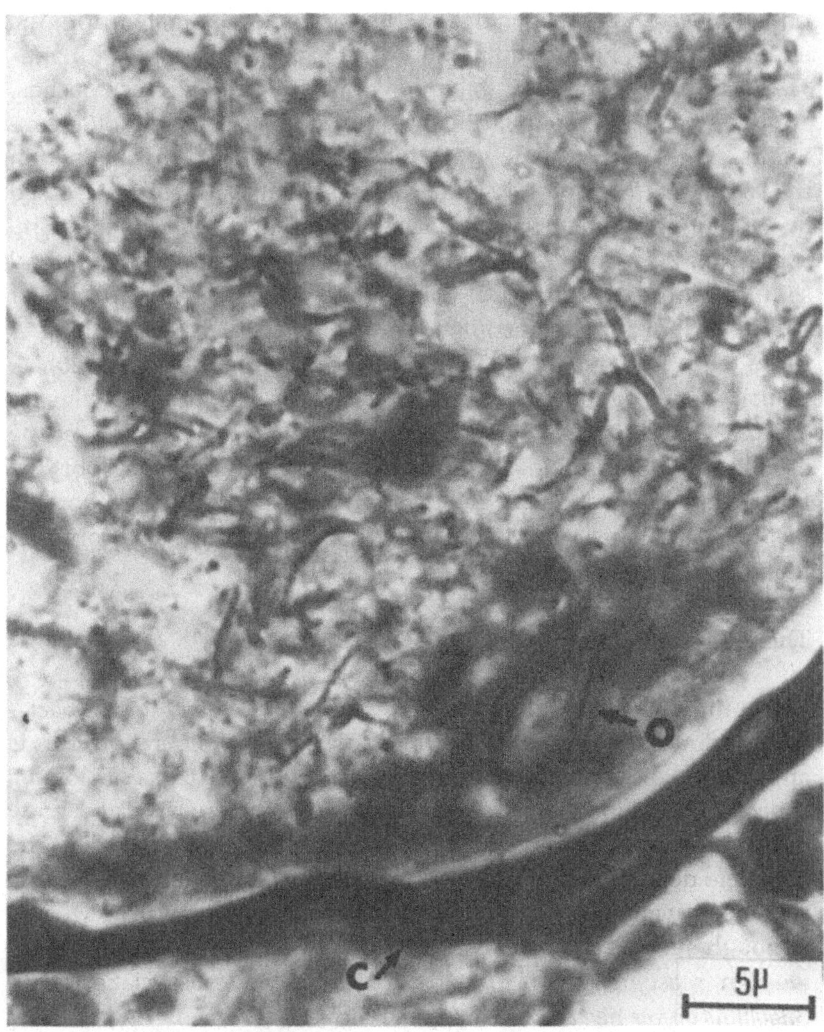

Figure 3
Longitudinal section of a mature oocyte of *S. oryzae* stained with toluidine blue (pH 5.4). Note the numerous bacteria in contact with the oosome. C = chorion, O = oosome.

unknown reasons, the bacteria are very rapidly eliminated from the male germ line and the rudiments of the testes never contain them.

In most insect species symbiotic with bacteria (e.g., cockroaches, aphids, and leafhoppers), the host cannot survive without symbionts. Since aposymbiotic strains of *Sitophilus* have been obtained in the laboratory (Nardon 1973), the role of symbiosis has been assessed by comparing symbiotic and aposymbiotic weevils (Nardon and Grenier 1988). Aposymbiotic strains have also been found in the natural setting (Mansour 1930, 1935; Schneider 1954; Musgrave and Miller 1956; Nardon and Grenier 1989). We have discussed the meaning of these wild aposymbiotic strains elsewhere (Grenier et al. 1986). A plausible explanation is that modern cereals are richer in growth substances than those which existed before mankind at the time of the establishment of symbiosis in weevils, and therefore can support the full development of aposymbiotic strains. This view is corroborated by the fact that it is easier to obtain aposymbiotic weevils on sorghum than on wheat, which has less riboflavin.

The Symbiotic Bacteria

Disrupting the symbiotic association results in the death of the bacteria (Nardon 1973). Unless we invoke spontaneous generation, this is an example of a symbiont that, over evolutionary time, has lost its capacity to live independently and can no longer survive and grow outside the host, even in the most sophisticated culture media. The complete dependence on the host may be the consequence of the loss of genetic material by mutation and/or gene transfer to the weevil, although such an interpretation remains to be proved. Nonetheless, we can reasonably conclude that the symbiotic association with *Sitophilus oryzae* has modified the symbiont; it is difficult to go further in an analysis, since we do not know the free-living form.

This case is not unique. Most intracellular bacteria harbored by insects cannot be cultured *in vitro*. However, certain yeast strains symbiotic with Anobiidae and Cerambycidae have been successfully cultured *in vitro* (see review in Nardon and Grenier 1989).

The intimacy between the insect and its symbiont is illustrated by their structural relation: in *S. oryzae* and other Rhynchophorinae (Nardon 1971; Nardon et al. 1985), the symbiotic bacteria lie free in the cytoplasm (figure 4). Such a situation is rare (Nardon 1988). We

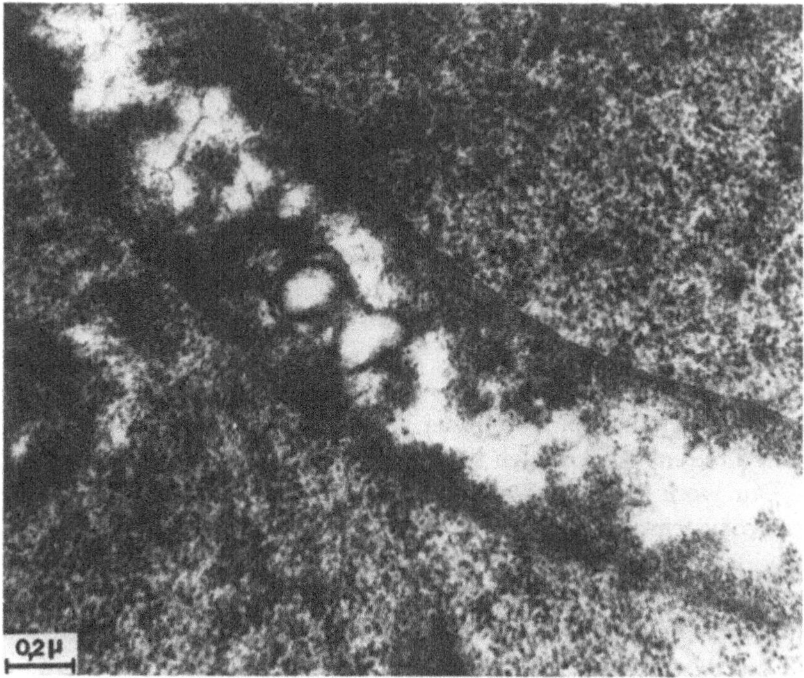

Figure 4
Symbiotic bacterium in the oocyte of *S. oryzae*. The bacterium is not included in a vacuole. Note the desmosome-like structure.

think that the phagocytotic membrane has probably been lost secondarily in order to favor exchange. This implies a full immunological compatibility between the two partners, acquired during the coevolutionary period.

The insect may influence its symbiont's metabolism and morphogenesis, without this necessarily implying genetic consequences and heritable effects. In the bedbug *Cimex*, the companion symbiont, immotile inside the cell, acquires six flagella very rapidly when it is experimentally extricated from the cell (Louis et al. 1973). This suggests that within the host cell, the gene products or proteins required for flagellar motility are repressed. Yeasts harbored by *Stegobium* and *Lasioderma* produce a reddish carotenoid pigment only *in vitro* (Jurzitza 1979). Although it is often difficult to distinguish changes due to metabolic interaction from those due to genetic modifications, the latter possibility cannot be rejected.

There have been very few studies of genome size in insect symbionts. On the basis of information from ultrastructural measurements, the *Euscelis* symbiont genome may be only about 150 kb long (Schwemmler 1983), which suggests the loss of numerous genes. In contrast, the size of aphid symbiont genomes, estimated by the reassociation kinetics of DNA dissociated by sonication, is 500 times larger (and more than 4 times larger than the *E. coli* genome) (Ishikawa 1987). *In vitro*, the symbionts of *Sitophilus oryzae* seem capable of synthesizing about thirty proteins (unpublished data), and the aphid symbionts several hundred. The genome size of *Sitophilus oryzae* symbionts is not precisely known, but Campbell and Unterman (1989) estimate it at more than 23 kb. If confirmed, the difference in size between genomes of leafhopper and aphid symbionts could represent different steps in their evolution inside the host.

Our work supports the view that symbiosis provides a new microecosystem for the symbiotic microorganism, which differs from its original medium. The host's cytoplasm provides a rich environment in which the symbionts are able to find nutrients necessary for growth, allowing the loss of genes coding for these factors. The mode of nutrition and energetic metabolism of these bacteria evolved in a way that would have been impossible outside the host. It is a specific consequence of symbiosis.

Let us now examine the symbiotic relationship from the weevil's point of view.

The Host Insect

For the weevil, the effects of symbiosis are even more apparent, and the presence of the symbiont has a range of consequences at different levels, all originating from a basic metabolic change. Certain modifications are not specific to the symbiosis and, in agreement with the classic neo-Darwinian concept, may have occurred autogenously by mutation (e.g., modifications of color, cell differentiation). However, other modifications, such as the production of certain vitamins, may be considered specific because they are prominent only in the presence of the symbiont.

Morphology, Cytology, Behavior

The symbionts have a broad and important influence on *Sitophilus oryzae* despite the fact that symbiosis is not obligatory for the insect

(Nardon and Grenier 1988). The differentiation of bacteriocytes and bacteriomes, both in the larva and in the adult, is under the influence of the bacteria (Nardon 1973, 1988). Specifically, except in the oocytes, the bacteria induce elevated ploidies of their host cells (32N to 128N; Nardon 1978a). This phenomenon resembles a "cell transformation" and suggests that the bacteria are able to elicit host DNA synthesis. Moreover, the aposymbiotic adult weevils are softer and paler than symbiotic ones, and (probably as a consequence of the decreased energetic metabolism) they no longer fly.

Effect on Nutrition, Growth, and Fertility

Although symbiotic and aposymbiotic weevils eat the same quantity of wheat, the symbiotic insects grow better (Grenier et al. 1986). The efficiency factor of conversion of digested food (ECD) is respectively 17.09 ± 0.20 and 15.95 ± 0.26 for symbiotic and aposymbiotic weevils (fresh weight). This is likely to be in part the consequence of the supply of growth factors to the host by the bacteria. Work with artificial diets shows that the bacteria can supply the insect with five vitamins: even if absent from the food, pantothenic acid, biotin, and riboflavin are provided in sufficient quantities to promote growth, while pyridoxin and folic acid are provided in lower quantities (Wicker and Nardon 1983). Moreover, the bacteria probably supply tyrosine or phenylalanine (Wicker and Nardon 1982). The modern cereals, in particular sorghum, possess high vitamin concentrations, which could explain the rearing of aposymbiotic weevils. Nevertheless, it is very difficult to grow aposymbiotic weevils on wheat or corn, so most "wild" strains are symbiotic. The elimination of bacteria reduces insect fertility about 30%, probably because of folic acid deficiency (Nardon 1973, 1987).

As a consequence of the supply of growth factors, development time is greatly accelerated in the presence of symbionts: at 27.5°C, it is 27.303 ± 0.153 days, versus 44.353 ± 0.125 for aposymbiotic insects. Furthermore, development time correlates directly with the number of bacteria: the more numerous the symbionts, the faster the development (Nardon 1978a,b; Nardon and Grenier 1988). Most important, we also demonstrated that the symbiont number is genetically controlled by chromosomal factors of the host (Nardon 1978b; Nardon and Wicker 1983). Such a system is both the consequence of symbiosis and the reason for its maintenance.

Effect of Symbiosis on Population Dynamics

As a consequence of increased fertility and diminished development time, symbiosis greatly affects population dynamics (Grenier et al. 1986). This is a crucial point for demonstrating the importance of symbiosis in the course of evolution. Biological parameters affecting the potential growth of an insect population (survival rates, sex ratio, age-specific fertility, development time) have been studied in two strains of *S. oryzae,* one symbiotic and one aposymbiotic. From these parameters, a Malthusian growth model was established which shows that the intrinsic rate of weekly increase is much higher in symbiotic than in aposymbiotic insects: 0.61 versus 0.46 at 27.5°C. We have experimentally verified the validity of the model. Furthermore, when a symbiotic female is introduced into a small group of aposymbiotic weevils, the symbiosis spreads very rapidly in the population (Nardon and Grenier 1990).

Interactions with Metabolism

The most important point is the significance of symbionts in the prevention of overaccumulation of sarcosine. Methionine, an amino acid, is in excess in wheat for *Sitophilus oryzae*. Owing principally to the bacteria in symbiotic weevils, methionine is reversibly sulfoxidized in a reaction that does not require energy. In the absence of symbionts, methionine is preferentially demethylated via a glycine-N-methyl-transferase-like activity, and sarcosine is produced. Since sarcosine can be neither incorporated into proteins nor excreted by the insect, it accumulates continuously during larval development. This undesirable accumulation shows that weevil mitochondria are deficient in the demethylating activity that recycles sarcosine to glycine (Gasnier-Fauchet and Nardon 1986, 1987). Remarkably, *the presence of symbionts allows the weevil to compensate for its mitochondrial deficiency and to save its energy,* since the formation of sarcosine needs ATP.

These observations are of great interest since, according to the Serial Endosymbiosis Theory (SET) of the origin of the eukaryotic cell, the evolution of symbiotic respiring bacteria has resulted in their transformation into mitochondria (Margulis 1981; Jacobs and Lonsdale 1987). If this is true, we may logically expect to find intermediate situations in which symbiotic bacteria replace mitochondria (as in

Pelomyxa palustris; see Whatley 1981) or assume some of their functions. It seems that *S. oryzae* could provide such an example, since a mitochondrial deficiency is compensated for by the activity of symbionts.

Another consequence of the symbiosis is the influence of bacteria on the index of methylation (SAM/SAH ratio), which is 3 times higher in aposymbiotic larvae than in the symbiotic larvae (Gasnier-Fauchet et al. 1986). This might have a crucial influence on genome expression.

To study the possible influence of activities of symbionts on mitochondrial metabolism, six different mitochondrial enzymatic activities have been compared in symbiotic and aposymbiotic larvae (Lefèbvre and Heddi 1990; Heddi and Lefèbvre 1990): cytochrome c oxidase, succinate cytochrome c reductase, isocitrate dehydrogenase, glycerophosphate cytochrome c reductase, pyruvate dehydrogenase, and keto-glutarate dehydrogenase. Due to the presence of bacteria, the specific enzymatic activities are significantly higher in symbiotic larvae: from +25% to +94%. Although we do not know how the bacteria increase mitochondrial enzyme activity, the simplest hypothesis implicates the supply of vitamins to the host, particularly pantothenic acid and riboflavin. The higher performance of energetic metabolism probably explains the fact that symbiotic weevils fly quite well, in contrast to the aposymbiotic ones.

Discussion and Conclusion

From these observations it is evident that the establishment of a symbiosis greatly affects the associated partners, creating a new organism whose special features are the results not simply of the addition, but especially of the interactions of two or more partners. In the family Rhynchophorinae, the genus *Sitophilus* is the only species living inside or on cereal grains. The other species live at the junction of the roots and the pseudo-stem of monocotyledons. When the symbiosis was established, about 100 million years ago, the wild cereals probably contained fewer growth factors than the highly selected cereals do now. The symbiosis most likely played a major role in the adaptation of *Sitophilus* to growth on cereal grains. Therefore, the acquisition of a symbiont fits well into the paradigm of natural selection; the coevolutionary partners are still subject to standard selection pressures.

Most of the resulting modifications in *Sitophilus oryzae* are not in themselves heritable, and, since they are absent in aposymbiotic animals, they are amenable to study. This is the paradoxical effect of symbiosis: at least at first,organisms may be heritably transformed without being themselves genetically affected. *Rather than any given characteristic, the symbiosis itself is transmitted to progeny.*

Undoubtedly this evolutionary mechanism has been underrated by neo-Darwinian biologists, since it corresponds to Lamarckian conceptions of evolution. The establishment and the maintenance of a symbiotic association over generations may be viewed as an example of the "inheritance of acquired characteristics" (Lamarck 1809). As Taylor (1986) remarked, these microbial genomes were not the type of "characteristics" that Lamarck had in mind. Although Lamarck was the first to present the theory of evolution in a coherent manner, his work is not well known (Szyfman 1982; Lovtrup 1987). In this context, it is sufficient to recall that he had an ingenious intuition: the possibility that organisms gain information from their environment and transmit it to their progeny. This is a process quite different from mutagenesis. We designate by "neo-Lamarckism" a theory of evolution that takes into account the possibility of acquiring exogenous characteristics or genes. Symbiosis is a neo-Lamarckian mechanism of evolution: we assert that the genome evolution required for organism evolution is not only a consequence of the accumulation of mutations.

As Lamarck predicted, the acquisition of new characteristics requires a very long time. We cannot expect to watch any bacteria or yeast spontaneously become symbiotically associated with any insect species. A certain degree of specificity must exist (Nardon 1988). The symbioses found in most curculionid beetles are undoubtedly quite ancient and exhibit numerous modalities. The integration of the symbiont required numerous reciprocal adjustments. It is impossible to assess precisely the consequences of symbiosis in our example, since we do not know the characteristics of the weevils before symbiosis was established. It is beyond doubt, however, that the acquisition and the subsequent integration of the symbiotic bacteria have greatly modified the insect. Some of the modifications, including the bacteriome differentiation and the acquisition of the ability to synthesize five vitamins via the bacteria, are strictly specific to the symbiosis. For other traits, such as the ability to fly, the role of symbiosis is not so clear. Flight is probably a primitive beetle trait. Posterior wings are absent in *Sitophilus granarius*, but present in *S. oryzae* and *S. zeamais*.

Nevertheless, probably as a secondary consequence of the removal of bacteria, aposymbiotic adults of *S. oryzae* and *S. zeamais* no longer fly when they are placed in the sunlight. This, too, demonstrates the metabolic importance of the symbionts.

Symbiosis confers a huge advantage to the weevil. Because of both higher fertility and more rapid development, once established, symbiotic insects rapidly replace asymbiotic ones in the population (Nardon and Grenier 1990). Furthermore, from a genetic point of view, the mechanism of transferring information from one generation to the next is twice as efficient in the case of symbiosis than in Mendelian heredity, since half the insects (the females) transmit bacteria to all the progeny. Even a mutation affecting the weevil genome would be greatly favored by symbiosis. *Therefore, symbiosis is not only in itself a powerful factor of evolution, but it could facilitate the evolution of the species by enhancing the population dynamics.*

What is an endocytobiont, from a genetic point of view? A new cluster of genes! In *Sitophilus oryzae,* the symbiotic bacterium supplies five vitamins and other growth factors, so we can estimate that the acquisition of these symbionts represents the simultaneous acquisition of at least twenty genes. The probability of simultaneously obtaining twenty genes by mutation is quite negligible. The main difference between a symbiont gene and a nuclear gene lies in the fact that the latter is present, but not necessarily functioning, in all the cells of the weevil, whereas the bacteria are present in only some of them. But we have to consider that in a larva the number of symbionts is between 1 million and 3 million. Thus, there are about the same number of symbionts in a larva as there are insect cells, and the bacteriome is quite similar to other insect organs. We must consider that *symbiosis is in fact the most important phenomenon capable of modifying a population,* and then we have to take into account its possible role in evolution. We could speculate that the eventual consequence of symbiosis is the acquisition of the symbiont DNA by the host via gene transfer. Therefore, symbiosis could be only a transitory step in evolution as *the most effective mechanism of gaining new genes in eukaryotes* (Nardon 1978a).

Nevertheless, as important as it may be, any mechanism for modifying an organism does not necessarily contribute to its evolution toward a new species. Therefore, one important question arises: Is symbiosis able to promote speciation? With respect to weevils, reciprocal hybrids between the two sibling species *Sitophilus oryzae* and *S.*

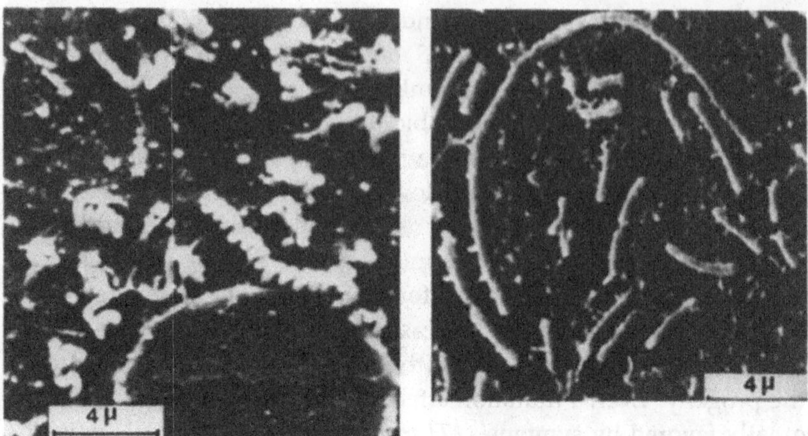

Figure 5
Comparison of the morphology of the symbiotic bacteria in the two sibling species *S. zeamais* (left) and *S. oryzae* (right). In *S. zeamais* the bacteria are always more or less curled and spiral.

zeamais are sterile. These species are distinguishable strictly by the morphology of their endocytobionts (figure 5; Nardon and Wicker 1981). We have no proof that the symbionts are directly implicated in the sexual isolation of these two *Sitophilus* species; however, it has been reported that nucleocytoplasmic incompatibility between insect strains of different geographical origin may be due to the presence of symbionts in one strain (rickettsias, in general). This has been observed in *Hypera postica* (Hsiao and Hsiao 1985), *Tribolium confusum* (Wade and Stevens 1985), *Drosophila paulistorum* (Daniels and Ehrman 1974), *Culex pipiens* (Fine 1978), and *Laodelphax striatellus* (Noda 1984). In *Sitophilus oryzae*, the transfer of the symbiotic genome into an aposymbiotic cytoplasm (by means of successive backcrosses) progressively leads to sterility. Thus, the bacteria seem to be implicated in the genome expression of their host. These facts strongly suggest that symbiosis may favor speciation.

Another important question concerns the possibility of explaining the origin of new higher taxa in the same way that we explain the origin of species. Is it correct to suppose that there is a unique factor promoting both genetic variation and evolutionary changes at any level of organization? Is the neo-Darwinian supposition correct—that the gradual process of mutation and selection explains all evolution, from prokaryote to man? Or is it possible to explain evolution only by successive symbioses, rejecting mutagenesis as a source of innovation? We think that it is quite impossible to explain innovation, adaptation, and complexification in terms of a single mechanism. We must remain prudent and avoid any formalism. Acquisition and further integration of a symbiont do not imply the rejection of mutation and natural selection as complementary mechanisms. Acquisition of the symbiont is a neo-Lamarckian process, whereas integration may require reciprocal adjustments in which mutagenesis and selection play a role. Therefore, in our minds, the endosymbiotic theory contributes to a synthesis between neo-Lamarckianism, neo-Darwinism, and neutralism (Kimura 1986). We consider symbiosis to be the most important and powerful factor in evolution, but not the only one. Not all insects have bacterial symbionts, yet all are diversified. The establishment of a new pluralistic theory of evolution would encompass the "symbiosis" among all the different concepts stated above.

Note

1. The authors prefer the term *symbiote(s)*. Although in French *symbiote* is correct, in English *symbiont* is proper, and represents an entire series of terms: *biont, holobiont,* etc.—Editor. According to K. F. Meyer (*J. Infect. Dis.* 36:1, 1925), M. Hertig et al. (*J. Parasitol* 23:326, 1937), and E. A. Steinhaus (*Principles of Insect Pathology*, McGraw-Hill, 1949, p. 125), the term *symbiont* is not correct in English. M. A. Brooks uses *symbiote,* the sole correct term from an etymological point of view and from a historical point of view (*symbiote* is the term first used in 1879 by De Bary).—Author.

References

Campbell, B. C., and Unterman, B. M. 1989. Purification of DNA from the intracellular symbiotes of *Sitophilus oryzae* and *Sitophilus* zeamais (Coleoptera: Curculionidae). *Insect Biochem.* 19(1): 85–88.

Daniels, S., and Ehrman, L. 1974. Embryonic pole cells and mycoplasmalike symbionts in *Drosophila paulistorum*. *J. Invertebr. Pathol.* 24: 14–19.

Fine, P. E. M. 1978. On the dynamics of symbiote-dependent cytoplasmic incompatibility in culicine mosquitoes. *J. Invertebr. Pathol.* 30: 10–18.

Gasnier-Fauchet, F., and Nardon, P. 1986. Comparison of methionine metabolism in symbiotic and aposymbiotic larvae of *Sitophilus oryzae* L. (Coleoptera: Curculionidae). II. Involvement of the symbiotic bacteria in the oxidation of methionine. *Comp. Biochem. Physiol.* 85B(1): 251–254.

Gasnier-Fauchet, F., and Nardon, P. 1987. Comparison of sarcosine and methionine sulfoxide levels in symbiotic and aposymbiotic larvae of two sibling species, *Sitophilus oryzae* L. and *Sitophilus zeamais* Mots (Coleoptera:Curculionidae). *Insect Biochem.* 17(1): 17–20.

Gasnier-Fauchet, F., Gharib, A., and Nardon, P. 1986. Comparison of methionine metabolism in symbiotic and aposymbiotic larvae of *Sitophilus oryzae* L. (Coleoptera:Curculionidae). I. Evidence for a glycine N-methyltransferase-like activity in the aposymbiotic larvae. *Comp. Biochem. Physiol.* 85B(1): 245–250.

Grenier, A. M., Nardon, P., and Bonnot, G. 1986. Importance de la symbiose dans la croissance des populations de *Sitophilus oryzae* L. (Coleoptére Curculionidae). Etude théorique et expérimentale. *Acta Oecologica, Oecol. Applic.* 7(1): 93–110.

Heddi, A., and Lefébvre, F. 1990. Energetic metabolism of mitochondria in hybrids of symbiotic and aposymbiotic larvae of *Sitophilus oryzae* (Coleopt., Curculionidae). In: Nardon, P., Gianinazzi-Pearson, V., Grenier, A. M., Margulis, L., and Smith, D. C., eds., *Endocytobiology IV*. INRA, Paris.

Hsiao, T. H. and Hsiao, C. 1985. *Rickettsia* as the cause of cytoplasmic incompatibility in the alfalfa weevil, *Hypera postica* (Gyllenhal). *J. Invertebr. Pathol.* 45(2): 244–246.

Ishikawa, H. 1987. Nucleotide composition and kinetic complexity of the genomic DNA of an intracellular symbiont in the pea aphid *Acyrthosiphon pisum*. *J. Mol. Evol.* 24: 205–211.

Jacobs, H. T., and Lonsdale, D. M. 1987. The selfish organelle. *Trends Genet.* 3(12): 337–341.

Jurzitza, G. 1979. The fungi symbiotic with anobiid beetles. In: Batra, L. R., ed., *Insect-Fungus Symbiosis: Nutrition, Mutualism and Commensalism*. Wiley.

Kimura, M. 1986. *The Neutral Theory of Molecular Evolution*. Cambridge University Press.

Lamarck, J. B. 1809. *La Philosophie Zoologique* (collection 10–18, Union générale d'éditions, Paris, 1968).

Lefèbvre, F., and Heddi, A. 1990. Energetic metabolism of mitochondria in symbiotic and aposymbiotic larvae and adults of *Sitophilus oryzae* (Coleopt., Curculionidae). In: Nardon, P., Gianinazzi-Pearson, V., Grenier, A. M., Margulis, L., and Smith, D. C., eds., *Endocytobiology IV*. INRA.

Louis, C., Laporte, M., Carayon, J., and Vago, C. 1973. Mobilité, ciliature et caractéres ultrastructuraux des microorganismes symbiotiques endo et exocellulaires de *Cimex lectularius* L.(Hemiptera, Cimicidae). *C. R. Acad. Sci. D. Paris* 277: 607–611.

Lovtrup, S. 1987. *Darwinism: The Refutation of a Myth*. Croom Helm.

Mansour, K. 1930. Preliminary studies on the bacterial cell mass (accessory cell-mass) of *Calandra oryzae*: the rice weevil. *Q. J. Microsc. Sci.* 73: 421–436.

Mansour, K. 1935. On the microorganism free and the infected *Calandra granaria*. *Bull. Soc. Roy. Entomol. Egypt* 19: 290–306.

Margulis, L. 1981. *Symbiosis in Cell Evolution: Life and Its Environment on the Early Earth*. Freeman.

Musgrave, A. J., and Miller, J. J. 1956. Some micro-organisms associated with the weevils *Sitophilus granarius* (L.) and *Sitophilus oryzae* (L.) (Coleoptera). II. Population differences of mycetomal micro-organisms in different strains of *S. granarius*. *Can. Entomol.* 88: 7–100.

Nardon, P. 1971. Contribution à l'étude des symbiotes ovariens de *Sitophilus sasakii*: Localisation, histochimle et ultrastructure chez la femelle adulte. *C. R. Acad. Sci.* 272D: 2975–2978.

Nardon, P. 1973. Obtention d'une souche asymbiotique chez le charançon Sitophilus sasakii: Différentes méthodes et comparaison avec la souche symbiotique d'origine. *C. R. Acad. Sci.* 277D: 981–984.

Nardon, P. 1978a. Etude des interactions physiologiques et génétiques entre l'hôte et les symbiotes chez le coléoptère Curculionide *Sitophilus oryzae*. Thèse Doctorat, INSA–Université Claude Bernard, Lyon.

Nardon, P. 1978b. Etude de l'action des rayons X sur les symbiotes ovariens et l'ovogenèse chez *Sitophilus oryzae* L. (Col. Curculionidae). *Bull. Soc. Zool. Fr.* 103: 295–300.

Nardon, P. 1987. Rôle des endocytobiotes dans la nutrition et le métabolisme des Insectes. In: Léger, C. L., ed., *La Nutrition des Crustacés et des Insectes*. CNERNA.

Nardon, P. 1988. Cell to cell interactions in insect endocytobiosis. In: Scannerini, S., Smith, D. C., Bonfante-Fasolo, P., and Gianinazzi-Pearson, V., eds., *Cell-to-cell Signals in Plant, Animal and Microbial Symbiosis*. Springer-Verlag.

Nardon, P., and Grenier, A. M. 1988. Genetical and biochemical interactions between host and its endocytobiotes in the grain weevils *Sitophilus* (Coleoptera, Curculionidae) and other related species. In: Scannerini, S., Smith, D. C., Bonfante-Fasolo, P., and Gianinazzi-Pearson, V., eds., *Cell-to-cell Signals in Plant, Animal and Microbial Symbiosis*. Springer-Verlag.

Nardon, P., and Grenier, A. M. 1989. Endocytobiosis in Coleoptera: Biological, biochemical and genetic aspects. In: Schwemmler, W., ed., *Insect Endocytobiosis: Morphology, Physiology, Genetics and Evolution*. CRC Press.

Nardon, P., and Grenier, A. M. 1990. Symbiosis as an important factor for the growth and the evolution of populations of *Sitophilus oryzae* L. (Coleoptera, Curculionidae). In: Nardon, P., Gianinazzi-Pearson, V., Grenier, A. M., Margulis, L., and Smith, D. C., eds., *Endocytobiology IV*. INRA.

Nardon, P., Louis, C., Nicolas, G., and Kermarrec, A. 1985. Mise en évidence et étude des bacteries symbiotiques chez deux charançons parasites du bananier: *Cosmopolites sordidus* (Germar) et Metamasius hemipterus (L.) (Coleoptère Curculionidae). *Ann. Soc. Ent. Fr.* 21(3): 245–258.

Nardon, P., and Wicker, C. 1981. La symbiose chez le genre *Sitophilus* (Coléoptère Curculionidae). Principaux aspects morphologiques, physiologiques et génétiques. *Ann. Biol.* 20: 327–373.

Nardon, P., and Wicker, C. 1983. Genetic control of symbiotes by the host in the insect *Sitophilus oryzae* L. (Coleoptera, Curculionidae). In: Schenk, H., and Schwemmler, W., eds., *Endocytobiology II: Intracellular Space as Oligogenetic Ecosystem*. Walter de Gruyter.

Noda, H. 1984. Cytoplasmic incompatibility in a rice planthopper. *J. Hered.* 75: 345–348.

Scheinert. W. 1933. Symbiose und Embryonalentwicklung bei Russel Käfern. *Z. Morphol. Oekol. Tiere* 27: 76–198.

Schneider H. 1954. Kunstlich symbiontenfrei gemachte Kornkäfer (*Calandra granaria*). *Naturwiss.* 41: 147.

Schneider, H. 1956. Morphologische und experimentelle Untersuchungen über die Endosymbiose der Korn und Reiskäfer (*Calandra granaria* und *C. oryzae*). *Z. Morph. Oekol. Tiere* 44: 555–625.

Schwemmler, W. 1983. Analysis of possible gene transfer between an insect host and its bacteria-like endocytobionts. *Intern. Rev. Cytol. Suppl.* 14: 247–266.

Szyfman, L. 1982. *Jean-Baptiste Lamarck et son époque*. Masson.

Taylor, F. J. R. 1986. An overview of the status of evolutionary cell symbiosis theories. *Ann. N.Y. Acad. Sci.* 503: 1–16.

Wade, M. J., and Stevens, L. 1985. Microorganism mediated reproductive isolation in flour beetles (Genus *Tribolium*). *Science* 227: 527–528.

Whatley, F. R. 1981. The establishment of mitochondria: *Paracoccus* and *Rhodopseudomonas*. *Ann. N.Y. Acad. Sci.* 361: 330–340.

Wicker, C., and Nardon, P. 1982. Development responses of symbiotic and aposymbiotic weevils *Sitophilus oryzae* L. (Coleoptera, Curculionidae) to a diet supplemented with aromatic amino acids. *J. Insect Physiol.* 28: 1021.

Wicker, C., and Nardon, P. 1983. Differential vitamin requirements of symbiotic and aposymbiotic weevils, *Sitophilus oryzae*. In: Schenk, H., and Schwemmler, W., eds., *Endocytobiology II: Intracellular Space as Oligogenetic Ecosystem*. Walter de Gruyter.

13 Cell Symbiosis, Adaptation, and Evolution: Insect-Bacteria Examples

Toomas Tiivel

Among the many and diverse organismal interactions, there occur countless symbioses in which microbial endosymbionts exist within the cells of another organism, a host symbiont. Some of these associations, such as the DNA-containing mitochondrial and plastid organelles (former endosymbionts of eukaryotic cells), are permanent. In other cases, where associations are cyclical, the symbiont may exist extracellularly as well as intracellularly. Symbioses present intriguing problems concerning the morpho-functional integrity of organisms, their autonomous existence, and aspects of their formation and development (Tiivel 1984a, 1989).

Intracellular Adaptation

The insides of all eukaryotic cells are more or less similar, so the evolutionary transition to intracellular life may be viewed as involving the solution of a common set of problems (Smith 1979; Moulder 1985). Numerous endocytobiotic systems, both mutualistic and parasitic, provide an amazingly broad spectrum of examples of how cells invade, survive within, and are integrated into the complex system of a quite different cell.

Sitte (1983) proposes four stages of increasing mutual dependence of partners in such associations. Slightly modified, they are as follows:

(1) *Non-obligatory*. The first prerequisite for assembling an endocytobiosis is ingestion coupled with avoidance of digestion. Both recognition signals and corresponding receptors are implied. Either the fusion of primary lysosomes with phagosomes is blocked, or the destructive impact of lysosomal hydrolases is otherwise prevented.

(2) *Obligatory*. Isolated partners can still be cultivated separately; however, in the natural habitat they are found exclusively in the combined system. Either the cytobionts are directly inherited, or there are special adaptations assuring re-infection of the progeny.

(3) *Established*. Mutual dependence of the partners is great enough to preclude separate cultivation even under optimal conditions. There exists a permanent exchange of metabolites, and gene transfer is possible.

(4) *Organellar*. The organelle state of endocytobiosis is characterized by an exchange of nucleoside phosphates among the partners, transfer of genetic information from the former microorganism to the nucleus of the host cell, the corresponding synthesis of the majority of organelle-specific proteins outside the organelle, and the transfer of these proteins into the organelle by specific mechanisms. Relics of the earlier stages of evolution are the double-membraned envelope and remnants of the former genome.

Adaptations involve the organism's ability to adjust itself to the particular features of its habitat or to complexes of certain features (Timofeeff-Ressovsky et al. 1977; Tiivel 1988). As the concept suggests, an adaptation always involves a fit to "something"; this something is, in its broader sense, the organism's habitat (Timofeeff-Ressovsky et al. 1977).

When two organisms enter into an endosymbiotic relationship, the host cytoplasm becomes the environment for the endosymbiont. After a period of coexistence, adaptation may occur, usually involving both partners. The integrating genetic material must "be able to change its behavior" after the contact with a new genome—the principle of change in functioning (Chaikovsky 1977; Tiivel 1988).

The Intracellular Contact of Symbionts

One of the characteristics of all eukaryotic cells is the abundant presence of intracytoplasmic membranes forming a kind of network in the cell (Seravin 1986). As a result, different parts of one cell may be separated and distinguished by a multitude of biochemical characteristics, which often exclude one another. This network is not stable; it is permanently renewing itself, and so endocytosis and exocytosis are possible. This feature of the eukaryotic cell may serve as a basis for solving problems of more stable aspects of intracellular life.

Many authors have pointed out that one of the main peculiarities of nearly all endocytobiotic associations in eukaryotes is that the microorganisms are segregated from the rest of the host cell by one or more membranes (Smith 1979; Sitte 1983; Tiivel 1984a; Moulder 1985; Tiivel 1989). The outer encircling membrane is of host origin, and appears to be progressively formed as if in response to endocytobiont presence. This is the case in some insects which have complicated endocytobiotic interconnections with microorganisms. The interdependence of the partners borders on a complete loss of their functional autonomy and the formation of compound organisms. The bacteria-like microorganisms have two endogenous peripheral membranes (a cytoplasmic membrane and a membraneous cell wall typical of Gram-negative bacteria), which, on the outside, are surrounded by an additional membraneous casing. Most probably derived from the host cell, this membraneous casing forms a kind of space around each cytobiont—a bacteriophoric (endocytobiotic) vacuole. Several characteristics of this vacuole and the encircling membranes vary in different associations. Variation occurs when microorganisms inhabit different host cells, as well as when there are several types (species?) of endocytobionts in one host cell, as for example in leafhoppers. Ultrastructural peculiarities of such vacuoles, which vary in size and internal structure during the life histories of the leafhopper host and its cytobionts, may reflect physiological or developmental changes (Tiivel 1984b, 1989). In a few weevils, encircling host membranes are absent, at least in some stages (Nardon 1988).

Apparently, endocytobionts become enclosed in host membranes when their relation to the host cell has stabilized (in contrast, to an aggressive parasitic infection in which surrounding host membranes are ruptured). In vacuoles, endocytobionts may retain the ability, or at least the potential, for independent growth (Smith 1979). The initial formation of bacteriophoric vacuoles is evidently a result of nonspecific "membrane reactions" of the eukaryotic cell to invading prokaryotes. There is a tendency toward the formation of peribacterial membranes, separating endocytobionts from host cytoplasm; this forms a vacuolar system that does not necessarily inhibit normal microbial activity or propagation (Tiivel 1988).

The formation of vacuoles by host cells may protect both the host cell and the cytobiont from immediate lysis, or may provide a place for cytobiont multiplication. Metabolic interconnections of the partners, and in some cases even genetic interactions, may take place across the vacuolar membranes. The vacuole may also create the space where host control of endoctyobionts is regulated—in some cases, processes of destruction and degeneration of endocytobionts have been documented (Tiivel 1984a).

In adapting to growth inside host cells, obligate intracellular cytobionts may have gained new features inimical to extracellular multiplication or have lost former essential ones; they may also have done both (Moulder 1985; Tiivel 1988).

Metabolic Interactions, Specificity, and Endosymbiont Control

A characteristic feature of intracellular cytobiosis is the one-way or reciprocal movement of substances between host and cytobiont (Smith 1979; Schwemmler 1983; Moulder 1985; Tiivel 1989). The more stable the association, the less random, and the more genetically controlled the synthesis and transport of substances is likely to be. Reciprocal gene transfer from the host to the cytobiont may take place, resulting in the synthesis by the cytobiont of some metabolites required by the host but not necessarily produced by the isolated microorganism (Schwemmler 1983; Nardon 1988).

The invasion of free-living organisms into the cells of a host requires recognition mechanisms between cytobionts and specific host cells; such mechanisms allow the symbiont to bypass the normal invasion responses of the host. Although the mechanism of "like-to-like" recognition is not completely understood, it is generally assumed that intracellular cytobionts and host cells recognize each other through the interaction of complementary structures on their cell surfaces. Whatever their original function, these structures may have been selected to bind more efficiently to host cells or, conversely, to cytobionts.

Successful endosymbioses require that the host be able to control the growth and reproduction of the endosymbionts. Several types of endosymbiont control may be exhibited by the host, such as control of endocytobiont reproduction, of the breakdown of endocytobionts in the host cell, and of complex transovarial passage of some microorganisms (Tiivel 1988, 1989).

An essential feature of the evolution of stable intracellular symbioses is the development of mechanisms for the control and regulation of the cytobionts by the host cell. Intracellular cytobionts have fine regulatory mechanisms, defined by the host's metabolism, that regulate the multiplication of the microorganisms (Gromov 1978; Smith 1979; Schwemmler 1983; Moulder 1985). At least some endocytobionts of insects act as a kind of trigger mechanism, enabling the host to direct the development of the cells that constitute its tissues (Tiivel 1984b).

Endosymbiont Transit

A general indication that symbionts have evolved suitable control mechanisms is the frequency with which microbes undergo morphological transformations upon entering host cells. Extracellular transit or infectious forms are morphologically distinguishable from intracellular reproductive or vegetative forms (Schwemmler 1983; Moulder 1985; Tiivel 1989). Special infectious forms of microorganisms may attach to and enter host cells with extraordinary efficiency and resist extracellular inactivation to an unusual degree (Smith 1979; Ossipov 1981). One of the most complicated examples is the transmission of the obligate, host-dependent endocytobionts of Homoptera, in which the developmental cycles of hosts and endocytobionts are strongly correlated. The progeny die if a certain quota of cytobionts fail to infect the developing egg (Schwemmler 1983; Tiivel 1984b).

Adaptations that facilitate transit vary greatly between cytobionts. Usually they occur as a result of splitting the conflicting demands of intracellular transit between two or more phenotypes, so that the cytobionts' survival as a species is ensured by the alternation of cell types and often by alternation of host cell types as well (Smith 1979; Tiivel 1988).

The Role of Compartmentalization of Cells

Considering the various adaptive modifications of eukaryotic cells and their prokaryotic endocytobionts, I suggest that the critical feature making these associations possible relates to the membrane system of eukaryotic cells and the compartmentalization of the latter. Indeed, the fact that an invader is surrounded by a host-derived membrane as soon as it enters the host cell and that there is a vacuole between this casing and the microorganism (the endocytobiotic vacuole) allows both partners to preserve their identities, to possess and control (sometimes genetically) transfer of metabolites in both directions by selective permeability, and to balance the number of endocytobionts by regulating their multiplication and their selective removal.

For both endocytobionts and hosts, adaptation to endocytobiosis is often associated with the acquisition of new genetic material, the specialization and alteration of both the structures and the functions of partners, and the production of diverse new forms of microorganisms, making it possible for the hosts to occupy new ecological niches. These adaptations are of decisive significance for the existence of a great number of species and their endocytobionts (Gromov 1978; Kordyum 1982; Schwemmler 1983). In a great variety of organisms, the endocytobiotic way of life (cellular associations) seems to be one of the sources of evolutionary innovation. Such associations direct the evolution of metabolic pathways and organelle systems, providing degrees of flexibility not found in single organisms (Taylor 1981). Preadaptations to endocytobiosis are not determined solely by the interaction of partners, but may be much more random; thus, we can speak of possible and impossible associations of organisms, and of certain prohibitions to endocytobiosis. Perhaps this is why stable endocytobioses are less abundant among plants and vertebrates—the number of prohibitions here are greater than in invertebrates, algae, and protozoa (Gromov 1978).

Several authors have pointed out that eukaryote-microbial cytobiosis represents a model for organelle evolution: it provides evidence for the necessary intermediate evolutionary steps and demonstrates the genetic integration of cell components (Chaikovsky 1977; Schwemmler 1979; Margulis 1981; Taylor 1983; Tiivel 1988). It has been proposed that a number of structural, functional, and evolutionary aspects of endocytobiosis in leafhoppers will allow the investigation of at least four major steps in the evolution of endocytobionts into organelles: cell-wall reduction, DNA alteration, reciprocally coordinated gene expression, and protein synthesis (Schwemmler 1980; Tiivel 1989). Perhaps the compartmentalization of the protoeukaryotic cell was a prerequisite for the establishment and existence of DNA-containing organelles precisely because it made these changes possible. As a result, the eukaryotic cell is a chimera consisting of different genetic parts.

References

Chaikovsky, J. V. 1977. Genetical integration of cellular structures as a factor of evolution. *J. General Biology (USSR)* 38: 823–235.

Gromov, B. V. 1978. Bacteria-intracellular symbionts of animals. *Progress in Microbiology (USSR)* 13: 50–72.

Kordyum, V. A. 1982. *Evolution and Biosphere*. Naukova Dumka, Kiev.

Margulis, L. 1981. *Symbiosis in Cell Evolution*. Freeman.

Moulder, J. W. 1985. Comparative biology of intracellular parasitism. *Microbiol. Rev.* 49: 298–337.

Nardon, P. 1988. Cell to cell interactions in insect endocytobiosis. In: Scannerini, S., Smith, D., Bonfante-Fasolo, P., and Gianinazzi-Pearson, V., eds., *Cell to Cell Signals in Plant, Animal and Microbial Symbiosis*. Springer-Verlag.

Ossipov, D. V. 1981. *Problems of Nuclear Heteromorphism in the Unicellular Organisms*. Nauka, Leningrad.

Schwemmler, W. 1979. *Mechanismen der Zellevolution: Grundriss einer modernen Zelltheorie*. Walter de Gruyter.

Schwemmler, W. 1980. Endocytobiosis: General principles. *Biosystems* 12: 111–122.

Schwemmler, W. 1983. Endocytobiosis as an intracellular ecosystem. In: Schenk, H. E. A., and Schwemmler, W., eds., *Endocytobiology II: Intracellular Space as Oligogenetic Ecosystem*. Walter de Gruyter.

Seravin, L. N. 1986. Evolution of eukaryotic cells. III. Some principles of morphofunctional organization of eukaryotic cells. *Tsitologija (USSR)* 28: 779–789.

Sitte, P. 1983. General organization of the eucyte and its bearings on cytosymbiosis and cell evolution. In: Schenk, H. E. A., and Schwemmler, W., eds., *Endocytobiology II: Intracellular Space as Oligogenetic Ecosystem*. Walter de Gruyter.

Smith, D. C. 1979. From extracellular to intracellular: The establishment of a symbiosis. *Proc. R. Soc. Lond. B* 204: 115–130.

Taylor, D. L. 1981. Evolutionary impact of intracellular symbiosis. *Ber. Deutsch. Bot. Ges.* 94: 583–590.

Taylor, F. J. R. 1983. Some eco-evolutionary aspects of intracellular symbiosis. *Int. Rev. Cytol. Suppl.* 14: 1–28.

Tiivel, T. 1984a. Ultrastructural aspects of endocytobiosis in leafhopper (Insecta: Cicadinea) cells. *Proc. Acad. Sci. Estonian SSR. Biology* 33: 244–255.

Tiivel, T. 1984b. Morphogenetic movements and insect development. In: Tiivel, T., Kull, K., Neuman, T., and Sutrop, U., eds., *Theory and Models in Life Science*. Tartu.

Tiivel, T. 1988. Adaptation to endocytobiosis in the evolution of cells. In: Kull, K., and Tiivel, T., eds., *Lectures in Theoretical Biology*. Valgus, Tallinn.

Tiivel, T. 1989. Leafhopper endocytobiosis. In: Schwemmler, W., ed., *CRC Handbook of Insect Endoctyobiosis: Morphology, Physiology, Genetics, Evolution*. CRC Press.

Timofeeff-Ressovsky, N. V., Vorontsov, N. N., and Yablokov, A. V. 1977. *A Short Survey of the Theory of Evolution*. Nauka, Moscow.

14 Symbiogenesis in Insects as a Models for Cell Differentiation, Morphogenesis, and Speciation

Werner Schwemmler

Physiochemistry of Bacteria, Plants, and Animals

It was in 1953 that graphs were first used to systematically evaluate chemical parameters such as the inorganic ions and organic molecules of insect hemolymph (Duchâteau et al. 1953). Sutcliffe (1963) classified the different hemograms and those of other animal cell fluids into three phylogenetic types. These were later expanded to include the physiological values of pH, osmotic pressure, and ion or molecule concentration ratios, and to extend their application to plants and bacteria (Schwemmler 1984).

Physiochemical type I is characterized by the predominance of inorganic ions over organic molecules; the reverse is true for type III. Type II represents a stage intermediate between types I and III (figure 1). In particular, the concentrations of sugar, Cl^-, and Na^+ decrease from type I to type III, while the concentrations of amino acids, other organic acids, PO_4^{3-}, K^+, Mg^{2+}, and Ca^{2+} increase. The chemical and physiological parameters appear to be correlated. The H^+ concentration increases with the concentration of organic acids from a pH of about 8 to about 6. At the same time, the replacement of many small inorganic ions with fewer larger organic molecules causes the osmotic pressure to drop from approximately 450 to 250 mosmol.

Thus, an intracellular or extracellular animal or plant fluid can be roughly assigned to one of three physiochemical types on the basis of pH values and osmotic pressures, or on the basis of ion and molecule concentration ratios. Of course, different parts of a single animal or plant may belong to different physiochemical types. In the crab *Astacus*, for instance, the hemolymph is type I whereas the muscle cells are type II/III (figure 2). There are also characteristic differences between animals and plants. The sugar concentration of type III plant

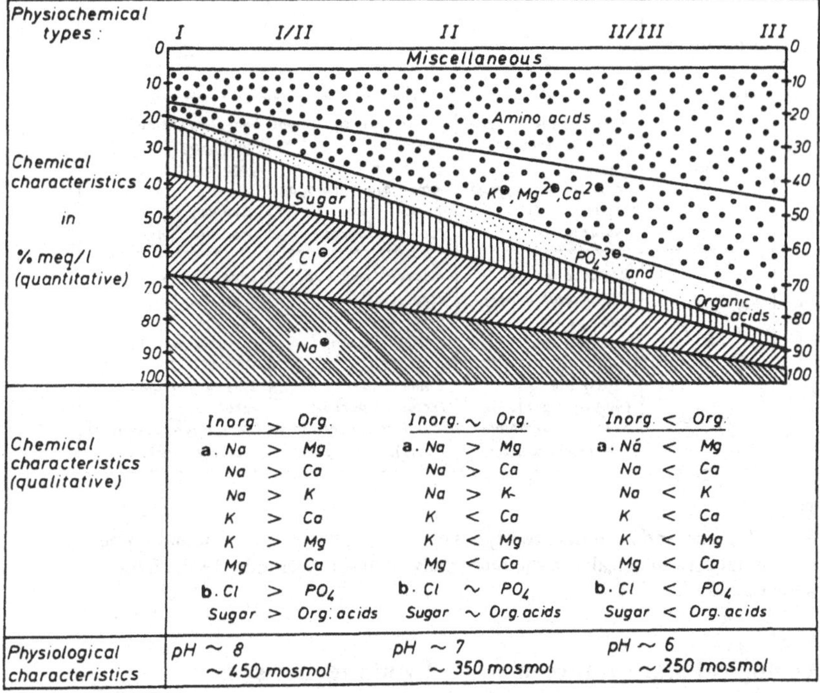

Figure 1
Characteristics of the three basic physiochemical types in the soluble procyte fraction and in the extra- or intracellular animal and plant fluids; given in physiochemical trends of evolution, in concentration ratios of inorganic ions and organic molecules, and in pH and osmotic values. (Source: Schwemmler 1984.)

phloem, for example, is higher than the amino-acid concentration, in contrast to the corresponding type of animal hemolymph. In plants, sugar is the decisive osmotic effector. The physiochemical type of extracellular animal and plant fluids is determined by the sum of the various physiochemical types of all the cells constituting the entire system. Of course, such superimposed regulatory mechanisms as anal pumps, Malphigian glands, and livers must be taken into account. Environment and nutrition also have a certain influence on the physiochemical type. Thus, the division into three physiochemical types is a handy but simplified representation. It is a typification, however, that can lead to concrete results, as I have attempted to demonstrate with the example of the Hemiptera (Schwemmler 1980a, 1983b).

Physiochemical type		I	II	III
Examples	Animals	Hemiptera (hemolymph) Euscelis, asymbiotic Corixa Crayfish (hemolymph) Intestinal cells (soluble part)	Hemiptera (hemolymph) Euscelis Triatoma Amphibians (blood) Muscle cells (soluble part)	Hemiptera (hemolymph) Palomena Graphosoma Cyanocytes, some (soluble part)
	Plants	Algae, anaerobic I/II* (soluble part) Root cells (soluble part)	Mosses (cell sap) II/III* Stem cells (soluble part)	Seed plants (cell sap) Leaf cells (soluble part)
	Microbes	Clostridia (soluble part) Primary symbionts I/II* (soluble part)	Eubacteria (soluble part) Auxiliary symbionts II/III* (soluble part)	Blue-green algae, aerobic Companion symbiont (soluble part)

Figure 2
Animal, plant, and bacterial prototypes of the three basic physiochemical types. (For data about physiological and chemical analysis see Schwemmler 1984; cf. figure 1 above.)

Ecological Significance of Insect Symbiogenesis

Hemiptera have developed various physiochemical types in connection with their habitats and nutrition. This is clearly shown by the cation ratios of some typical species (figures 1, 2). The waterbug *Corixa* feeds on algae of cell sap type I/II and fish of hemotype I (Schwemmler 1984), and has a cation ratio that places it in hemotype I. The waterbug *Neotrephes* also leads an aquatic life, but feeds preferentially on cell sap type II salamander eggs and salamander larvae of hemotype II, and only occasionally on crustaceans of hemotype I. In agreement with its nutrition, this bug has developed hemotype II. Corresponding to their terrestrial habitat and nutrition, the leafbugs *Palomena* and *Graphosoma* have hemotype III, but they feed on the fruits of angiosperms, which have fruit sap type III. The predatory bugs *Triatoma* and *Rhodnius* occupy relatively moist terrestrial habitats and feed on mammalian blood of type I. On the basis of their cation ratios, they belong to hemotype II. The scale insect *Rastrococcus* and the leafhopper *Euscelis* occupy a moist terrestrial transition zone, feed on angiosperms of phloem type III, and have hemotype II. *Euscelis* instars that have been artificially freed of their intracellular symbionts have the pH and osmotic characteristics of hemotype I.

It is interesting to compare the cation ratio of the hemolymphs of the above-mentioned Hemiptera, the consumers, with the cation ratios of their food producers (figure 3), in order to determine whether the type of hemipteran symbiosis is dependent on the relationship between its hemolymph type and the physiochemical type of its nutritional source (Grassé 1949). *Corixa* and *Neotrephes*, whose cation ratios correspond in principle to those of the food producers on which they feed, do not harbor symbionts, aside from the usual extracellular intestinal microbiota. In contrast, *Graphosoma* and *Palomena* differ from their producers with respect to one cation ratio, and live in a loose symbiotic relationship with extracellular bacteria which are housed in special folds or crypts of the intestine. Inheritance is maternal. The bacteria are transmitted to the next generation externally (on eggs smeared with feces) and by coprophagia. (In Greek, *kopros* means "feces" and *phagein* means "to eat.") *Triatoma* and *Rhodnius* differ from their food producers with respect to two cation ratios and harbor vitally necessary bacterial symbionts both intracellularly in the intestinal epithelium and extracellularly in intestinal folds. These symbionts are also passed on by smearing the eggs with feces. *Rastrococcus* differs from its producers in about four cation ratios and also harbors essential bacterial and fungal endocytobionts, which are present in the fatbody of the insect. They are passed on to the next generation inserted between the egg coat and the egg cells. *Euscelis* differs from its nutritional source with respect to at least four cation ratios. The symbiont-free leafhoppers probably differ with respect to all six ratios. Here we also find the strictest form of obligate endocytobiosis. The endocytobionts live in special host cells (bacteriocytes) arranged in two symbiotic organs (bacteriomes) and are transmitted by maternal inheritance via intra-ovarian infection.

Microbial symbioses are involved when the Hemiptera differ from their food producers with respect to cation ratios. The greater the difference, the more integrated the symbiosis. Thus, the degree of endosymbiosis seems directly related to both the difference in cation ratios between consumer and food producer and the level of symbiosis (figure 4):

• In the case of complete agreement between the cation ratios (i.e., agreement between the physiochemical types of consumer and producer), there is no significant symbiosis (endosymbiosis degree 0).

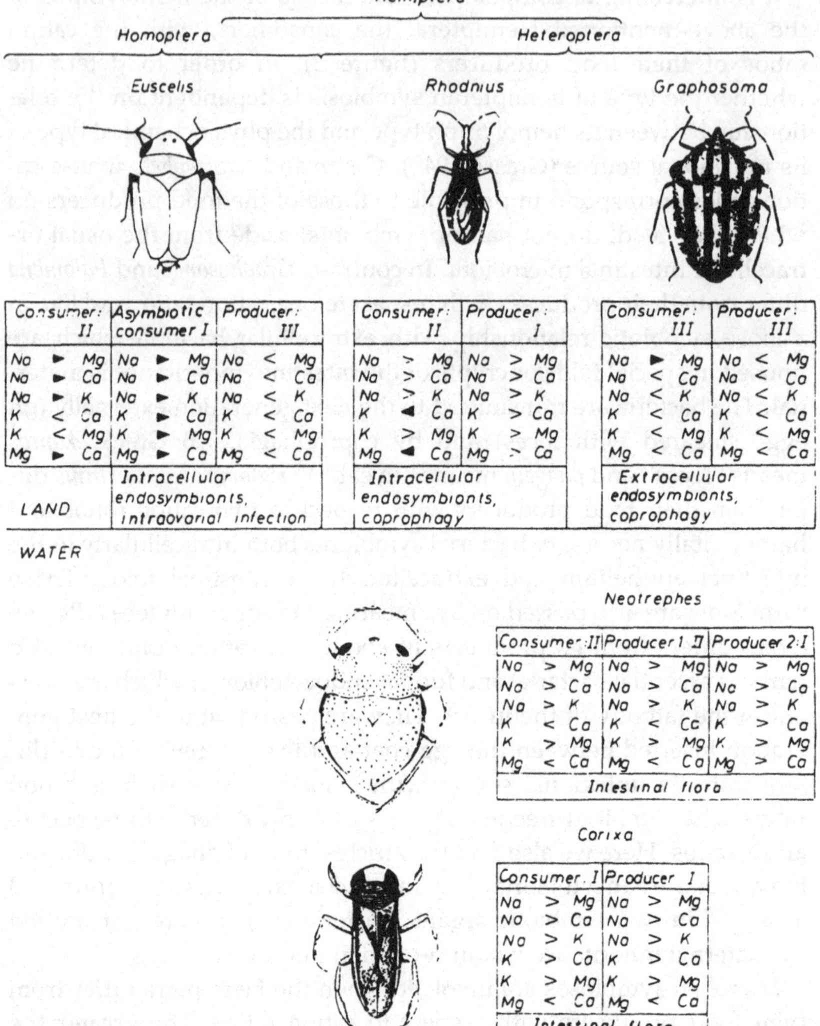

Figure 3
Comparison of the cation ratios (see figure 1) in the hemolymph of insect hosts (consumers) with those in their food (substrate producers) with reference to the degree of symbiosis in typical representatives of the Homoptera and Heteroptera (superorder Hemiptera). A solid triangle indicates a difference in the cation ratios of consumer and producer. The greater the number of differences in the cation ratios, the higher the degree of symbiosis (Schwemmler 1984). *Coprophagy* means the eating of feces (and other excrement); the symbionts are transmitted by the eggs smeared with feces, and the larvae become infected after hatching by eating the feces. The Roman numerals after consumer and producer indicate the physiochemical type.

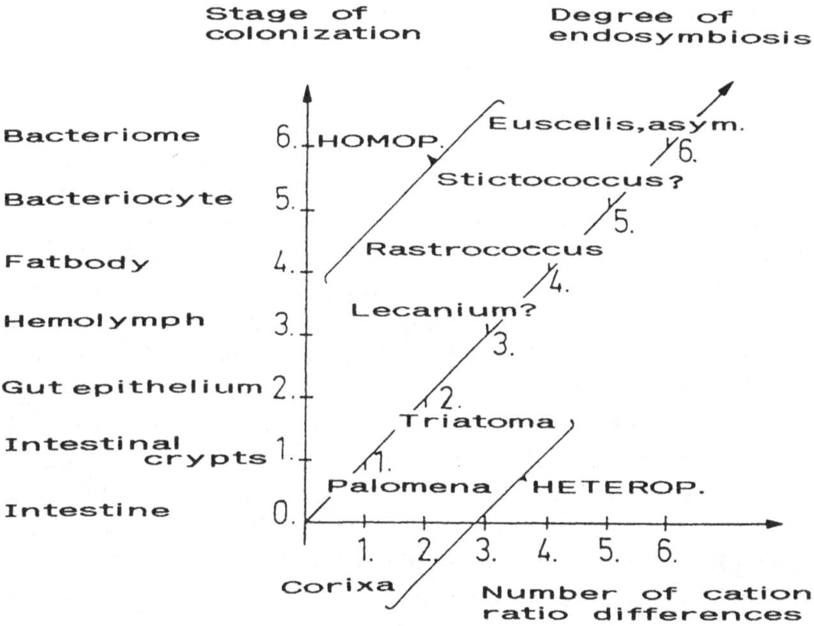

Figure 4
Diagram of the degree of endosymbiosis: the integration level of symbionts is plotted against the difference of cation ratios between the insect host and its food for typical Homoptera and Heteroptera. With increasing difference in cation ratios, the level of integration increases and thus, with it, the degree of endosymbiosis. The date for *Lecanium* and *Stictococcus* are incomplete. (Source: Schwemmler 1984; see figure 3 above.)

• When there are differences in up to three cation ratios, equivalent to the difference between type I and II or between type II and III, an exosymbiosis is present (endosymbiosis degree < 3).

• When more than three cation ratios are different, equivalent to the difference between type I and III, an intracellular symbiosis (endocytobiosis) is present (endosymbiosis degree > 3).

Buchner's (1965) assertion that symbionts serve only to compensate one-sided nutrition and to remove host waste products was in need of revision (Schwemmler 1980a, 1983a). Apparently, the symbionts, as extra- or intracellular decomposers, mediate chemically and physiologically between the host consumer and its food producer. This physiochemical correlation between host and symbiont must be a product of a long period of coevolution.

Speciation of Insects by Symbiogenesis

The predecessors of the mandibulates were the last completely aquatic ancestors of the insects (Henning 1969). During the early Paleozoic, about 600 million years ago, this group diverged into the aquatic Crustacea and the aquatic-terrestrial Tracheata. At the base of the Cambrian period, about 500 million years ago, the Myriapoda and the Insecta emerged from the latter group. The waterbugs (Hydrocorixae, subgroup of Hemiptera) are still (or again) aquatic today. The *Corixa* of physiochemical type I (and perhaps the hemipteran ancestors) feed almost exclusively on algae which are also of type I/II (Schlee 1969; figure 2 above), and no symbiosis has been found in this group (figure 5). The Peloridiidae, about 400 million years old, are considered the ancestral form of either the leafhoppers or the Homoptera, to which they belong (Buchner 1965). They feed on moist forest mosses of physiochemical type II/III. *Hemiodoecus fidelis* (physiochemical type I/II),[1] a representative of the Peloridiidae, harbors only *a*-endocytobionts. These bacteria are an essential presence in the entire hemipteran group, and are thus called primary symbionts. Individual species of Hemiptera may have up to six different types of endocytobionts. According to Müller (1949), primary endocytobionts are those which are able to exist without other types of symbionts. The primary symbionts are harbored in the host's primary bacteriocytes.

The oldest fossil leafhoppers date from the late Paleozoic or from the early Mesozoic, about 200 million years ago. They are the direct ancestors of the extant forms, and may have fed on the ferns which developed at that time. These ferns probably had the physiochemical type II/III (see note 1). The ancestral forms of the leafhoppers (Procicadoidae, Profulgoroidae) are thought to have harbored the essential auxiliary symbionts *t*, *f*, and *X* in the so-called auxiliary bacteriocytes. Auxiliary endocytobionts exist only in association with primary endocytobionts (Müller 1949).

Modern leafhoppers feed primarily on the phloem of angiosperms, whose evolution began about 130 million years ago (Mesozoic). These insects have taken up facultative companion symbionts, including *B* and *W*. The acquisition of these symbionts is thought to have occurred in relatives of leafhoppers, the scale insects (Coccidae), during the early Cenozoic, about 60 million years ago. The companion symbionts were in each case taken up into companion bacteriocytes of the host.

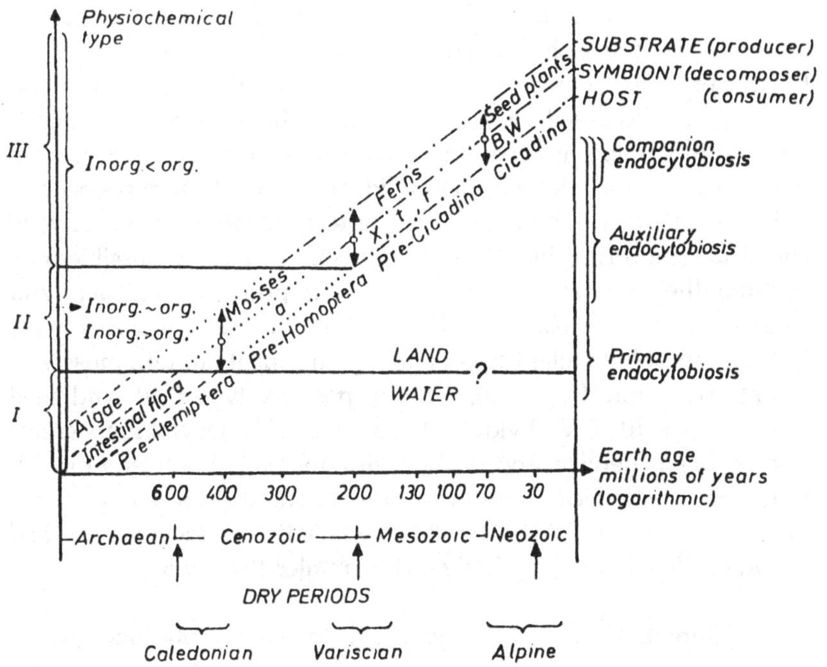

Figure 5
Probable relationship in the evolution of the leafhopper as host-consumer and its food or substrate-producer and symbiont decomposer: hypothetical aquatic Hemiptera seem to have evolved into moisture-loving, peloridiid-like forms. Either the aquatic or the semiaquatic forms then developed into the modern, purely terrestrial leafhopper species. This occurred through integration of primary, auxiliary, and companion endocytobionts, thereby adapting to new producers—mosses, then ferns, and finally angiosperms. The symbionts must in each case have mediated between the more primitive physiochermical type of the host and the higher physiochemical type of its food. (Source: Schwemmler 1984.)

During the evolution of the primary, auxiliary, and companion bacteriocytes, the global climate was arid and there were geological folding processes (figure 5, Seyfert and Sirkin 1973). The facts and conclusions discussed here suggest that the bacteriocytes developed from aquatic hemipteran predecessors of the physiochemical type I, via moisture-loving, peloridiidean forms of the physiochemical type I/II to the modern purely terrestrial forms of the physiochemical type II/III. This development was apparently possible only through the incorporation of intracellular primary, auxiliary, and companion endocytobionts. Only through these symbioses were leafhoppers able to adapt to changing plant nutrition, i.e., mosses of the physiochemical type I/II, ferns of possibly type II/III, and seed plants of type III. The phylogenetic function of endocytobionts seems to have been chemical and physiological mediation between the differing physiochemical types of the animal and its food source. This, however, is precisely the function which the endocytobionts had already fulfilled ontogenetically (Schwemmler 1979, 1984).

Cell Differentiation and Morphogenesis by Insect Symbiogenesis

The leafhopper endocytobiosis, one of the most integrated mutual dependencies of host and symbiont, is a good model for analysis of these physiochemical relationships. The egg cell of *Euscelidius* (closely related to *Euscelis*: Jassidae, Cicadina, Homoptera, Hemiptera) seems especially well suited for such studies.

Probably as a result of gene transfer into the host genome, the molecular weight of the endocytobiont DNA is probably between 2.2×10^7 and 2.6×10^7 daltons, only slightly higher than the molecular weight of *Euscelis* mitochondria (approximately 1×10^7 daltons). As a result, endocytobionts, like mitochondria, are not capable of growth or reproduction outside their host cell (Schwemmler 1987). If, however, the endocytobionts of one generation are prevented from infecting the leafhopper eggs and thus from passing to the next generation, only head-thorax embryos (organisms lacking abdomens) result. Similarly, if the egg mitochondria are inactivated, "headless" abdomens may sometimes develop, which grow *in vitro* like tumors. It is obvious from these data that leafhoppers have evolved to their extant form by integrating at least part of the genome of their mitochondria and endocytobionts (figure 6).

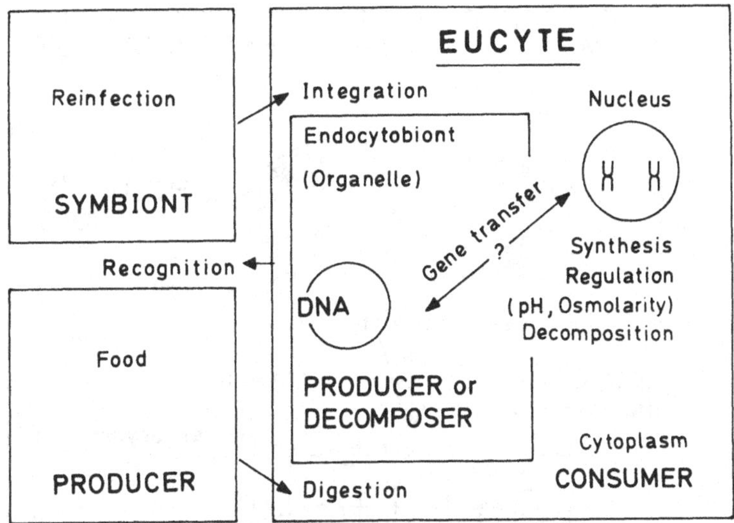

Figure 6
Representation of the eukaryotic leafhopper egg cell with its semi-autonomous DNA-organelles and endocytobionts as a central regulating cybernetic system. The DNA-organelles and endocytobionts work as decomposers of catabolites and waste products of the host cell or the food producer, as synthesizers of anabolites for the nucleo-cytoplasm or consumer, and as regulators of pH, of osmotic pressure, and probably of the intracellular clock. (Source: Schwemmler 1983.)

There are other close structural and functional analogies between leafhopper endocytobionts and leafhopper mitochondria (Schwemmler 1989). Apparently, both mitochondria and endocytobionts synthesize host anabolites (vitamins, amino acids, morphogenetic substances, etc.) using the host catabolites (e.g., uric acids, urea, carbon-bodies), and thereby regulate the pH, the osmolarity, and probably the intracellular clocks of the host by gene transfer. Moreover, endocytobionts and mitochondria control the interaction of cytoplasmic glycolysis and respiration in the host (figure 7). This kind of interaction seems to stabilize cell rhythm and initiate embryonic development; if disturbed, it is apparently capable of inducing tumor-like growth (Schwemmler et al. 1989).

Leafhopper symbiogenesis can serve as a model for molecular analysis not only of insect speciation and morphogenesis, but also of (de-)differentiation of eukaryotic cells. A new interdisciplinary research field, endocytobiology, is currently being established from the fusion of symbiosis and cell research.

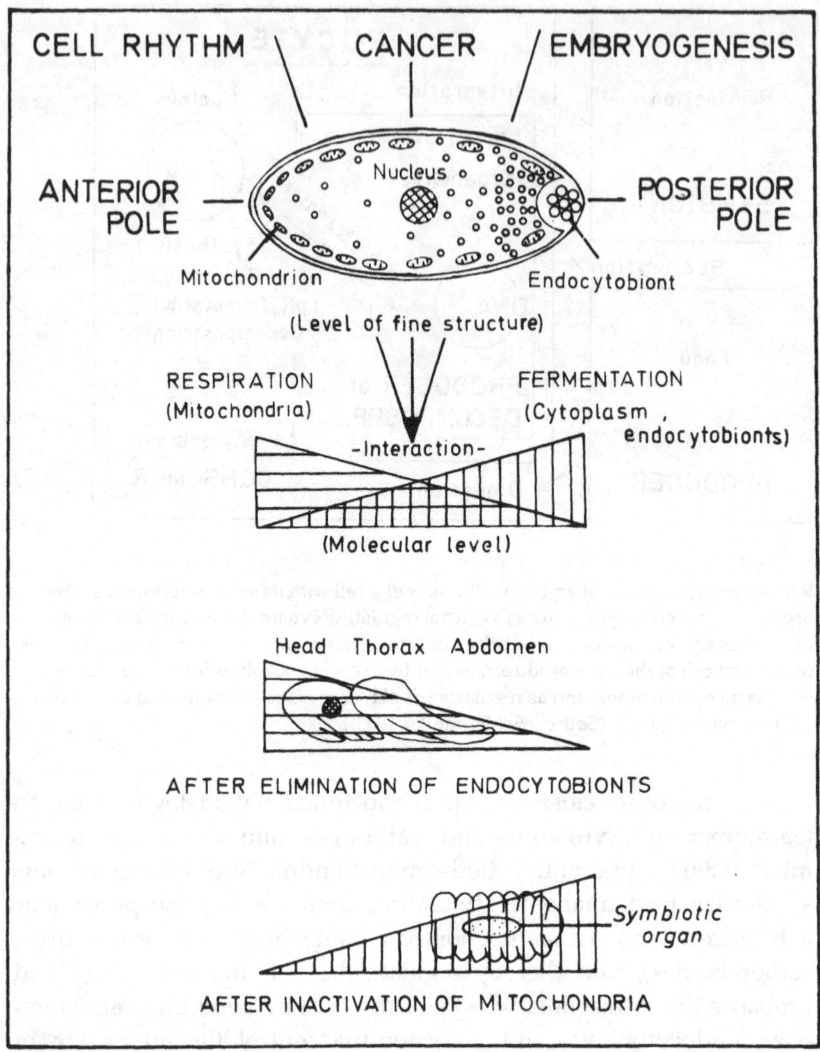

Figure 7
The leafhopper egg cell of *Euscelidius variegatus* (Jassidae, Cicadina, Homoptera, Hemiptera) is a suitable model for molecular analysis of the influence of symbiogenesis on morphogenesis and cell differentiation: the initiation of egg development is controlled by the interaction between an ooplasmic/endocytobiotic glycolysis gradient with maximum activity at the posterior pole and a mitochondrial respiration gradient with maximum activity at the anterior pole, following the third day after oviposition. This interaction also appears to stabilize the intracellular clock, and may induce tumor-like growth when disturbed. (Source: Schwemmler 1987.)

Study of Symbiogenesis by Endocytobiology

The term *symbiosis* was introduced to the literature by De Bary in 1879. It originally included not only mutualism (beneficial interdependence of different species) but also parasitism, a form of association which is beneficial to one partner and detrimental to the other. At present the term *symbiosis* is often equated with the concept of mutualism. (The term is also used in this sense in figure 8.) The phylogenetically younger partner is generally considered to be the host, and the older partner the symbiont.

Around the turn of the century, the symbiotic theory was formulated. According to this, plastids (Schimper 1883), mitochondria (Altmann 1890), and later also the nucleus (Mereschkovskii 1910) were proposed to be the final products of a long process of symbiotic integration. When DNA was discovered in plastids (Ris 1961) and in mitochondria (Nass and Nass 1963), symbiosis research received new impetus to investigate questions of cell evolution. This is reflected in the book *Symbiosis in Cell Evolution*, by Lynn Margulis (1981), founder of the modern endosymbiont hypothesis. She postulates that in addition to mitochondria and plastids, the eukaryotic motility apparatus (undulipodium or euflagellum) is the result of a symbiotic integral process (serial endosymbiosis theory; see Taylor 1974). The founder of systematic symbiosis research, Paul Buchner (1965), investigated the function of endosymbionts and assessed the significance of their contribution to the breakdown of nutrients for the host and to the provision of vitamins to a host with a limited diet. Later, this concept was developed further for a number of intracellular endosymbionts to account for physiological and biochemical studies which showed that endocytobionts take on functions of DNA-containing cell organelles (Schwemmler 1980a,b).

There was thus a need for a clear terminology distinguishing the various types of symbiosis (figure 8). The term *cytobiosis*, introduced by Taylor (1980), includes not only intracellular but also epicellular and intercellular symbiosis, and it is not clear whether it completely excludes extracellular symbiotic relationships. The term *endosymbiosis*, frequently used by Buchner, is equally vague. The term *cytosymbiosis*, coined by Sitte (Kleinig and Sitte 1984), also presents difficulty, especially because it does not include cytoparasitism according to the definition given above. However, the term *endocytobiosis* fits well; Eberhard Schnepf was the first to use it, in 1975. In 1979 I elaborated

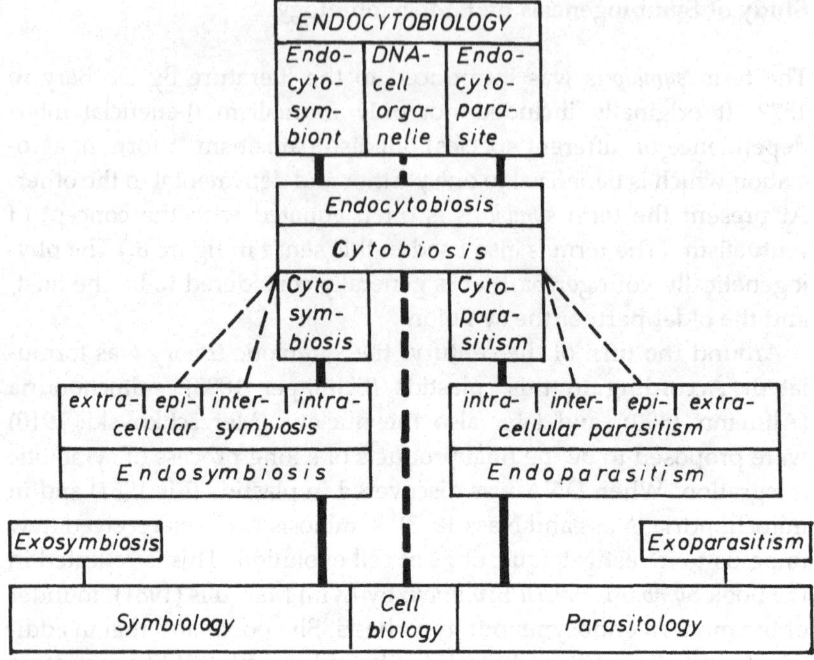

Figure 8
Derivation and terminology of the new interdisciplinary research field, endocytobiology. According to this, endocytobiosis means the living together of a pro- or eukaryotic host with a pro- or eukaryotic, DNA-autonomous, intracellular symbiont or parasite; in endocytobiology not only structures, functions, genes, and evolution of endocytobionts and parasites are analyzed, but also organelles which are only DNA-semiautonomous, and which have devloped from DNA-autonomous endocytobionts. (Source: Schemmler 1989.)

the term to mean both intracellular symbionts (endocytosymbionts) and intracellular parasites (endocytoparasites), thus taking into account the ease of transition between symbiosis and parasitism. I also derived the term *endocytobiology* from *endocytobiosis* (Schwemmler 1979). In 1980 and 1983, at the first and second International Colloquia on Endocytobiology in Tübingen, Schenk and I proposed the official introduction of the term and its definition, in order to establish it internationally at the conference of the New York Academy of Sciences in 1986 (Lee and Fredrick 1987). The term *endocytobiology* has now become generally accepted among symbiologists, parasitologists, and cytologists who investigate intracellular symbioses. This term also includes the study of DNA-containing cell organelles, in accor-

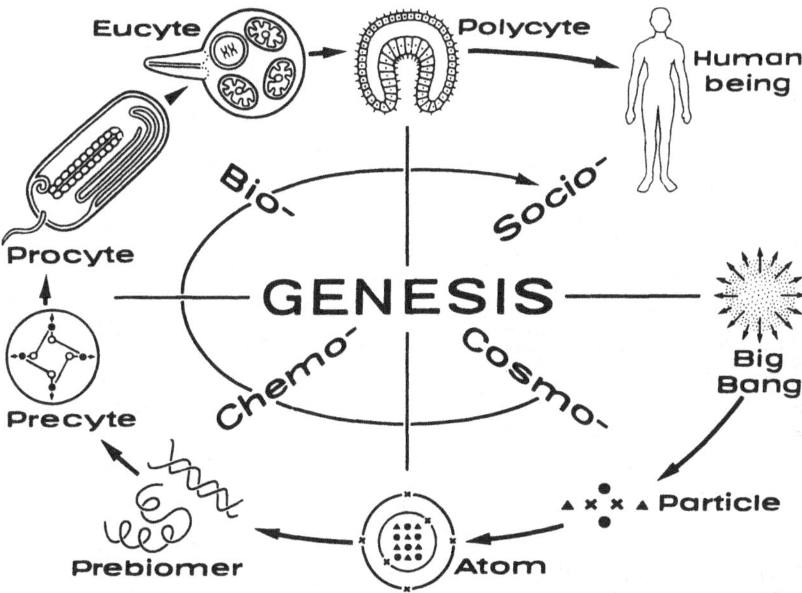

Figure 9
Evolution in the broad sense from the Big Bang to humanity. It may be regarded as a coherent process which may be divided into physical, chemical, biological, and cultural phases. The most important stages in the course of this process are the elementary particles, the atoms, cells, multicellular systems (polycytes), and human beings.

dance with the Endocytobiological Cell Theory (an expansion of the serial endosymbiosis theory). International cooperation of scientists in this field resulted in the establishment of the International Society of Endocytobiology and the interdisciplinary international journal *Endocytobiosis and Cell Research*. These institutions, as a forum for communication, may stimulate future research and thus contribute to achieving understanding of the problems of symbiogenesis as a major mechanism in evolutionary innovation (figure 9).

The General Process of Evolution

It is widely believed that biological evolution is a continuous process with smooth transitions (figure 10A), an idea first proposed by Darwin (1859) and further developed by neo-Darwinists, including Dobzhansky, Mayr, Huxley, Rensch, and Maynard Smith (Rieppel 1989). This process is characterized by the gradual adaptation of slightly heterogenous populations to their environment through natural

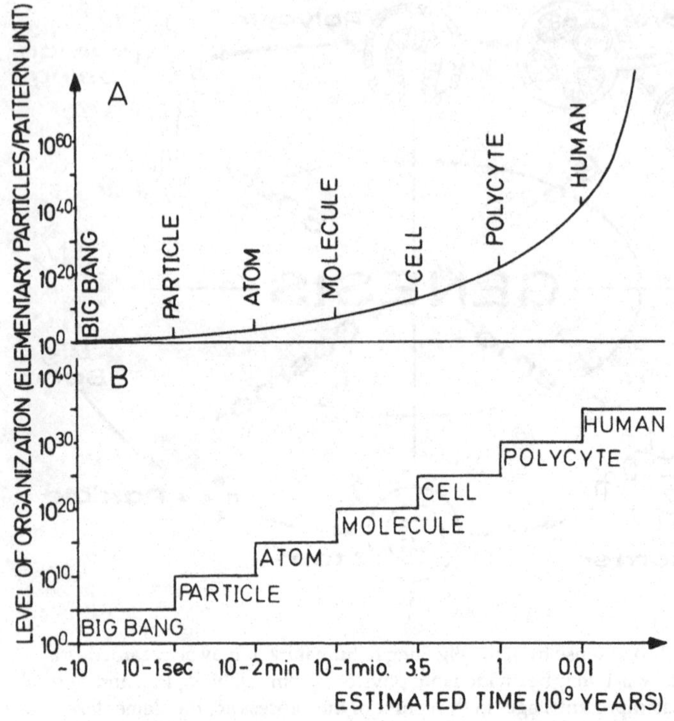

Figure 10
Alternative hypotheses on the nature of the general evolutionary process from Big Bang to humanity. (A) Continuous change: gradualism. (B) Punctuated equilibrium: building-block mechanism (during biological evolution: symbiogenesis). (Source: Schwemmler 1989.)

selection. Adaptive changes (mutations) must spread through the population by a slow process if they are to be preserved in subsequent generations. This mode is called *gradualism*, in the terminology of Eldredge and Gould (1972).

Some scientists are of the opinion that biological evolution has also occurred in periodic leaps, e.g., by symbiogenesis: various living systems of the same organizational level have fused, forming more complex biological systems, which, because of selection, spread across the earth in a relatively short time and in small populations (figure 10B; Goldschmidt 1940; Margulis 1981; Schwemmler 1984, 1989). The basic principle of a unified theory of symbiogenesis is presented below; the details of arguments for or against these ideas are to be found in the more recent literature (Schwemmler 1989).

Evolution of Cellular Metabolism

Phenomenon

In the course of evolution, prokaryotic cells (procytes) have produced four basic types of metabolism, each of which now exists in both anaerobic and aerobic forms, and which made possible the exploitation of increasingly rich sources of energy. Energy is obtainable in the process of fermentation by cleavage of diphosphosugars; in photergy by a light-driven proton pump and a protein-bound carotene derivative (bacteriorhodopsin); in photosynthesis using a light-driven chlorophyll system; and in respiration from a respiratory chain which consists of redox systems, including the cytochromes.

Hypotheses

Two alternative hypotheses have been proposed for the phylogeny of the four metabolic types (fermentation, photergy, photosynthesis, and respiration). A decisive difference between these two hypotheses is the proposed time at which the chlorophyll system evolved. According to the conversion hypothesis (Broda 1975), chlorophyll arose prior to the emergence of cytochromes. The cleavage hypothesis (Schwemmler 1984, 1989) postulates that the chlorophyll system differentiated after at least the first parts of the cytochrome system were in place.

Reconstruction

At present, the evidence needed to assess the rivaling hypotheses includes chemical fossils, microfossils, consideration of ecological niches, and homologies in metabolism and in sequences of proteins and nucleic acids (Schwemmler 1989). The data clearly favor the cleavage hypothesis, according to which the order of the increasing redox potential of the respiratory chain's components is also the order of their evolution. By this reasoning, the chlorophyll system cannot have developed before the first cytochromes.

It is assumed that the original bacteria possessed only fermentation coupled to a sulfur metabolism. Only later were the first redox systems of the respiratory chain incorporated into their cell membranes.

The development of the first cytochromes was apparently followed by a divergence into three major lineages: the urkaryotes, which are thought to be the ancestors of the nucleocytoplasm system of nucleated cells (see next section), the original archaeobacteria, and the eubacteria. It is assumed that the archaeobacteria are just as closely related to the urkaryotes as to the eubacteria (Woese and Fox 1977; Woese 1981). The development of the archaeobacteria into eubacteria is thought to have involved successive incorporation of further members of the cytochrome system into the respiratory chain. The values of their redox potentials suggest that integration of the last redox system, cytochrome oxidase, was preceded by that of the chlorophyll system, which closed the circuit (Dickerson 1980). The development of respiring organisms would then have led to a secondary loss of chlorophylls (Broda 1975).

Discussion

The basic concept developed by Woese (1981) is that the most primitive bacteria diverged into three lineages: the urkaryotes, the archaeobacteria, and the eubacteria. This is generally accepted as a reasonable basis for further discussion. However, there is still disagreement about further classification into subgroups on the basis of sequence analysis (Woese and Fox 1977) and comparison of metabolism or structure (Lake et al. 1985). A possible explanation for the differences in opinion could be the phenomenon of horizontal gene transfer. It is likely that, in the course of bacterial evolution, genes for special metabolic functions were exchanged in the form of small rings (plasmids) of DNA. (See pages 96–100 of this volume.) In such cases, sequence homologies could no longer be used as evidence for phylogenetic relationships, because the homologous sequences are of exogenous origin (Dayhoff 1972).

A series of important questions concerning the evolution of bacteria remain unanswered. It is still uncertain whether fermentation is actually the oldest metabolic type or whether it instead constitutes a later reductive adaptation to special organic substrates. It also was inferred that the anaerobic sulfur bacteria and their metabolic type are the most ancient (König 1986; Stackbrandt 1986).

Evolution of DNA-Containing Cell Organelles

Phenomenon

The further evolution of eukaryotic cells, the eucytes, was characterized by the formation of morphologically separate reaction spaces (compartments) with different functions, including heredity (nucleus), fermentation (cytoplasm), aspects of motility (kinetid[2]), respiration (mitochondria), and photosynthesis (chloroplasts). These spaces are genetically more or less independent cell organelles or complexes of organelles. The nucleocytoplasm complex is genetically the most autonomous. Mitochondria and chloroplasts are only semiautonomous organelles, whereas only transfer RNA has been clearly identified as endogenous to the kinetid (Hartmann 1975).[3]

Hypotheses

There are two major alternative hypotheses concerning the origin of the cell organelles containing DNA or RNA. The classical cell hypothesis claims that the eucyte evolved directly from one single procyte as an elemental organism. Its separate metabolic activities (fermentation, photergy, respiration, and photosynthesis) became compartmentalized by segregation of duplicated nuclear genes. The alternative, the serial endosymbiosis theory and its expansion the endocytobiological cell theory, posits that the corresponding cell organelles are descendants of intracellular symbionts which were specialized for one of the four basic metabolic types. The essential difference between the two hypotheses is thus the derivation of the DNA and RNA found in the organelles: in the first case, they are endogenous; in the second, exogenous.

Reconstruction

The experimental data consist of cell fossils, genetic autonomy, analogies with endocytobionts, and homologies with bacteria (Schwemmler 1989). From these data, it is nearly certain that mitochondria and plastids have descended from endocytobionts; however, the origin of the motility apparatus of eukaryotes has not yet been satisfactorily clarified, because of the lack of relevant sequence analyses (Margulis 1981; Schwemmler 1984; Hinkle, this volume). On the basis of DNA

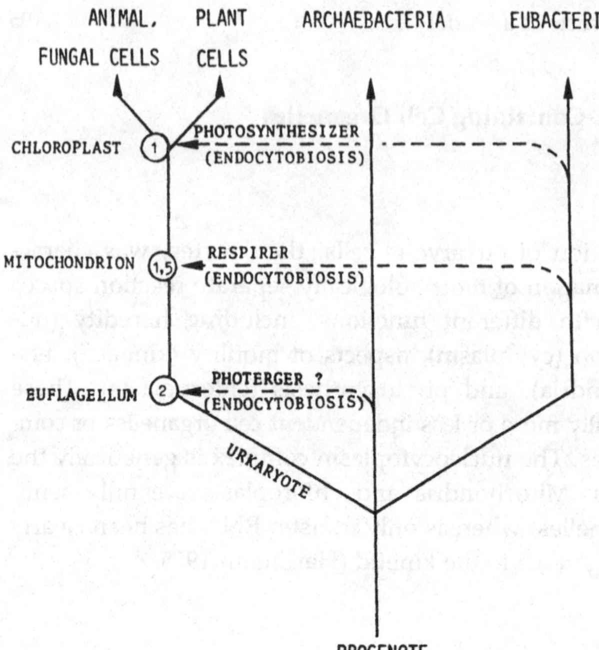

Figure 11
A model for eucyte evolution. At least three endocytobioses occurred, producing the precursors of animal, fungal, and plant cells. The existence of photergic endocytobiosis, giving rise to the eucytic motility organelles, is still under discussion (see text). The numbers in circles are estimated dates in billions of years.

and RNA sequence analysis (Dayhoff and Schwartz 1980; Woese and Fox 1977; Woese 1981; Doolittle and Bonen 1981), it is now generally accepted that mitochondria and plastids arose by the assimilation and incorporation first of respiring and then of photo-assimilating, intracellular eubacterial symbionts into archaeobacterial urkaryote hosts (figure 11). Precursors arose in this way first for animal and fungal cells, then for plant cells.

Discussion

The origin and the nature of the motility apparatus, including the undulipodium (eukaryotic flagellum), are at present the most disputed points in eucyte evolution. Margulis (1981) has suggested that the 9(2) + 2 microtubular apparatus arose from symbioses between spirochete-like endocytobionts and a *Thermoplasma*-like archaeobacterial urkaryote. Immunofluorescence studies appear to indicate homologies between eukaryotic microtubules and tubulin-like proteins from spirochetes, but not between microtubules and bacterial flagellin. The presumed flagellar symbionts may have been photergic (Schwemmler 1984, 1989). Under certain circumstances, photergic genes which were conserved within the host genome appear to activate. Certain green algae form protein-bound pigments known as rhodopsin (Foster et al. 1984). Rhodopsin, in conjunction with a neighboring eye spot (stigma) and the undulipodium,[4] generates light-orientated motion. Certain cells in the retinas of vertebrates also form membrane structures in which rhodopsin, acting as a visual pigment, is embedded. Homologies between these pigments and the sensory pigments of halo-rhodopsin suggest a common origin of both proteins (Martin et al. 1986; Blanck and Oesterhelt 1987; Dencher 1988). Eucytes could thus have obtained the genes for rhodopsin either by horizontal gene transfer or by assimilation and integration of a photergic endocytobiont. If symbiosis was involved in the origin of eukaryotic motility, there are reasons to presume that motility arose before mitochondrial endocytobiosis (Schwemmler 1984, 1989; Margulis 1988).

Another unsolved problem of eucyte evolution is the origin of the nucleus. According to Giesbrecht and Drews (1981), the eucytic cell's nucleus is not homologous to the mitochondria and plastids, but should probably be regarded as a composite system that developed from more or less independent karyomers, a process which can still be observed in the early oogenesis of many organisms. Each karyomer would represent a single chromosomal apparatus of a procyte.

The approach to cellular evolution presented here has revealed the origins of six systematic groups: archaeobacteria, eubacteria, protists, fungi, plants, and animals. Since sequence analysis provides highly convincing evidence for phylogenetic relationships, it is unlikely that much will change in the current basic concepts, with possible exceptions concerning the origin of the nucleus and of the eucytic euflagellar apparatus.

Mutation and Selection as the Major Mechanism of Continuous Evolution

There are a number of different mechanisms which drive the small steps of biological evolution. These include isolation, recombination, genetic drift, mutation, and selection. For the relatively small, continuous steps of biological evolution, the term *microevolution* has been coined; mechanisms of microevolution are correspondingly called *micromechanisms*.

Natural selection favors organisms that, because of beneficial and heritable traits, are better adapted to their environment (Darwin 1859). According to many geneticists, the microevolutionary process —the combination of random mutation and natural selection—is adequate to explain the gradual formation of new races or species.

Meso-evolutionary mechanisms, such as cell or nuclear fusion, polyploidization, and gene transfer, have been discussed elsewhere (see Haynes, this volume).

Symbiogenesis as the Major Mechanism of Discontinuous Evolution

The evolution of one basic type of organism from another, such as the development of the eucyte from the procyte, is a macroevolutionary step for which the existence of macromechanisms must be postulated (Stanley 1981; Illies 1983). Triality, periodicity, phase, and building-block principles have been suggested as constituting this type of macromechanism (Schwemmler 1984, 1989). According to the building-block principle, each higher evolutionary type is composed of different representatives of the next lower type (figure 12). Applied to the total process of cellular evolution, associations of various types of eucytes were formed from various types of procytes, and multicellular plants, fungi, and animals may have formed from different types of eucytes. At this level, evolution appears to occur in jumps (figure 10B). This can be explained by the formation of symbioses (symbiogenesis) between different, previously autonomous, components (Schwemmler 1972). Symbiogeneses are still occurring, as has been demonstrated with endocytobiosis of leafhoppers. With respect to the question of the major mechanisms of the biological evolutionary process, both gradualism (microevolution) and symbiogenesis (macro-/mega-evolution) make valid contributions—as is so often the case when apparently alternative hypotheses are presented. (See pages 9 and 46–47 above.) The combination of the two explanations makes further progress toward a unified theory of evolution possible.

Conclusions

Bacteria, animals, and plants can be classified into three distinct physiochemical composition groups according to pH, osmotic values,

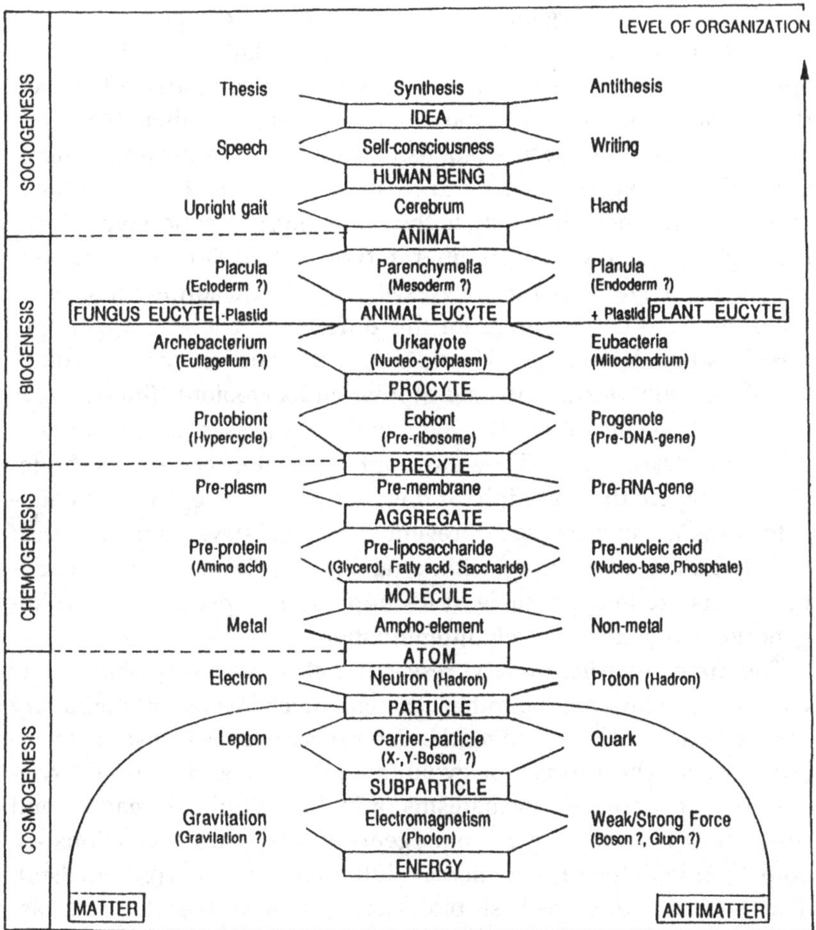

Figure 12
The total process of evolution, including biogenesis, according to the unified theory of evolution (compare with the grand unified theory of nature: Hawking 1988). Evolution is presented as a causal process, in which phases of smooth transitions (gradualism) and of apparent stagnation in the increase of complexity (open blocks; microevolution) alternate with phases of evolutionary, discontinuous changes (rectangles; macroevolution). The latter changes are caused primarily by incorporation of already existing structural elements (symbiogenesis), forming new, more complex systems. (Sources: Schwemmler 1972, 1985, 1989.)

and the ratios of inorganic-ion concentrations and organic-molecule concentrations in their intracellular or extracellular sap. The widespread existence of these types allows a direct comparison between the physiochemical composition of consumers and their food producers. Examples of such relationships are found in the order Hemiptera. Bacterial symbioses tend to be absent when the insect consumer and the food producer belong to the same physiochemical type. If the insect and its food organism differ in type, extracellular or intracellular microbial symbioses usually exist. The conclusion is drawn that endosymbiosis compensates for this difference. The leafhopper species (Cicadina, Homoptera, Hemiptera) proliferate within their ecological niches only through the aid of their endocytobionts (intracellular bacteria-like symbionts). The symbionts are passed on to the next generation via the egg. They move into special insect organs (bacteriomes), where they multiply. If the infection of the egg by the endocytobionts is experimentally prevented, normal development ceases; only head-thorax embryos without abdomens develop. Thus, endocytobionts are indispensable in leafhoppers for speciation, morphogenesis, and oogenesis (cell differentiation).

The new interdisciplinary research field of endocytobiology is concerned with, among other things, mechanisms of hereditary intracellular symbiosis and mechanisms of symbiosis formation (symbiogenesis). The process of evolution was and is driven and controlled by the various mechanisms of biological micro-, macro-, and mega-evolution. During the emergence of a species, transitions are smooth only in the case of microevolutionary processes (gradualism). The transition from one basic biological type to another (for example, from procytes to eucytes) is not a continuous process but occurs in jumps (punctuated equilibrium). Therefore, symbiogenesis must be considered as the generally applicable mechanism for speciation, morphogenesis, and cell differentiation in cellular and multicellular evolution. The process of symbiogenesis is ongoing, as can be seen with the example of leafhopper endocytobiosis. Analyses of cases of extant symbiogenesis are a basis for any reconstruction of biological evolution.

Notes

1. Since experimental data are lacking for this system, the physiochemical classification is hypothetical.

2. The author prefers the term *euflagellar complex.*—*Editors*

3. Hall et al. (1989) recently reported the discovery of kinetosomal DNA in *Chlamydomonas reinhardtii* (see Hinkle, this volume).—*Editors*

4. The author prefers the term *euflagellum*.—*Editors*

References

Altmann, R. 1980. *Die Elementarorganismen und ihre Beziehung zu den Zellen*. Veit, Leipzig.

Blanck, A., and Oesterhelt, D. 1987. The halo-opsin gene. II: sequence, primary structure of halorhodopsin and comparison with bacteriorhodopsin. *EMBO J*. 6: 265–273.

Broda, E. 1975. *The Evolution of the Biogenetic Processes*. Pergamon.

Buchner, P. 1965. *Endosymbiosis of Animals with Plant Microorganisms*. Wiley.

Darwin, C. 1859. *On the Origin of Species by Means of Natural Selection, or the Preservation of Favoured Races in the Struggle for Life*. John Murray, London.

Dayhoff, M. O. 1972. *Atlas of Protein Sequence and Structure*, volume 5. National Biochemical Research Foundation, Washington.

Dayhoff, M. O., and Schwartz, R. M. 1980. Prokaryote evolution and the symbiotic origin of eukaryotes. In: Schwemmler, W., and Schenk, H., eds., *Endocytobiology I: Endosymbiosis and Cell Research, a Synthesis of Recent Research*. Walter de Gruyter.

De Bary, A. 1879. *Die Erscheinung der Symbiose*. Karl J. Trübner, Strassburg.

Dencher, N. A. 1988. Rhodopsin-like membrane proteins in the archaebacterium *Halobacterium halobium*: A clue to cellular evolution. *Endocyt. C. Res*. 5(1): 1–16.

Dickerson, R. E. 1980. Cytochrome c and the evolution of energy metabolism. *Scientific American* 3: 98–110.

Doolittle, W. F., and Bonen, L. 1981. Molecular sequence data indicating an endosymbiotic origin for plastids. In: Fredrick, J., ed., *Origins and Evolution of Eukaryotic Intracellular Organelles*. New York Academy of Sciences.

Duchâteau, G., Florkin, M., and Leclerq, J. 1953. Concentration des bases fixes et types de composition de la base totale de l'hémolymphe des insectes. *Arch. Intern. Physiol*. 61(4): 518–549.

Eldredge, N., and Gould, S. J. 1972. Punctuated equilibria: An alternative to phyletic gradualism. In: Schopf, T. J. M., ed., *Models in Paleobiology*. Freeman.

Foster, K. W., Saranak, J., Patel, N., Zarilli, G., Okabe, M., Kline, T., and Nakanishi, K. 1984. A rhodopsin is the functional photoreceptor for phototaxis in the unicellular eukaryote *Chlamydomonas*. *Nature* 311: 756–759.

Giesbrecht, P., and Drews, G. 1981. Die "Kernstrukturen" der Bakterien und ihre Beziehung zu denen der "Mesokaryoten." In: Metzner, H., ed., *Die Zelle: Struktur und Funktion. 3.* Wissenschaftliche Verlagsgesellschaft, Stuttgart.

Goldschmidt, R. 1940. *The Material Basis of Evolution.* Yale University Press.

Grassé, P. P. 1949. *Traité de Zoologie (Anatomie, Systématique, Biologie)*, volume IX. Masson.

Hall, J. L, Ramanis, Z., and Luck, D. J. L. 1989. Basal body/centriolar DNA: Molecular genetic studies in *Chlamydomonas. Cell* 59: 121–132.

Hartmann, H. 1975. The centriole and the cell. *J. Theor. Biol.* 51: 501–509.

Hawking, S. 1988. *A Brief History of Time: From Big Bang to Black Holes.* Bantam.

Henning, W. 1969. *Die Stammesgeschichte der Insekten.* Kramer, Frankfurt.

Illies, J. 1983. *Der Jahrhundertirrtum: Würdigung und Kritik des Darwinismus.* Umschau, Frankfurt.

Kleinig, H., and Sitte, P. 1984. *Zellbiologie.* Gustav Fischer, Stuttgart.

König, H. 1985. Biologie der Archäbakterien. *Biuz* 3: 71–82.

Lake, J. A., Clark, M. W., Henderson, E., Fay, S. P., Oakes, M., Scheinemann, A., Thorner, J. P., and Mah, R. A. 1985. Eubacteria, halobacteria and the origin of photosynthesis: The photocytes. *Proc. Natl. Acad. Sci.* 82: 3716–3720.

Lee, J., and Fredrick, J. F., eds. 1987. *Endocytobiology III.* New York Academy of Sciences.

Margulis, L. 1981. *Symbiosis in Cell Evolution.* Freeman.

Margulis, L. 1988. Serial endosymbiotic theory (SET): Undulipodia, mitosis, and their microtubule systems preceded mitochondria. *Endocyt. C. Res.* 5: 133–162.

Martin, R. L., Wood, C., Baehr, W., and Applebury, M. L. 1986. Visual pigment homologies revealed by DNA hybridization. *Science* 232: 1266–1269.

Mereschkovskii, C. 1910. Theorie der zwei Plasmaarten als Grundlage der Symbiogenesis. *Biol. CB* 30: 378, 321, 353.

Müller, H. J. 1949. Zur Systematik und Phylogenie der Zikaden-Endosymbiosen. *Biol. Zbl.* 68: 343–368.

Nass, S., and Nass, M. M. K. 1963. Intramitochondrial fibers with DNA characteristics. II. Fixation and electron staining reactions. *J. Cell Biol.* 19: 593–611.

Rieppel, O. 1989. *Unterwegs zum Anfang. Geschichte und Konsequenzen der Evolutionstheorie.* Artemis, Zürich.

Ris, H. 1961. Ultrastructure and molecular organization of genetic systems. *Can. J. Genet. Cytol.* 3: 95–120.

Schimper, A. F. W. 1883. Über die Entwicklung der Chlorophyllkörner und Farbkörper. *Bot. Z.* 41: 105–114.

Schlee, D. 1969. Morphologie und Symbiose: Ihre Beweiskraft für die Verwandschaftsbeziehungen der Coleorrhyncha. Phylogenetische Studien IV: Heteropteroidea (Heteroptera, Coleorrhyncha) als monophyletische Gruppe. *Stuttgarter Beitr. Naturkunde* 210: 1–27.

Schwemmler, W. 1972. Endosymbiosebildung: Mechanismus der Evolution. *Naturw. Rdschau* 25(9): 350.

Schwemmler, W. 1979. Endocytobiose und Zellforschung. *Naturwissenschaften* 66: 366.

Schwemmler, W. 1980a. Endocytobiosis: general principles. *Biosystems* 12: 111–122.

Schwemmler, W. 1980b. Endocytobiology: A modern field between symbiosis and cell research. In: Schwemmler, W., and Schenk, H., eds., *Endocytobiology I: Endosymbiosis and Cell Research, A Synthesis of Recent Research.* Walter de Gruyter.

Schwemmler, W. 1983a. Endocytobiosis as an intracellular ecosystem. In: Schenk, H., and Schwemmler, W., eds., *Endocytobiology II: Intracellular Space as Oligogenetic Ecosystem.* Walter de Gruyter.

Schwemmler, W. 1983b. Analysis of possible gene transfer between an insect host and its bacteria-like endocytobionts. *Int. Rev. Cytol.* 14: 247–266.

Schwemmler, W. 1984. *Reconstruction of Cell Evolution: A Periodic System.* CRC Press.

Schwemmler, W. 1987. Endocytobionts and mitochondria as determinants of leafhopper egg cell polarity. In: Fredrick, J. F., and Lee, J., eds., *Endocytobiology III.* New York Academy of Sciences.

Schwemmler, W. 1989. *Symbiogenesis: A Macro-Mechanism of Evolution. Progress Towards a Unified Theory of Evolution.* Walter de Gruyter.

Schwemmler, W., Schirpke, B., Behrends, B., Dencher, N. A., and Schröder, W. H. 1989. Investigations on the initiation of insect embryogenesis by ions (*Euscelidius variegatus* KBM: Cicadina, Homoptera). *Cytobios* 57: 19–32.

Seyfert, C. K., and Sirkin, L. A. 1979. *Earth History and Plate Tectonics.* Harper and Row.

Stackbrandt, E. 1986. Das hierarchische Systemder Eubakertien. Probleme und Lösungsansätze. *Forum Mikrobiol.* 5: 225–260.

Stanley, S. M. 1981. *New Evolutionary Time Table: Fossils, Genes and the Origin of Species*. Basic Books.

Sutcliffe, D. W. 1963. The chemical composition of hemolymph in insects and some other arthropods in relation to their phylogeny. *J. Comp. Biochem. Physiol.* 9: 121–135.

Taylor, F. J. R. 1974. Implications and extensions of the serial endosymbiosis theory of the origin of eukaryotes. *Taxon* 23: 229–258.

Taylor, F. J. R. 1980. The stimulation of cell research by endosymbiotic hypotheses for the origin of eukaryotes. In: Schwemmler, W., and Schenk, H., eds., *Endocytobiology I: Endosymbiosis and Cell Biology, A Synthesis of Recent Research*. Walter de Gruyter.

Whittaker, R. H., and Margulis, L. 1978. Protist classification and the kingdoms of organisms. *Biosystems* 10: 3–18.

Woese, C. R. 1981. Archaebacteria. *Scientific American* 6: 98–122.

Woese, C. R., and Fox, G. E. 1977. The concept of cellular evolution. *J. Mol. Evol.* 10: 1–6.

15 Luminous Bacteria Symbiotic with Entomopathogenic Nematodes

Kenneth H. Nealson

Symbiosis as a source of evolutionary innovation can be easily demonstrated in the abundant associations that occur between prokaryotic symbionts which possess unusual biochemical abilities and eukaryotic hosts which acquire and use these abilities. Examples include nitrogen-fixing bacteria symbiotic with leguminous plants, sulfur chemolithotrophic or methanotrophic bacteria symbiotic with marine invertebrates (Vetter, this volume), and symbiotic luminescent bacteria of marine fishes (McFall-Ngai, this volume). I describe here a symbiosis involving luminous bacteria, *Xenorhabdus luminescens*, and soil nematodes, *Heterorhabditis bacteriophora*. This symbiotic duo is pathogenic for a variety of different insects (Gaugler 1988), and is one of several such systems that is under investigation for possible use in the biological control of insect pests (Gaugler 1988; Gaugler and Kaya 1990; Poinar 1979). Although I will deal here with the bacterium *X. luminescens* and its heterorhabditid host, *H. bacteriophora*, it should be stressed that there are at least four other species of *Xenorhabdus* (Akhurst and Boemare 1988), each of which is found specifically associated with other nematode hosts. Though all these systems share many of the traits that will be described here, there are also some differences. For a complete discussion, the reader is referred to a recent volume discussing many aspects of the entomopathogenic nematodes: Gaugler and Kaya 1990.

Life Cycle

The life cycle of the symbiosis/parasitism, as exemplified by *X. luminescens* and *Galleria mellonella*, is illustrated in figure 1. Although the bacteria interact with the nematodes throughout the life cycle, these interactions are quite different at different stages. The free-

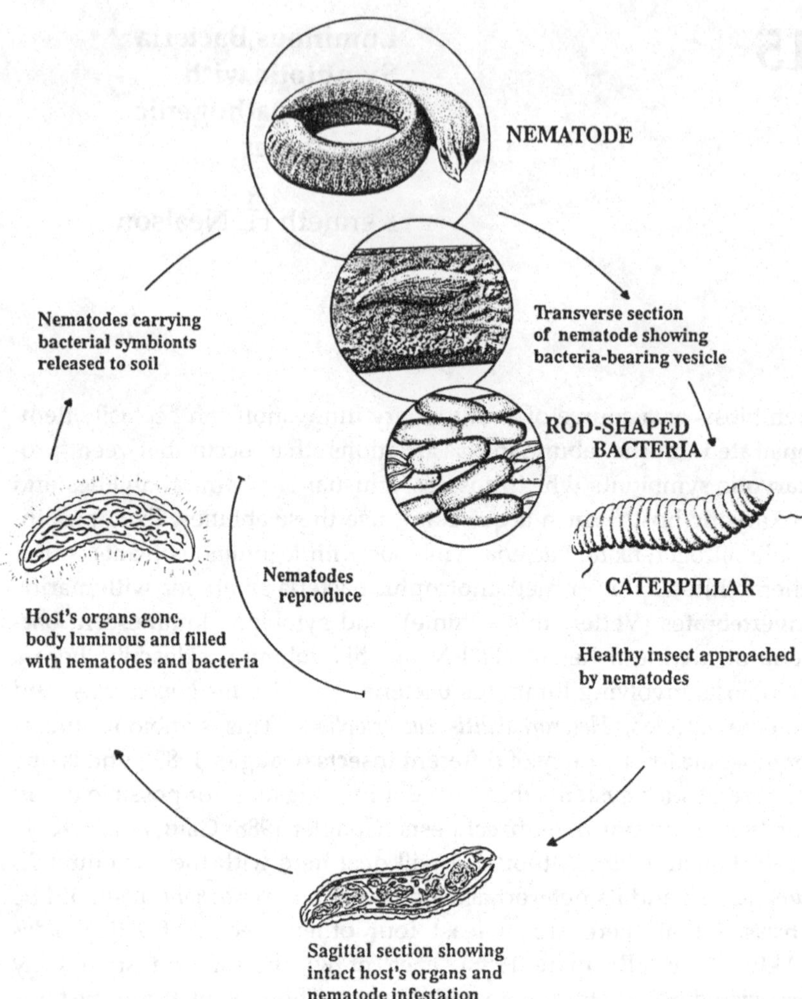

Figure 1
The bacteria /nematode/insect life cycle. This figure depicts the various phases of the nematode-bacterial symbiosis. Free-living nematodes, *Heterorhabditis bacteriophora*, carry the luminous bacteria *Xenorhabdus luminescens* as symbionts. After the location and penetration of a suitable insect (such as *Galleria mellonella*) by the nematode, the bacteria are released into the hemocoel, where they proliferate, leading to the death of the insect, the luminescence of the insect, and the pigmentation of the insect carcass. After reproducing inside the insect carcass, the nematodes emerge to begin the cycle again. Drawing by Christie Lyons.

living nematodes found in the soil as third-instar infective juveniles carry the nongrowing symbiotic bacteria either in the intestine or in specialized vesicles (Akhurst 1983; Bird and Akhurst 1983; Poinar 1966). The infective juveniles, which forage in search of insect prey, are covered with a waxy cuticle that precludes either food uptake or waste elimination (Poinar 1979). They are remarkably hardy, being capable of survival for months or longer in dilute Ringer's solution (Milstead 1981). They also have a rather specific and limited host range, which provides the specificity needed for the development of useful biological control systems. Akhurst and Boemare (1988) showed that in nature each species of entomopathogenic nematode always carries a single species of *Xenorhabdus*.

Another phase of the life cycle occurs when the foraging nematode locates a suitable insect prey, enters it via virtually any external opening (anus, mouth, spiracle, etc.), and penetrates into the hemocoel of the insect by boring through the wall (Bedding and Molyneux 1982). Once in the hemolymph, the infective juvenile dissolves its waxy cuticle and eliminates the bacterial symbionts from its intestine (Poinar 1976, 1979). The bacteria, which are known to have an LD_{50} of only a few cells for some insects (Akhurst 1982a; Milstead 1979; Poinar et al. 1980), begin to multiply; the death of the insect occurs within hours of the infection (Akhurst 1982a; Boemare and Akhurst 1988).

The interactions that occur between nematode and bacteria are not altogether clear in these early stages. However, it is postulated that the bacteria do not survive well outside the nematode (Poinar 1979). It is well known that bacteria that are simply ingested by the insects cannot gain entry to the hemocoel, and thus are effectively nonpathogenic (Milstead 1979; Poinar 1979); this suggests that the nematode is important both for the long-term survival of the bacteria and for its transport to the insect prey. Furthermore, the nematode protects the bacteria during the early stages of the infection by inactivating the insect's immune system (Goetz et al. 1981).

There is also some specificity involved with the association (Akhurst 1982a; Akhurst and Boemare 1988). Since the nematodes can be grown with bacteria as their sole source of food, natural specificity can be circumvented in the laboratory by growing the nematodes on pure cultures of other strains or species of *Xenorhabdus* or non-*Xenorhabdus* bacteria. The nematodes can also be grown axenically on a nonbacterial medium (Akhurst 1982a; Akhurst and Boemare 1988; Boemare 1983; Dunphy et al. 1985). Laboratory studies with nematodes that have been given bacteria other than their usual symbionts indicate that different *Xenorhabdus* strains endow a given nematode with different levels of virulence (Dunphy et al. 1985), which suggests that the natural host specificity may be important for the virulence of this symbiosis in nature.

Both bacterial growth and nematode reproduction begin shortly after infection. The bacteria grow rapidly (doubling in 2½–3 hours) within the insect cadaver (Goetz et al. 1981; Poinar et al. 1980), producing both lipopolysaccharide endotoxins (Dunphy and Webster 1988) and exotoxins, in the form of proteases and lipases, which presumably participate in the breakdown of insect proteins and lipids. The bacteria are in turn ingested by the nematode, which completes its life cycle in the insect cadaver. Axenically maintained nematodes are unable to complete their life cycle in the insect cadaver; bacteria are needed to produce appropriate nutrient conditions (Boemare 1983; Poinar and Thomas 1966). Laboratory studies have shown that nematode reproduction can occur with other (non-*Xenorhabdus*) bacteria, but it is much less efficient (Akhurst 1982a). During the bacterial growth and nematode reproduction phase of the growth cycle, not only does the insect die, but the insect carcass undergoes some notable changes as a result of the infection: it becomes bioluminescent and darkly pigmented, and, despite being a dead, bacteria-laden carcass, it does not putrefy. These changes are believed to be the results of bacterial metabolites: (1) bioluminescence is due to bacterial luciferase, (2) pigmentation is due to the production of an anthraquinone pigment, and (3) antibiotics produced by the bacteria inhibit putrefaction of the cadaver, presumably by limiting the growth of other bacteria.

The bacterial contributions mentioned above are thought to be intimately linked to the success of the symbiotic pathogenesis (Gaugler and Kaya 1990). Bioluminescence and pigmentation are postulated to be valuable for the attraction of prospective prey organisms to the insect cadaver when emergence of infective-stage nematodes occurs (Poinar et al. 1980). The antibiotics are presumably used to inhibit non-*Xenorhabdus* bacteria from interfering with nematode reproduction (Akhurst 1982b; Akhurst and Boemare 1990). Extracellular bacterial enzymes are thought to establish the appropriate nutrient regime for nematode reproduction (Popiel et al. 1989) and possibly to assist in the killing of the insect by rendering its immune system inactive in conjunction with the activity of the nematodes (Goetz et al. 1981; Schmidt et al. 1988).

The nematodes grow and reproduce via a complex series of events: the larvae molt and become sexually mature; the females lay eggs, which hatch to yield second-stage larvae; these larvae feed, increase in size, and molt to become the third-instar infective juvenile larvae. At this point, they cease feeding on the luminous bacteria, and instead take them on as symbionts, form a waxy cuticle over their entire body, and emerge in search of new prey (Poinar 1979). Up to 500,000 nematodes per gram of infected insects can be produced (Boemare and Akhurst 1988).

Properties of the Bacteria

Taxonomically, the genus *Xenorhabdus* is grouped with the Enterobacteriaceae (Akhurst 1983; Akhurst and Boemare 1988, 1990; Ehlers et al. 1988; Farmer et al. 1989; Grimont et al. 1984). While *X. luminescens* shares many properties with other enterobacterial genera, it is a rather unusual organism with regard to several properties: bioluminescence (Schmidt et al. 1989), formation of red pigment (Richardson et al. 1988), production of antibiotics (Akhurst 1982b, Paul et al. 1981), formation of intracellular protein crystals (Boemare et al. 1983, Couche et al. 1987), and strong protease and lipase activity (Boemare and Akhurst 1988; Schmidt et al. 1988). Several different taxonomic approaches, including numerical taxonomy (Akhurst 1983; Grimont et al. 1984), DNA/DNA hybridization (Farmer et al. 1989), and 16S rRNA sequence analyses (Ehlers et al. 1988), have yielded similar conclusions (Akhurst and Boemare 1990).

The bioluminescent systems of two *X. luminescens* strains are similar to previously described bacterial *lux* systems. Colepicolo et al. (1989) concluded that although the luciferase of *X. luminescens* was similar to other bacterial luciferases with regard to substrate requirements, the regulation of the luminous system was substantially different. Virtually none of the factors that are known to control light emission in the marine luminous bacteria—autoinduction, catabolite repression, iron repression, regulation by ionic strength (Nealson et al. 1990)—appear to be involved in the *X. luminescens* system. Schmidt et al. (1989) purified the luciferase from *X. luminescens* (strain Hb) and demonstrated that the separated subunits of the enzyme could be

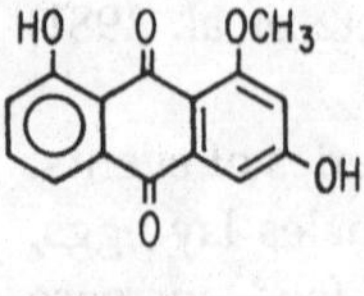

Figure 2
Structure of the red anthraquinone pigment produced by *Xenorhabdus luminescens*, strain HK (Richardson et al. 1988).

complemented by luciferase subunits from the marine species *V. harveyi* to give active enzyme, which suggests that these systems were quite similar. Further studies (Frackman et al. 1990) have shown that the cloned *lux* genes of *X. luminescens* are similar both in gene arrangement and in nucleotide sequence to the bacterial lux genes of other luminous bacteria (Meighen 1988). Since *X. luminescens* is ecologically quite different from the other (marine) luminous bacteria, it is perhaps not surprising that the regulation of the *lux* system appears to be quite different.

One of the diagnostic characteristics of the genus *Xenorhabdus* is the ability to produce pigment (Grimont et al. 1984). For one species, *X. luminescens* (strain Hk), the red pigment was purified and identified as an anthraquinone of the structure shown in figure 2 (Richardson et al. 1988). The structure, which has the basic polyketide backbone of compounds like erythromycin, is unusual as a product of the enterobacteria, being much more commonly produced by actinomycetes (actinobacteria; Malpartida et al. 1987). The structures of other pigments have not been identified. The genes for pigment formation have been cloned and expressed in *E. coli* (Frackman et al. 1989), and hybridization analysis indicates that similar genes are present in other *X. luminescens* strains, but not in *X. nematophilus* (P. Ragudo, unpublished data).

Another property common to virtually all *Xenorhabdus* species is the ability to produce compounds with potent and broad-spectrum antibiotic activity (Akhurst l982b; Boemare and Akhurst, 1988; Paul et al. 1981). From five organisms examined to date, four different classes of compounds have been identified (figure 3); this suggests that the presence of an inhibitor is important for the symbiosis.

Intracellular protein crystals were first observed in *Xenorhabdus* species by Boemare et al. (1983). Some cells have more than one type of crystal. The crystals have been characterized both by electron microscopy (Boemare et al. 1983) and biochemically (Couche and Gregson

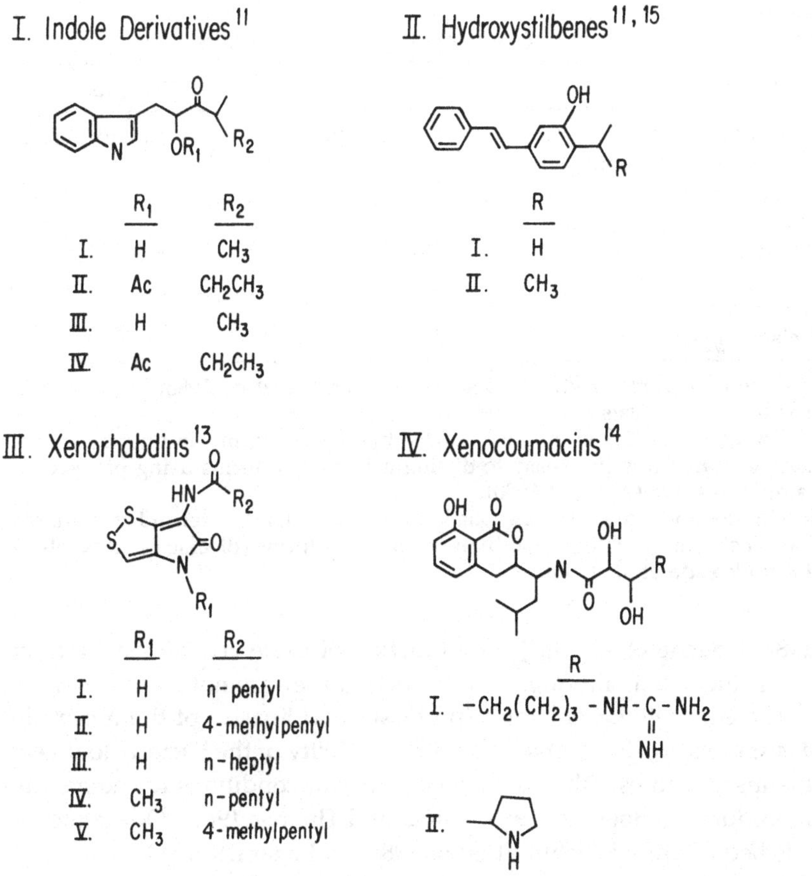

I. Indole Derivatives[11]

	R_1	R_2
I.	H	CH_3
II.	Ac	CH_2CH_3
III.	H	CH_3
IV.	Ac	CH_2CH_3

II. Hydroxystilbenes[11, 15]

	R
I.	H
II.	CH_3

III. Xenorhabdins[13]

	R_1	R_2
I.	H	n-pentyl
II.	H	4-methylpentyl
III.	H	n-heptyl
IV.	CH_3	n-pentyl
V.	CH_3	4-methylpentyl

IV. Xenocoumacins[14]

	R
I.	$-CH_2(CH_2)_3-NH-C(=NH)-NH_2$
II.	2-pyrrolidinyl (N–H)

Figure 3
Structures of four antibiotics produced by various species and strains of *Xenorhabdus* (Gregson and McInerney 1986; Paul et al. 1981; Rhodes et al. 1984; Richardson et al. 1988).

Table 1
Properties of primary and secondary forms of *X. luminescens*.

Property	Primary	Secondary
Bioluminescence	Bright (100%)	Dim (0.1–1%)
Antibiotic production	+	−
Pigment production	+	−
Intracellular crystal	+	−
Protease	+	−
Lipase	+	−
Dye uptake[a]	+	−
Relative growth rate[b]	Slow	Fast

Data compiled from Akhurst 1980; Bleakley and Nealson 1988; Boemare and Akhurst 1988; Nealson et al. 1990.
a. Tetrazolium is taken up and reduced by primary forms, but not by secondary variants. It is thus easy to distinguish the two forms using plates containing this dye (Akhurst 1980).
b. On any given medium, especially minimal media, the secondary variants have a shorter doubling time than the primary forms (Bleakley and Nealson 1988; Nealson et al. 1990).

1987; Couche et al. 1987). The function of these crystals has not yet been elucidated, although it appears that they are not insecticidal.

The extracellular enzymes (proteases and lipases) of the *Xenorhabdus* species are likely involved with the ability of the bacteria to digest the insect and establish the proper nutrient conditions for nematode reproduction (Boemare 1983; Poinar and Thomas 1966). One protease has been purified (Schmidt et al. 1988) and characterized as an alkaline metalloprotease of molecular weight 61,000. Using a variety of protease assays in conjunction with gel electrophoresis, Boemare and Akhurst (1988) showed that different strains of *Xenorhabdus* produced a wide variety of different types and numbers of proteases. Similar studies with lipase have not been done.

A final property that all *Xenorhabdus* species seem to share is the tendency to form spontaneous variants (Akhurst 1980)—or phases, as they are called by Akhurst and Boemare (1990). In *X. luminescens* (strain Hm), conversion of the primary to the secondary form resulted in the loss of the unique properties usually associated with the primary forms (Bleakley and Nealson 1988; table 1 here). In fact, at first glance, the secondary forms appear to be contaminants, and are difficult to differentiate from other enterobacteria. This problem was discussed by Grimont et al. (1984), who considered that many so-called secondary variants were actually contaminants. Although this may be true in some cases, two reports (Akhurst and Boemare 1988; Bleakley and Nealson 1988) have presented compelling evidence in support of the secondary forms' being true spontaneous variants.

There is some disagreement about the properties that are lost during conversion of primary to secondary forms, and about the ability of the secondary forms to revert back to primary. This phenomenon may be due to differences between *Xenorhabdus* species or even between strains (Hurlbert et al. 1989). In our studies of *X. luminescens* (Bleakley and Nealson 1988), the conversion of secondary forms back to primary was not observed, although it has been reported in the literature for *X. nematophilus* (Akhurst and Boemare 1990). Hybridization analysis of DNA (from both primary and secondary forms) using both labeled *lux* genes and genes for pigment production indicate that the genes are present in both the primary and secondary forms, and that no gene rearrangements appear to have taken place. Furthermore, transfer of plasmid-borne *lux* genes into a secondary (dark) form resulted in restoration of bioluminescence, indicating that gene expression was possible in the secondary form (Frackman and Nealson 1990; Ragudo and Nealson, unpublished).

Neither the factors that cause formation of the secondary forms nor the factors that stabilize them or cause their reversion back to primary are known. The issue is an important one, however, as the secondary isolates are defective both with regard to killing insects and with regard to allowing the nematode to complete its life cycle and reproduce effectively (Akhurst 1982a; Boemare and Akhurst 1988).

Symbiosis and Evolutionary Innovation

Innovation via symbiosis is an obvious point of discussion for this system. In a single event, the nematode acquires several characteristics, some of which are rather unique for this group of organisms. Genetically this is quite an accomplishment, the extent of which may be approximated in terms of structural gene acquisition. For instance, the structural genes for bioluminescence number five: two for luciferase and three for the aldehyde reductase. On the basis of an analogy to the polyketide pathway in the *Streptomyces coelicolor* (Malpartida et al. 1987), at least eight genes are needed for the synthesis of the red pigment. Depending on the antibiotic synthesized, from five to fifteen genes are needed for biosynthesis, and one gene each is probably required for protease, lipase, and intracellular crystal formation. Thus, discounting the regulatory genes, one is left with twenty or more genes that have been acquired via symbiosis.

There are undoubtedly bacterial contributions to the symbiosis that are not yet evident (involved with pathogenesis, nematode reproduction, etc.), but just considering the ones mentioned above it is difficult not to be impressed with the evolutionary aspect of this organismal association. Several completely new functions have been obtained and are being put to use by the host nematode. Although the nematode may well have evolved these functions itself, there are no known extant examples; this suggests that symbiosis is an efficient and effective pathway for evolutionary innovation, at least in this system.

However, innovation by acquisition is not a one-way process, and the host must undergo substantial structural and/or functional modifications of its own. Some of these modifications appear to be remarkably complex. First, with regard to the bacterial symbionts, the nematode must develop the ability to cease feeding so that the luminous bacteria can be safely stored as gut symbionts, must develop a structure or system for housing the symbiotic bacteria, and must develop a system for expelling the bacteria from its gut once it has entered the insect hemolymph. Second, the nematode must itself be or become resistant to the rather potent broad-spectrum toxins produced by the bacteria. Third, the nematode must develop some system of recognition so that the correct bacteria can be obtained and maintained as symbionts. Finally, the nematode must somehow develop a system that either selects for the primary form or induces conversion of secondary forms to primaries, in order to ensure maintenance of a virulent symbiosis.

It is not at all clear how many genes are involved in such host adaptations, but it is clear that this symbiosis, like virtually all others, involves rather impressive responses (both functional and structural) on the part of the host to the acquisition of new functions. In this sense, innovation by acquisition is just the beginning, which leads to—in some way, by some mechanism—much more of the same. It brings up many questions regarding the factors that lead to and select for innovation—questions that might well be answered by studies of systems like the nematode/bacterial symbioses described here.

References

Akhurst, R. J. 1980. Morphological and functional dimorphism in *Xenorhabdus* spp., bacteria symbiotically associated with the insect pathogenic nematodes *Neoaplaectana* and *Heterorhabditis*. *J. Gen. Microbiol.* 121: 303–309.

Akhurst, R. J. 1982a. Bacterial Symbionts of Insect Pathogenic Nematodes. Ph.D. thesis, University of Tasmania.

Akhurst, R. J. 1982b. Antibiotic activity of *Xenorhabdus* spp., bacteria symbiotically associated with insect pathogenic nematodes of the families *Heterorhabditidae* and *Steinernematidae*. *J. Gen. Microbiol.* 128: 3061–3065.

Akhurst, R. J. 1983. Taxonomic study of *Xenorhabdus*, a genus of bacteria symbiotically associated with insect pathogenic nematodes. *Int. J. Syst. Bacteriol.* 33: 38–45.

Akhurst, R. J., and Boemare, N. E. 1988. A numerical taxonomic study of the genus *Xenorhabdus* (Enterobacteriaceae) and proposed elevation of the subspecies of X. *nematophilus* to species. *J. Gen. Microbiol.* 134:1835–1845.

Akhurst, R. J., and Boemare, N. E. 1990. Biology and taxonomy of *Xenorhabdus*. In: Gaugler, R., and Kaya, H., eds., *Entomopathogenic Nematodes in Biological Control*. CRC Press.

Bedding, R. A., and Molyneux, A. S. 1982. Penetration of insect cuticle by infective juveniles of *Heterorhabditis* spp. *Nematology* 28: 354–359.

Bird, A. F., and Akhurst, R. J. 1983. The nature of the intestinal vesicle in nematodes of the family *Steinernematidae*. *Int. J. Parasitol.* 13: 599–606.

Bleakley, B., and Nealson, K. H. 1988. Characterization of primary and secondary forms of *Xenorhabdus luminescens* strain Hm. *FEMS Microb. Ecol.* 53: 241–250.

Boemare, N. 1983. Recherches sur les complexes nemato-bacteriens éntomopathogenes: étude bacteriologique, gnotobiologique et physiologique du mode d'action parasitaire de *Steinernema carpocapsae*. Ph.D. thesis, Université des Science, Montpellier.

Boemare, N., and Akhurst, R. J. 1988. Biochemical and physiological characterization of colony form variants in *Xenorhabdus* spp. *J. Gen. Microbiol.* 134: 751–761.

Boemare, N., Louis, C., and Kuhl, G. 1983. Etude ultrastructurale des cristaux chez *Xenorhabdus* spp., bacteries infectées aux nematodes entomophages Steinernematidae et Heterorhabditidae. *C. R. Soc. Biol.* 177: 107–115.

Colepicolo, P., Cho, K.-W., Poinar, G. O., and Hastings, J. W. 1989. Growth and luminescence of the bacterium *Xenorhabdus luminescens* from a human wound. *Appl. Environ. Microbiol.* 55: 2601–2606.

Couche, G. A., and Gregson, R. P. 1987. Protein inclusions produced by the entomopathogenic bacterium *Xenorhabdus nematophilus* subsp. *nematophilus*. *J. Bacteriol.* 169: 5279–5288.

Couche, G. A., Lehrbach, P. R., Forage, R. G., Cooney, G. C., Smith, D. R., and Gregson, R. P. 1987. Occurrence of intracellular inclusions and plasmids in *Xenorhabdus* sp. *J. Gen. Microbiol.* 133: 967–973.

Dunphy, G. B., and Webster, J. M. 1988. Lipopolysaccharides of *Xenorhabdus nematophilus* and their haemocyte toxicity in nonimmune *Galleria mellonella* larvae. *J. Gen. Microbiol.* 134: 1017–1028.

Dunphy, G. B., Rutherford, T. A., and Webster, J. M. 1985. Growth and virulence of *Steinernema glaseri* influenced by different subspecies of *Xenorhabdus nematophilus*. *J. Nematol.* 17: 476–482.

Ehlers, R.-U., Wyss, U., and Stackebrandt, E. 1988. 16S rRNA cataloguing and the phylogenetic position of the genus *Xenorhabdus*. *Syst. Appl. Microbiol.* 10: 121–125.

Farmer, J., Jorgensen, J., Grimont, P., Akhurst, R., Poinar, G., Pierce, G., Smith, J., Carter, G., Wilson, K., and Hickman-Brenner, F. 1989. *Xenorhabdus luminescens* (DNA hybridization group 5) from human clinical specimens. *J. Clin. Microbiol.* 27: 1594–1600.

Frackman, S., Anhalt, M., and Nealson, K. H. 1990. Cloning, organization, and expression in bioluminescence genes of *Xenorhabdus luminescens*. *J. Bacteriol.* 172: 5767–5773.

Frackman, S., and Nealson, K. H. 1990. The molecular genetics of *Xenorhabdus*. In: Gaugler, R., and Kaya, H., eds., *Entomopathogenic Nematodes in Biological Control*. CRC Press.

Frackman, S., Ragudo, P., and Nealson, K. H. 1989. Bioluminescence, pigment and protease genes of *Xenorhabdus luminescens*. *Abstr. Ann. Mtg. Amer. Soc. Microbiol.* H-93: p 185.

Gaugler, R. 1988. Ecological considerations in the biological control of soil-inhabiting insects with entomopathogenic nematodes. *Agric. Ecosyst. Envir.* 24: 351–360.

Gaugler, R., and Kaya, H. 1990. *The Entomopathogenic Nematodes in Biological Control*. CRC Press.

Goetz, P., Boman, A., and Boman, H. 1981. Interactions between insect immunity and an insect-pathogenic nematode with symbiotic bacteria. *Proc. R. Soc. Lond.* B212: 333–350.

Gregson, R. P., and McInerny, B. V. 1985. *Xenocoumacins*. Australian patent PCT/AU85/00215.

Grimont, P. A. D., Steigerwalt, A. G., Boemare, N., Hickman-Brenner, F. W., Deval, C., Grimont, F., and Brenner, D. J. 1984 . Deoxyribonucleic acid relatedness and phenotypic study of the genus *Xenorhabdus*. *Int. J. Syst. Bacteriol.* 34: 378–388.

Hurlbert, R., Xu, J., and Small, C. 1989. Colonial and cellular polymorphism in *Xenorhabdus luminescens*. *Appl. Environ. Microbiol.* 55: 1136–1143.

Malpartida, F., Hallam, S., Kieser, H., Motamedi, H., Hutchinson, C., Butler, M., Sugden, D., Warren, M., McKillop, C., Bailey, C., Humphreys, G., and Hopwood, D. 1987. Homology between *Streptomyces* genes coding for synthesis of different polyketides used to clone antibiotic biosynthetic genes. *Nature* 325: 818–821.

Meighen, E. A. 1988. Enzymes and genes from the *lux* operons of bioluminescent bacteria. *Ann. Rev. Microbiol.* 42: 151–176.

Milstead, J. 1979. *Heterorhabditis bacteriophora* as a vector for introducing its associated bacterium into the hemocoel of *Galleria mellonella* larvae. *J. Invert. Pathol.* 33: 324–327.

Milstead, J. 1981. Influence of temperature and dosage on mortality of seventh instar larvae of *Galleria mellonella* caused by *Heterorhabditis bacteriophora* and its bacterial associate *Xenorhabdus luminescens*. *Nematology* 27: 167–171.

Nealson, K. H., Schmidt, T. M., and Bleakley, B. 1990. Physiology and biochemistry of *Xenorhabdus*. In: Guagler, R., and Kaya, H., eds., *Entomopathogenic Nematodes in Biological Control*. CRC Press.

Paul, V. J., Frautschy, S., Fenical, W., and Nealson, K. H. 1981. Antibiotics in microbial ecology: Isolation and structure assignment of several new antibacterial compounds from the insect-symbiotic bacteria *Xenorhabdus* spp. *J. Chem. Ecol.* 7: 589–597.

Poinar, G. O. 1966. The presence of *Achromobacter nematophilus* in the infective stage of a *Neoaplectana* sp. *Nematology* 12: 105–108.

Poinar, G. O. 1976. Description and biology of a new insect parasitic *Rhabditoid, Heterorhabditis bacteriophora*. *Nematology* 21: 463–470.

Poinar, G. O. 1979. *Nematodes for Biological Control of Insects*. CRC Press.

Poinar, G. O., and Thomas, G. M. 1966. Significance of *Achromobacteria nematophilus* in the development of the nematode, DD136 (*Neoaplectana* sp.: Steinernematidae). *Parasit.* 56: 385–390.

Poinar, G. O., Thomas, G. M., Haygood, M., and Nealson, K. 1980. Growth and luminescence of the symbiotic bacteria associated with the terrestrial nematode *Heterorhabditis bacteriophora*. *Soil Biol. Biochem.* 12: 5–10.

Popiel, I., Grove, D. L., and Fricdman, M. J. 1989. Infective juvenile formation in the insect-parasitic nematode *Steinernema fetiae*. *Parasit*. 99: 77–81.

Rhodes, S. H., Lyons, G. R., Gregson, R. P., Akhurst, R. J., and Lacey, M. J. 1984. Canadian Patent 1214130; U.S. Patent 4672130.

Richardson, W. H., Schmidt, T. M., and Nealson, K. H. 1988. Identification of an anthraquinone pigment and a hydroxystilbene antibiotic from *Xenorhabdus luminescens*. *Appl. Environ. Microbiol*. 54: 1602–1605.

Schmidt, T. M., Bleakley, B., and Nealson, K. H. 1988. Characterization of an extracellular protease from the insect pathogen *Xenorhabdus luminescens*. *Appl. Environ. Microbiol*. 54: 2793–2797.

Schmidt, T. M., Kopecky, K., and Nealson, K. H. 1989. Bioluminescence of the insect pathogen *Xenorhabdus luminescens*. *Appl. Environ. Microbiol*. 55: 2607–2612.

16

Symbiosis and the Evolution of Novel Trophic Strategies: Thiotrophic Organisms at Hydrothermal Vents

Russell D. Vetter

Trophic Strategies of Eukaryotes

Prokaryotes employ a variety of trophic strategies to exploit different energy sources, whereas the known energy sources available to eukaryotic organisms are much more limited. Phototrophic organisms (plants and other organisms that contain plastids or photosynthetic symbionts) use solar energy. Heterotrophic organisms (animals, fungi, and other groups lacking functional plastids or photosynthetic symbionts) rely on the energy in reduced carbon compounds. Although there is no apparent *a priori* reason why eukaryotic organisms could not exploit chemical energy in reduced metal, nitrogen, sulfur, inorganic carbon compounds, until recently no evidence existed that eukaryotes exploited these other energy sources. The discovery in the late 1970s of ecosystems at hydrothermal vents has revolutionized our understanding of the trophic opportunities available to eukaryotes. Hydrothermal-vent communities are abundant oases of life in the otherwise depauperate realm of the deep sea (Corliss et al. 1979). These communities are clustered around fissures in the rock where geothermal fluids containing sulfides, methane, ammonia, hydrogen, and reduced metals flow out from buried aquifers. One of the early explanations for the abundant life surrounding these vents was that somehow these ecosystems were driven by the energy contained in the geothermal fluids. Further examination of invertebrate animals in these communities revealed that many of the species living around hydrothermal vents contained endosymbiotic, chemolithotrophic, sulfur-oxidizing bacteria capable of using energy in hydrogen sulfide for fixation of carbon to bacterial and invertebrate cell material (Cavanaugh et al. 1981; Felbeck 1981). Since then, symbioses between sulfur-oxidizing bacteria and various invertebrates have been dis-

Figure 1
A community of invertebrates clustered around a hydrothermal vent at the Galapagos Spreading Center. The large tube worm *Riftia pachyptila* and the vent mussel *Bathymodiolus thermophilus* contain symbiotic sulfur-bacteria that use the potentially toxic hydrogen sulfide of the vent waters to drive autotrophic carbon fixation. (See note 1.)

Figure 2
The giant white clam *Calyptogena magnifica,* also found at vent sites, forms dense clusters along fissures in the basalt where sulfide-containing ventwater is escaping. (See note 1.)

covered in a wide variety of other sulfide-rich environments, including salt marshes, mangrove swamps, sewage outfalls, anoxic basins, hydrocarbon seeps, and brine seeps (Fisher 1990). Subsequent study of the anatomy, physiology, and biochemistry of the host organisms has confirmed that, in at least some cases, the animals rely on chemolithoautotrophic processes for most or all of their carbon and energy requirements (Anderson et al., 1987; Conway et al. 1989). The most specialized organisms lack a gut and cannot live without nutritional input from their chemoautotrophic symbionts. The adaptations of bacterial symbionts and bacterial hosts involve far more than those of animals feeding on sulfur bacteria; these associations represent a fundamentally different trophic strategy for eukaryotes. To emphasize the sulfur-based autotrophic strategy for eukaryotes, I use the term *thiotrophic metabolism*.

In the typical animal cell, mitochondria carry out most of the oxidative catabolic metabolism of reduced carbon compounds via coupled electron transport and the reduction of molecular oxygen to water (figure 3). Believed to be the descendants of prokaryotic symbionts, mitochondria contain a genome distinct from the nucleus and organized similarly to that of prokaryotes (Margulis and Bermudes 1985; Sogin 1989). Much of the mitochondrial genome is believed to have been transferred to the nucleus (Gray and Doolittle 1982; Nagley and Devenish 1989), and many mitochondrial proteins are transcribed from genes located on nuclear chromosomes (Attardi and Schatz 1988; Hartl et al. 1989).

The phototrophic eukaryotic cell contains plastids which harvest the electromagnetic energy of sunlight for ATP production and for the reduction of CO_2 into cell carbon for cell growth and mitochondrial ATP production. Whereas many of the chloroplast genes are believed to have been transferred to the nucleus (Nagley and Devenish 1989), the chloroplast (like the mitochondrion) retains a separate genome. Genes have moved between the chloroplast and the mitochondrion in some species (Moon et al. 1988; Stern and Palmer 1984; Nugent and Palmer 1988; Pritchard et al. 1989).

Analogously, the thiotrophic cell (figures 3, 4) contains an organelle or a symbiotic bacterium that uses the chemical energy contained in reduced sulfur compounds to generate ATP, fix carbon for biosynthetic processes, and fuel mitochondrial energy production. The thiotrophic eukaryote, as defined here, is obligately dependent on chemical energy and is capable of autotrophic growth. Because none

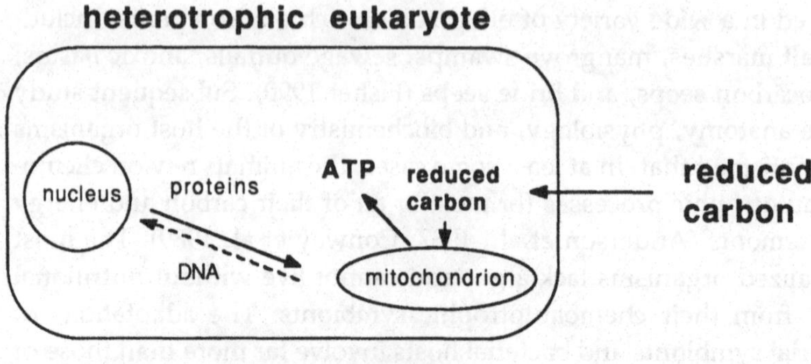

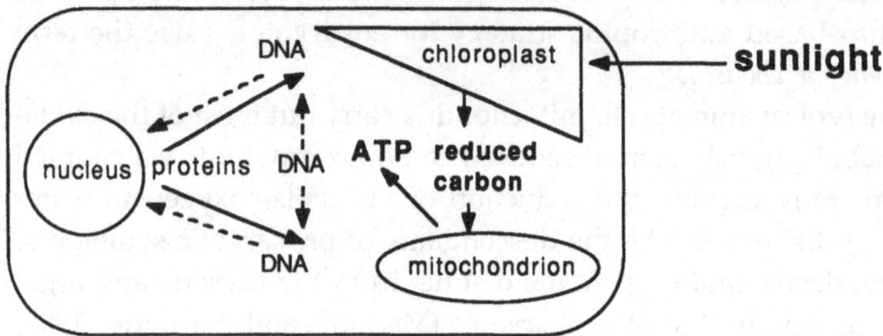

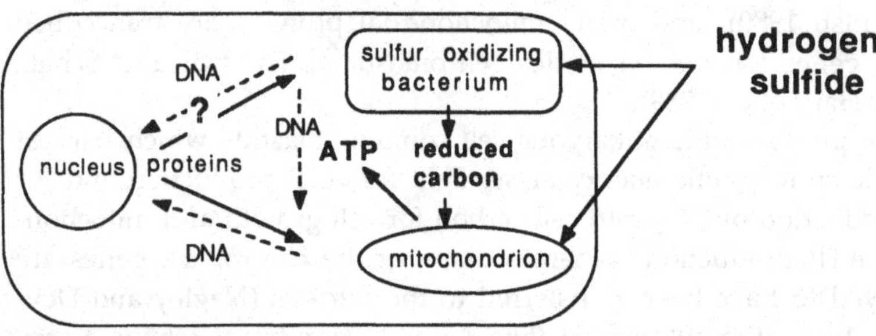

Figure 3
Cellular organization of different trophic stategies among the eukaryotes. (See note 2.)

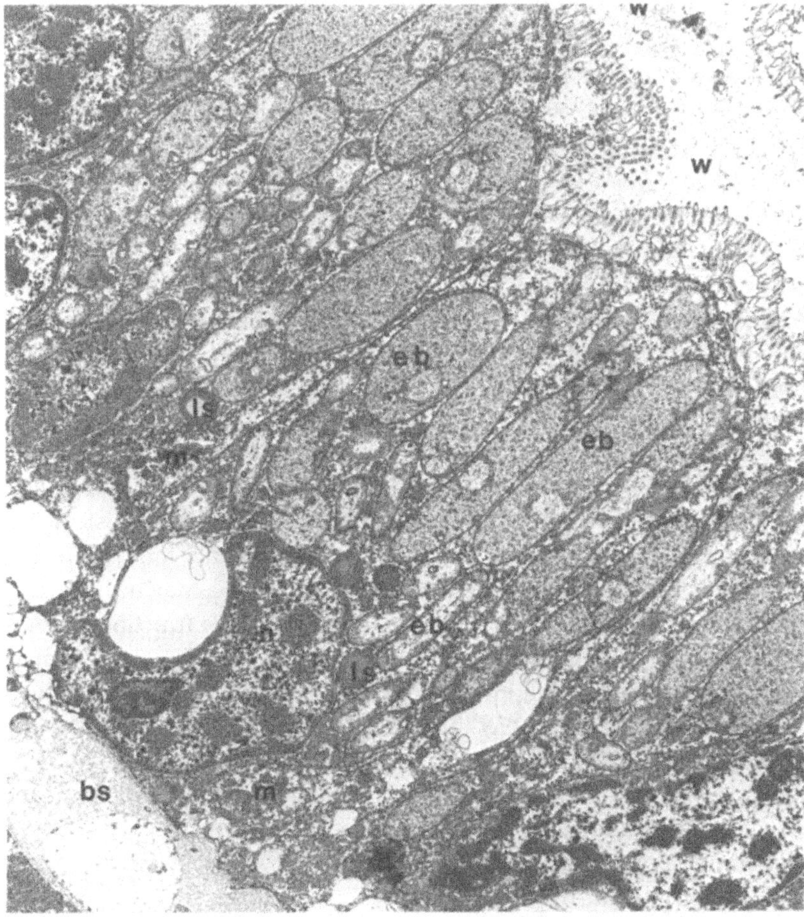

Figure 4
A bacteriocyte from the thiotrophic bivalve *Solemya reidi*. This electron micrograph contains many of the features typical of the bacteriocytes of many different thiotrophic symbioses. The bacteriocyte region of the gill is one cell layer deep. The basal portion of the bacteriocytes are in contact with the blood space (bs), and the apical portion is overlain by a thin extension of the epithelial cells but is in close contact with seawater (w). The apical portion is filled with endosymbionts (es), contained within vacuoles. The bacteriocyte typically contains numerous lysosomes (ls), which degrade endosymbionts and the other cell components. The epithelial cells that flank the bacteriocytes have abundant mitochondria (m). (Electron micrograph courtesy of D. Wilmot.)

of the symbiotic sulfur bacteria has been cultured in the laboratory, it is unclear whether the symbionts can exist outside the host or are dependent organelles. The extent to which genes have moved between the symbiotic bacteria and the host's mitochondria and nucleus is under study. An important distinction between phototrophic and thiotrophic organisms is that in a phototrophic eukaryote oxygen is generated by the plastids and consumed by mitochondria, whereas in thiotrophic organisms oxygen is consumed by both bacterial symbionts and mitochondria during sulfide oxidation. Nitrate can substitute for oxygen in some free-living sulfur-oxidizing bacteria, and there are indications that in some of the thiotrophic symbioses nitrate may be used (Cary et al. 1989; B. Javor, unpublished data).

Not all organisms that form symbiotic associations with chemoautotrophic, sulfur-oxidizing bacteria are thiotrophic as defined here. For example, the hydrothermal-vent polychaete *Alvinella pompejana* has a dense coat of sulfur bacteria on its dorsal epithelial surface. However, the animal feeds through its mouth, and food particles have been observed in the gut. The epibiotic bacteria may be a source of nutrition for the animal, may serve a protective function in the detoxification of sulfide, or may aid in the metabolism of animal waste. Alternatively, the interaction may simply be phoretic, rather than metabolic. Even where sulfide-oxidizing bacteria are endosymbiotic (e.g., in lucinid bivalves), since the animals may also maintain a digestive system, the quantitative nutritional importance and obligate nature of the interaction requires demonstration (Cary et al. 1989).

The anatomy, the physiology, the metabolic pathways involved in sulfide metabolism, the potential alternative energy sources (such as dissolved organic carbon), and the stable isotopic compositions have been studied in sufficient detail to allow the conclusion that two species are obligately thiotrophic: *Riftia pachyptila*, the large vestimentiferan tube-worm found at hydrothermal vents, and *Solemya reidi*, a protobranch bivalve associated with organic-rich, sulfide-rich, reducing sediments. Both appear as thiotrophs, but each of the symbiotic associations carries out sulfide metabolism by a different pathway and each displays distinct biochemical and physiological adaptations. Here I briefly review the metabolic schemes and the evidence for sulfide-dependent autotrophy in the two animals. To understand the complexities of thiotrophic metabolism, it is necessary to appreciate two properties of hydrogen sulfide. First, sulfide is one of the most potent inhibitors of respiratory electron transport. It acts as an

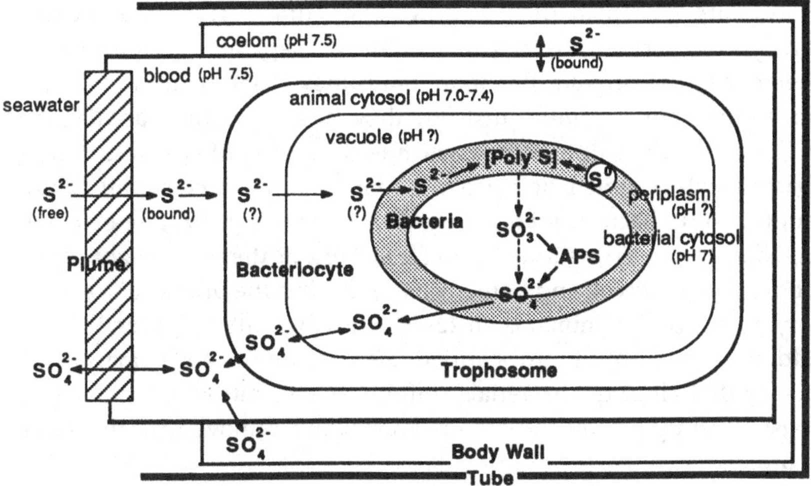

Figure 5
Sulfide metabolism in *Riftia pachyptila,* a sulfide specialist that concentrates and transports sulfide and oxygen to bacteria but does not actively oxidize sulfide except as a peripheral defense. The bacteria are buried deep in the animal, without access to seawater, and depend on the animal's vascular system for the delivery of nutrients and the removal of waste products. The bacteria do not oxidize other externally supplied forms of sulfur, including thiosulfate and sulfite.

inhibitor of cytochrome oxidase, the terminal electron transport site of mitochondria and aerobic sulfur-oxidizing bacteria (Beauchamp et al. 1984; Vetter and Bagarinao 1989). Second, sulfide in the presence of cell constituents such as heme iron can rapidly and spontaneously undergo chemical oxidation (National Research Council 1979), dissipating significant free energy as heat. Thus, any study of thiotrophic metabolism must consider how animals protect mitochondrial function, how bacteria protect their electron transport systems, and how the animal gathers and simultaneously supplies sulfide and oxygen to the bacteria without poisoning its own tissues and without seriously diminishing the energy content of the sulfide. In this chapter I will highlight adaptations to thiotrophic metabolism; a more detailed account is contained in a recent review by Somero et al. (1989).

Thiotrophic Metabolism in *Riftia pachyptila*

Riftia pachyptila (figure 5) is the largest and best studied of the organisms found at hydrothermal vent sites. *Riftia* and similar tubeworms are placed in the family Vestimentifera. The status of this family is uncertain, but it either defines a separate phylum or is a member of the phylum Pogonophora (Jones 1985). Adult *Riftia* have no mouth, gut, or anus. Instead, they have a highly vascularized organ called the trophosome, which is composed of specialized cells (bacteriocytes) which house dense populations of chemolithoautotrophic sulfur bacteria (van der Land and Norevang 1977; Jones 1985). The bacteria can constitute up to 35 percent of the wet weight of the trophosome (Powell and Somero 1986). All but the bright red respiratory plume of the animal is enclosed in a white tube (figures 1, 5). In addition, the trophosome is buried deep within the body cavity. This means that all of the bacteria's sulfur, oxygen, nitrogen, and carbon must be supplied by the blood. Wastes are removed by the same route.

The blood vascular system of *Riftia* contains a large extracellular hemoglobin with an apparent molecular weight of 1700 kD. A smaller, 400-kD hemoglobin is found predominantly in the coelomic fluid. The blood constitutes 30 percent of the wet weight of the animal and serves as a reserve and transport system for sulfide and oxygen. These hemoglobins, in which the same molecule binds both sulfide and oxygen, are very unusual. Bound-sulfide concentrations in the blood are as high as 8.7 mM; free concentrations are usually below a few hundred μM (Arp et al. 1987; J. J. Childress, personal communication). The oxygen is bound to the heme; the sulfide is bound to the globin portion of the molecule, probably through reduction of a disulfide bridge to form a cysteine persulfide (Arp et al. 1987). Because the heme moeity is unusual in that it does not readily form ferric hemoglobin or ferric hemoglobin sulfide derivatives, the animal simultaneously transports sulfide and oxygen to the trophosome. The sulfide-binding hemoglobin carries out several functions: it prevents sulfide poisoning of animal tissues; it maintains a bound (nonreactive) pool of sulfide and oxygen; it buffers external variations in the supply of both substrates and delivers them at constant and noninhibitory levels (Powell and Somero 1983; Powell and Somero 1986; Fisher et al. 1988).

Perhaps due to the efficiency of the sulfide-binding hemoglobin, the animal tissue of *Riftia* appears to be as susceptible to sulfide poisoning as other invertebrates. The cytochrome oxidase of *Riftia* mitochondria is inhibited by low concentrations of sulfide (Hand and Somero 1983; Powell and Somero 1986), and the mitochondria are only marginally able to oxidize sulfide before it reaches the cytochrome oxidase (J. O'Brien, unpublished data). An identified component, perhaps analogous to the hematin granules of the intertidal lugworm *Urechis caupo,* forms a peripheral sulfide-oxidizing layer in the skin of *Riftia* and protects the worm from sulfide in the water (Powell and Somero 1986; Powell and Arp 1989). The cellular concentration of free sulfide is unknown. It is also unclear if it is the high affinity of bacteria for sulfide or a high-affinity cellular form of the binding protein that protects mitochondria from sulfide poisoning.

The bacteria of *Riftia* display several adaptations to their endosymbiotic existence. The first adaptation is that they oxidize hydrogen sulfide (Wilmot and Vetter 1989; D. Wilmot, unpublished data): sulfide, but not thiosulfate, sulfite, or a number of heterotrophic substrates, stimulated oxygen consumption and was oxidized by isolated *Riftia* bacteria. Only sulfide stimulated CO_2 fixation (Fisher et al. 1989). Unlike free-living bacteria which oxidize other substances, the *Riftia* microbes oxidize only sulfide. Because sulfide is the only sulfur compound that is taken up by the animal from seawater and because the animal tissue oxidizes little sulfide to other forms of sulfur, such as thiosulfate, the metabolic specialization of the bacteria is consistent with the host's physiology. The bacteria are the primary sulfide-oxidizing component of the symbiosis and must protect the host from sulfide poisoning. Therefore, both the sulfide-oxidizing capacity and the resistance to sulfide toxicity of the bacteria are important properties. Both sulfide-dependent and thiosulfate-dependent oxygen-consumption rates have been measured for *Riftia* symbionts and for two free-living, sulfur-oxidizing bacteria, *Thiobacillus neapolitanus* and *Thiomicrospira crunogena.* The *Riftia* symbionts were able to oxidize sulfide and were resistant to sulfide inhibition at sulfide concentrations greater than 1 mM, the two free-living sulfur-oxidizing bacteria were inhibited at concentrations as low as 150 μM. When exposed to sulfide, *Riftia* bacteria sequester large amounts of elemental sulfur in their periplasmic space. Elemental sulfur can constitute over 20 percent of the wet weight of freshly isolated bacteria (Wilmot and Vetter 1989; D. Wilmot, unpublished data). High sulfide-oxidizing capacity, resistance to sulfide toxicity, and ability to rapidly store large amounts of elemental sulfur appear to be bacterial adaptations to the need of the host for a rapid and flexible capacity to respond to changes in external sulfide concentrations and avoid poisoning.

In summary, *Riftia* relies primarily on the sulfide-binding hemoglobin and subsequent bacterial oxidation to resist sulfide toxicity. The animal produces few partial oxidation products and does not take them up from solution. The bacteria, sulfide specialists with a high affinity and a specific activity for sulfide oxidation, are able to shunt excess sulfide into elemental sulfur deposits. Although feeding or absorption of dissolved carbon is not a plausible supplemental feeding strategy for *Riftia*, it is very difficult to prove experimentally by a mass balance approach that *Riftia* are net autotrophs (net consumption of CO_2 in the intact animal). Experiments done with shipboard flow-through pressure respirometry are extremely difficult, and the health of animals recovered from 2600 m is not known with certainty. Nevertheless, values indicating net CO_2 consumption have been achieved for short periods in on-deck, whole-animal pressure-respirometry experiments (J. J. Childress, personal communication). The carbon-and-sulfur-stable isotopic compositions of *Riftia* also suggest that host biomass is derived from chemosynthetic carbon-fixation pathways (Rau 1981; Fry et al. 1983).

Thiotrophic Metabolism in *Solemya reidi*

Solemya reidi inhabits organic sediments near pulp mills and sewage outfalls, where high sulfide concentrations result from microbial sulfate reduction in the anoxic sediments (figure 6). Although the larval form has an open gut, the adult, obligately symbiotic form is gutless (Gustafson and Reid 1988; Reid 1980). In basic body plan and physiology, *Solemya*, a protobranch mollusc, is very different from the tubeworm *Riftia pachyptila*, and this has important implications for how the symbiosis is organized (figure 7). *Solemya* does not contain circulating hemoglobin; instead, it contains a copper respiratory pigment, hemocyanin (Mangum et al. 1987). The gills, which contain the bacteria, are filamentous and lamellar, so that the bacteria are never more than one cell layer away from seawater (Felbeck 1983; Vetter 1990). This open body plan means that there is much more exposure of the animal to external sulfide. Because the animal is mobile and lives in organic sediments, there is a potential nutritional contribution from dissolved organic matter (Felbeck 1983).

The blood of *Solemya* does not bind sulfide or promote the oxidation of sulfide (Somero et al. 1989). Therefore, the animal tissues must be protected by a different mechanism. The blood of freshly collected

Solemya typically contains low or undetectable concentrations of sulfide (<10 μM) despite sulfide concentrations in porewaters surrounding the animals of between 20 and 800 μM (Vetter et al. 1989; Vetter, unpublished data). The blood may contain little sulfide either because the animals are impermeable to sulfide or because sulfide that enters is rapidly oxidized. The blood of *Solemya* contains high concentrations of thiosulfate, which suggests that the animal is detoxifying sulfide as it diffuses in (Anderson et al. 1987; Vetter et al. 1989).

How is this detoxification of sulfide to thiosulfate carried out? Unlike the mitochondria of *Riftia*, the mitochondria of *Solemya* show a stimulated respiration rate in the presence of sulfide. Oxygen consumption by mitochondria increases with sulfide concentrations up to 20 μM, and concentrations as high as 50 μM are metabolized. The sulfide is oxidized by the mitochondrial electron-transport chain, and the energy is partially harvested for the production of ATP (Powell and Somero 1985). This oxidation is a specific metabolic process which is inhibited by sodium azide and boiling. Unlike spontaneous chemical oxidation, mitochondrial sulfide oxidation produces one end product: thiosulfate (O'Brien and Vetter 1990). This appears to be the primary mechanism by which aerobic respiration is protected in the animal.

Thiosulfate, the oxidation product typical of animal sulfide detoxification (Vetter and Bagarinao 1989), is a reduced sulfur compound that can be oxidized by bacteria to sulfate to provide energy for growth. The bacteria in *Solemya* use the thiosulfate produced by animal detoxification as one of the substrates for their metabolism (Anderson et al. 1987). In addition, they may oxidize sulfide that has escaped animal oxidation and is bound to intracellular hemoglobin in the thin lamellar gills (Doeller et al. 1988). When external sulfide concentrations are high, the *Solemya* bacteria shunt excess thiosulfate or sulfide into elemental sulfur deposits. During periods of low external sulfide, the bacteria are able to draw on these reserves (Vetter 1985, 1990; Anderson et al. 1987). The blood reserves of thiosulfate and the bacterial reserves of elemental sulfur may allow the maintenance of stable bacterial metabolism in the presence of fluctuating external concentrations of sulfide. The molluscs maintain a type of chemostat where CO_2 fixation and not bacterial cell division is maximized. We are currently studying the metabolic properties of freshly removed *Solemya* bacteria in greater detail to determine if they show special adaptations to their endosymbiotic existence.

Figure 6
The gutless protobranch mollusc *Solemya reidi* inhabits organic-rich sediments containing high concentrations of hydrogen sulfide. These animals are 5–6 cm long and live in burrows in sediments surrounding the Los Angeles sewage outfall pipe in Santa Monica Bay. The endosymbionts are housed in the gills of the clam.

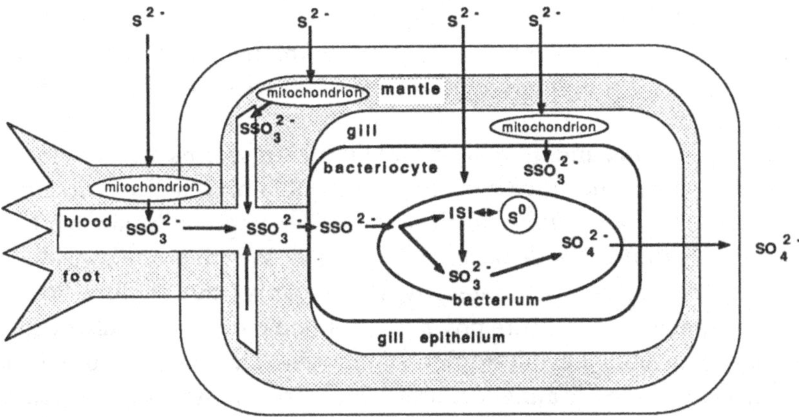

Figure 7
Sulfide metabolism in *Solemya reidi*. *Solemya* pumps water through the mantle cavity so the whole surface of the animal is exposed to sulfide-containing sediments and pore water. The animal is capable of sulfide oxidation without the aid of bacteria. The partial oxidation product, thiosulfate, is released to the blood. Bacteria are housed in the thin lamellar gill, where they can oxidize thiosulfate present in the blood or sulfide that penetrates directly from the seawater. The blood is unable to bind sulfide, and sulfide concentrations are low in the blood.

Net autotrophy has been demonstrated in the *Solemya* symbiosis at external sulfide concentrations between 50 and 100 mM (Anderson et al. 1987). The carbon, sulfur, and nitrogen isotope compositions of *Solemya reidi* and the related species *Solemya velum* also demonstrate complete dependence on chemoautotrophic metabolism (Conway et al. 1989; R. D. Vetter, unpublished data).

The Evolution of Thiotrophic Metabolism

How can symbiosis generate evolutionary novelty? Is present evolutionary theory sufficient to explain the genetic diversity observed today? Thiotrophic metabolism is clearly a novel trophic strategy for an animal, one that requires extensive adaptations on the part of the animal and on the part of its bacterial symbionts. Evolutionary consideration of thiotrophic metabolism begins with the question: "Can the sequential accumulation of single mutations account for the different thiotrophic strategies we observe in modern *Riftia* and *Solemya*?" We can approach this question by considering the likely habitat conditions and ancestral morphologies in these and related nonsymbiotic species to gain insights into how these thiotrophic animals might have evolved through intermediate forms that were favored by natural selection.

Riftia pachyptila and all other vestimentiferan worms are gutless in the adult form and have a trophosome containing sulfur-oxidizing bacteria. They are related to the far smaller pogonophoran worms that also harbor obligate bacterial symbionts. Whether one considers the Vestimentifera a separate phylum or part of the phylum Pogonophora, at least one entire phylum is defined morphologically by the existence of obligate thiotrophic symbioses. The Vestimentifera in many ways resemble annelid worms. The early life stages of *Ridgeia*, a vestimentiferan similar to *Riftia*, have recently been described by Southward (1988). No larval forms have been found, but newly settled juveniles (0.27 mm in length) have two tentacles, a mouth, a ciliated feeding ring, and a gut that is continuous from mouth to anus. The lumen of the gut contains a morphologically mixed population of bacteria. The cells lining the midgut contain vacuoles with both partially digested and intact bacteria (possibly the symbionts). In a slightly larger juvenile (1.8 mm) with fourteen tentacles, the trophosome is clearly developed from endodermal cells of the gut. The trophosome contains bacteria and has become vascularized. The gut is still open from mouth to anus. At the 100-tentacle stage, the juvenile resembles the adult in basic anatomical features, containing a fully developed trophosome and a closed gut. Southward notes that the smallest juveniles of *Ridgeia* bear a superficial resemblance to certain polychaete larvae. Thus, despite both the gross morphological changes in the adult form and the obligate thiotrophic metabolism, the juvenile forms are superficially similar to heterotrophic species which phagocytose bacteria cells from the lumen of the gut during normal feeding. An open gut also occurs in the juvenile stages of *Riftia* development (Jones and Gardner 1988).

Phagocytosis and lysosome fusion are common elements of digestion in most animals. The uptake of a mixed bacterial population, with subsequent lysis of all but one species, is a common route of symbiont infection (Muscatine and McNeil 1989). Resistance to lysis and recognition factors are often encoded for by plasmids in the symbiont (Vierny and Iaccarino 1989). The establishment of pure cultures of chemoautotrophic bacteria can be explained in terms of discrete, gradual adaptations on the part of both the free-living bacteria and the host, resulting in a decrease in cell lysis after ingestion.

The phylogenetic relation between pogonophorans and annelid worms is also apparent in hemoglobin structure. Annelids and pogonophorans have large extracellular hemoglobins which are unusual in that they are held together in part by inter-subunit disulfide bridges (Waxman 1975; Arp et al. 1987). Hemoglobins typically lack disulfide bridges of any kind. The presence of disulfide bridges could be viewed as a preadaptation, in that easily reducible disulfide bridges probably serve as the sulfide-binding mechanism via thiol-disulfide exchange (Arp et al. 1987).

It is possible to envision many of the intermediate steps whereby ancestral annelid worms, which include deposit-feeding species that inhabit sulfide containing environments, would have a gut microbiota that included sulfide-oxidizing bacteria. If these bacteria either performed a sulfide detoxification function or produced more carbon for the animal by fixation rather than lysis, the inhibition of lysis might be favored. Because the symbionts require the simultaneous availability of sulfide and oxygen, close interaction with the oxygen and sulfide-binding capacities of the blood would have been of tremendous advantage to sulfide-oxidizing bacteria. The symbiosis could be favored even if the bacteria only carried out a detoxification function for the host. If the host began to rely more and more on carbon fixation by the bacteria, the animal would move from inceased sulfide tolerance due to bacteria to actual nutritional dependence upon sulfide. Evolutionary pressures to maximize carbon fixation would serve to optimize substrate delivery and product removal. Since carbon input and hence sulfate-reduction rates are low in the deep sea, the nearby submarine hydrothermal vents would represent a rich source of energy relative to the sulfide levels in deep-sea sediments.

Now let us consider *Solemya reidi*, which differs considerably from *Riftia* in anatomy and physiology. *Solemya* lacks circulating hemoglobin, and the bacterial symbionts are never farther than 10–15 μm away from the surrounding seawater (Cavenaugh 1983; Doeller et al. 1988). Consequently, the potential intermediate steps in the evolution of the *Solemya*-type symbiosis probably differ from those proposed for *Riftia*. Because *Solemya* species generally inhabit eutrophic habitats, such as salt marshes, mangrove swamps, and sewage outfalls, and since virtually all their body surfaces are exposed to sulfide-containing water flowing through the mantle cavity, it is worth considering how other organisms from high-sulfide environments handle the problem of sulfide toxicity. Sulfide-resistant forms of cytochrome oxidase have not been found (Vetter and Bagarinao, in press). On the basis of studies of sulfide detoxification in crabs (Vetter et al. 1987), fish (Bagarinao and Vetter 1989), and *Solemya* (O'Brien and Vetter 1990), mitochondrial oxidation of sulfide to thiosulfate seems to be the most common way to prevent sulfide inhibition of respiration. This detoxification mechanism results in the accumulation of thiosulfate in the blood of all of these organisms. Indeed, high thiosulfate concentrations develop in the blood of all organisms experimentally exposed to sulfide (Vetter et al. 1987, 1989; Bagarinao and Vetter 1989). In addition, we have found no active excretion of thiosulfate, nor is thiosulfate rapidly cleared from blood (Vetter et al. 1987; R. D. Vetter, unpublished data). Thus, the mitochondrial oxidation of sulfide and the presence of thiosulfate in the blood appear to be common in sulfide-tolerant animals and probably existed in the ancestor of *Solemya* before the establishment of a thiotrophic symbiosis. Because thiosulfate is a charged molecule and the undissociated H_2S form of sulfide is membrane permeable, the diffusion of sulfide into an organism and its subsequent oxidation to an impermeable form results in the net concentration of thiosulfate. Autotrophic sulfur-oxidizing bacteria are extremely common in the aerobic zone above sulfidic sediments; thus, the existence of sulfur-oxidizing bacteria on the surface of *Solemya* and their subsequent internalization via phagocytosis would be expected. Because the carbon produced would be based on an animal waste product, the evolution of such a symbiosis could develop gradually by increased dependence on symbiotic bacterial carbon. The loss of a gut in adult forms and obligate dependence on sulfide nutrition probably occurred as later events.

The exact steps in the evolution of thiotrophic metabolism in *Riftia* and *Solemya* will probably never be known, but the above scenarios suggest that the intermediate steps in the evolution of obligate thiotrophy could be beneficial to at least one of the partners. Therefore, the development of thiotrophic symbioses can be reconciled with contemporary views of how natural selection operates at the level of individual fitness. But what about the broader aspects of evolutionary theory?

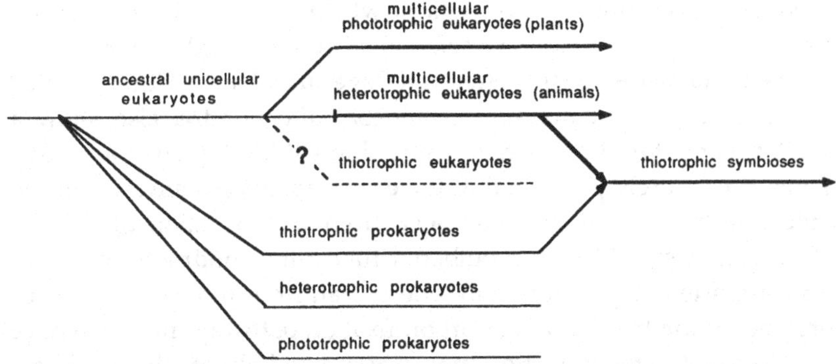

Figure 8
The evolution of different eukaryotic trophic strategies. Plants and animals arose from single-cell eukaryotes, which were the result of symbiotic events. There is no evidence for a similar progression for thiotrophic metabolism. Even if single-cell or simple multi-cell thiotrophic eukaryotes exist, thiotrophic invertebrates appear to have arisen from multicellular animals as a result of symbioses between invertebrates and chemoautotrophic prokaryotes. Thiotrophic metabolism resulted from a fusion of extant but unrelated genetic lineages and results in an emergent symbiotic phenotype with metabolic properties not found in either lineage.

The Evolutionary Significance of Thiotrophic Symbioses

When considering the evolution of different trophic strategies among the eukaryotes, it is useful to consider which potential strategies are absent in the modern world as well as which trophic strategies are common (figure 8). Animals developed from single-celled heterotrophs which contained a nucleus and mitochondria. This type of cellular organization was most likely the result of symbiotic associations. Likewise, plants developed from single-celled algae: photosynthetic eukaryotes which contained a nucleus, mitochondria, and chloroplasts. Why do we not see a similar progression among the eukaryotic thiotrophs, or among cells exploiting other chemoautotrophic means of nutrition? Are there single-celled eukaryotes with the chemosynthetic equivalent of a chloroplast? Were there such organisms when sulfide-rich habitats were more common, and are they represented in the fossil record? Could the primitive mitochondrion oxidize sulfide? All of these are important questions concerning the evolution of thiotrophic metabolism.

While mitochondria are the site of sulfide toxicity, it has recently been shown that mitochondria can oxidize low levels of sulfide-producing ATP from this oxidation, which uses the same electron-transport system as carbon metabolism (Powell and Somero 1985; O'Brien and Vetter 1990; Vetter and Bagarinao 1989). Although ATP production was not measured, amoeba mitochondria also show a similar respiratory burst when exposed to sulfide (Lloyd et al. 1981). Thus, it is entirely possible that the early prokaryotic ancestor of the mitochondrion could generate ATP from the oxidation of sulfide. However, it should be remembered that mitochondria as yet have been reported to generate only energy, and not reducing power or organic carbon from sulfide. Carbon-fixation pathways may also have been present in the mitochondrial ancestor. In light of what we know about sulfide-oxidizing mitochondria and thiotrophic symbiosis, it is quite possible that single-celled eukaryotes which contain sulfide-oxidizing bacteria or organelles do or did exist. Nevertheless, it is clear that thiotrophic metabolism in most live protists or animals did not evolve from single-celled thiotrophic ancestors in a manner analogous to the animal and plant lineages. Why did evolution not favor a thiotrophic lineage? How have animals responded to this unexploited resource? These questions challenge contemporary views of evolution.

Neo-Darwinian evolutionary theory would tell us that once an organism passes a branch point, convergent evolution may drive the formation of analogous structures (the bat wing and the bird wing), but that the genetic bases of the two structures are not related. The corollary to this is that the potential avenues of adaptation available to an organism are contained in the present genome. Is this the case for thiotrophic organisms? To answer this question we must identify the genetic unit of selection in thiotrophic animals. In a typical eukaryote, the nuclear genome undergoes recombination during sexual reproduction and the ancillary genomes in the mitochondrion and chloroplast undergo binary division. They usually, but not always, are passed maternally. Natural selection operates on the phenotypes expressed by the combined genomes. At what point do we consider the symbiotic bacterium and the thiotrophic host as a single unit of selection, a symbiotic phenotype (see Law, this volume)? One criterion for determining the unit of selection for thiotrophic symbioses is whether speciation occurs after the establishment of obligate thiotrophic symbioses. Although a modest fossil record has been discovered, indicating the presence of hydrothermal-vent communities dating from the early Late Cretaceous (Haymon and Koski 1985), the record is incomplete and no studies have addressed questions of speciation. The lucinid clams offer some insight into speciation and evolution (Reid and Brand 1986). The lucinid family contains hundreds of species, many of which co-occur in similar habitats. The most probable explanation for speciation, on the basis of the deduced extent of genetic relatedness of the symbionts within this family and the greater genetic distance between lucinid symbionts and the symbionts of unrelated hosts (Distel et al. 1988), is that speciation occurred after obligate thiotrophic metabolism was established. This is also the most likely scenario for the vestimentiferan worms, the solemyid and vesicomyid clams, and any group in which all members are morphologically similar, contain genetically related symbionts, and do not contain any nonsymbiotic species. The alternative hypothesis, that each species became symbiotic after speciation, is improbable in view of the similarity of the adaptations on the part of related hosts and the differences between unrelated hosts. Whatever we use as a criterion for determining when a single selective unit, a symbiotic phenotype, is formed from two symbiotic partners, it seems clear that the significance of thiotrophic symbioses is that it is possible to acquire a novel metabolic function from a different evolutionary lineage (figure 7). Thus, we must accept the possibility that analogous metabolic function can have a homologous genetic basis. In this way, symbiosis is a bridge to a different evolutionary lineage. Let us examine how novel genes of thiotrophic symbionts might be incorporated into the host genome.

Symbiosis as a Source of Genetic Novelty

One of the great questions of molecular evolution is "What is the source of new genes?" Though internal processes, such as gene duplication and subsequent divergence, can explain some metabolic innovations (such as different cellular and vascular forms of hemoglobin), other processes are surely at work. Selection operates at the level of present individual fitness regardless of future fitness. Consequently, there must be evolutionary barriers to some adaptations irrespective of their future selective value. The more new genes involved in a metabolic pathway, the greater the barrier. For a process such as chemoautotrophic sulfur metabolism to evolve in clams solely on the basis of the host's genome, the barrier would probably be prohibitive. Symbiosis may be viewed as a means of overcoming an evolutionary barrier, allowing the selective benefits of the new metabolic process to be exploited while the genetic basis of the metabolic capacity is being transferred. The acquisition of a novel metabolic function via symbiosis could be favored for two reasons. First, if it confers an increase in fitness, the advantage is conferred relatively quickly on an evolutionary time scale. Second, and perhaps more important, it allows the two genomes to coexist in close proximity for a sufficient length of time for the host to "learn the language" (acquire the genes) for the metabolic function.

If we consider the evolution of an organelle such as the mitochondrion, the advantage of a long-term symbiotic association is that the genes for a metabolic pathway in the mitochondrion might not have to be transferred to the nucleus in any particular order, and each gene does not have to have selective value by itself because mitochondrial function is maintained regardless of the location of individual genes. Because there are multiple copies of the mitochondrion in each cell, even gene transfers lethal to the donor mitochondrion will not be necessarily be lethal to the cell. The transfer of a complete metabolic pathway must be proportionally more difficult, in that the mitochondrion and the chloroplast still exist and their genes have not been entirely transferred to the nuclear chromosomes. However, individual genes or smaller portions of metabolic pathways can be transferred experimentally with relative ease (Nagley and Devenish 1989). Even if these genes are not used for their original purpose, they can be modified and used for other purposes by the host. Transient symbiotic associations in the past may be one explanation for the metabolic diversity that we observe today. Transfer of genes can also occur between mitochondria and chloroplasts and between mitochondria and symbiotic bacteria (Moon et al. 1988; Nugent and Palmer 1988; Stein et al. 1990). To the extent that mitochondria, chloroplasts, and perhaps some thiotrophic symbionts are maternally inherited (Gustafson and Reid 1988), incorporation into the nuclear genome is not even a requirement for the stable acquisition of novel metabolic capacities.

How easy is the transfer of genes? In any symbiosis it is surely a rare event. However, there is ample evidence for the transfer of chloroplast and mitochondrial genes into the nucleus (Margulis and Bermudes 1985; Nagley and Devenish 1989) and for the transfer of genes between chloroplasts and mitochondria (Levings and Brown 1989). In amoebae which actively digest prey by lysosome-phagosome fusion, the occurrence might be common (Pritchard et al. 1989) because the cell resembles a living recombinant-DNA experiment. In a typical recombinant-DNA experiment the DNA is cut into pieces, the cells are permeabilized, and the source and recipient genomes are brought into close proximity along with the required restriction and splicing enzymes. In the bacteriocyte there are large numbers of bacteria with weakened or nonexistent outer membranes, in close proximity to the host's mitochondria and nucleus. In addition, the bacteria are dividing and undergoing lysosomal digestion (Fiala-Medioni et al. 1990). Any DNA fragments that escape from the lysosome to the cytosol would be subject to transfer to the mitochondrion and the nucleus. Transportation of DNA fragments in the cytosol into the mitochondrion is particularly likely because DNA fragments can bind to cytosolic apoproteins destined for importation into mitochondria (Vestweber and Schatz 1989).

Preliminary evidence for the incorporation of the gene for the small subunit of the carbon-fixation enzyme, ribulose 1,5 bisphosphate carboxylase, into mitochondria has been recently reported for *Solemya reidi* (Stein et al. 1990). Subsequent transfer of genes incorporated into mitochondria to future generations is relatively easy because of maternal inheritance. Incorporation into the nuclear genome, either from the original bacterial donor or from a mitochondrial intermediate, is more difficult, but the latter seems to have occurred for mitochondria and chloroplasts. In thiotrophic symbioses, transfer of genes would be most probable for single-celled organisms that reproduce by binary division. It would be more difficult for those symbioses in which symbionts are passed along with the egg, a situation that has been suggested for *Solemya reidi* (Gustafson and Reid 1988). When symbiont incorporation occurs after fertilization, it is even more difficult to imagine how gene transfers occurring in somatic cells, such as bacteriocytes, could be incorporated into the germ-cell lines; however, there is still much to be learned about thiotrophic symbioses.

Symbiosis as a Source of Evolutionary Innovation

What then is the significance of symbiosis in modern evolutionary theory? Symbiosis is a way of bridging lineages with different metabolic capabilities. It is perhaps the only way to acquire a complex metabolic pathway, such as chemoautotrophy, once an organism has passed an evolutionary branch point. Before 1979 no one expected or predicted the occurrence of thiotrophic clams. Since then, we see that the incorporation of a bacterial genome made this mode of nutrition possible. Aside from the physiological adaptations required by both partners, the evolutionary significance of thiotrophic symbioses is that the organisms now have a new symbiotic phenotype. Freed from the competitive interactions of their ancestral forms (e.g., competition for food particles), they now compete for entirely different limiting resources (e.g., sulfide). Competition among thiotrophic organisms for the sulfide in hydrothermal-vent water and the relative resistance of different species to heat, pressure, sulfide toxicity, and hypoxia (figures 1, 2) have driven the evolution of thiotrophic hydrothermal-vent organisms in new directions, away from their ancestral phenotypes. Thus, the acquisition of novel metabolic pathways, through symbiosis, can even lead to evolutionary innovation at the community and ecosystem levels.

Notes

1. The thiotrophic invertebrates illustrated in figures 1 and 2 compete for the energy in sulfur in a manner more analogous to the competition between light-dependent plants than to the competition between the filter-feeding heterotrophic invertebrates that they resemble. A clear appreciation of the trophic basis of the symbiotic phenotype is required to interpret the ecology and competitive interactions of the different species.

2. In a *heterotrophic eukaryotic cell,* genes have apparently moved from the mitochondrion to the nucleus, and most mitochondrial proteins are now encoded by genes in the nucleus and imported into the mitochondrion after synthesis. The mitochondrion acts as the major site of oxygen-consuming energy-generating metabolism. The mitochondrion takes in reduced forms of carbon and puts out ATP for other cellular functions. In a *phototrophic eukaryotic cell* the chloroplast has undergone similar gene transfer to the nucleus as well as to the mitochondrion. The chloroplast harvests the electromagnetic energy of sunlight to fix carbon for the cell. Chloroplasts produce oxygen while mitochondria consume oxygen. The *thiotrophic eukaryotic cell* contains symbiotic bacteria which harvest the energy in hydrogen sulfide for the fixation of carbon. The mitochondria oxidize carbon in a manner similar to the heterotrophic cell and also generate some ATP directly from sulfide oxidation. Both the mitochondria and sulfide-oxidizing bacteria consume oxygen. The extent of genetic transfer is unknown, and the ability of the bacteria to remain viable in a free-living state is also unknown.

References

Anderson, A. E., Childress, J. J., and Favuzzi, J. A. 1987. Net uptake of CO_2 driven by sulphide and thiosulphate oxidation in the bacterial symbiont-containing clam *Solemya reidi*. *J. Exp. Biol.* 133: 1–31.

Arp, A. J., Childress, J. J., and Vetter, R. D. 1987. The sulfide-binding protein in the blood of the vestimentiferan tube-worm, *Riftia pachyptila*, is the extracellular hemoglobin. *J. Exp. Biol.* 128: 139–158.

Attardi, G., and Schatz, G. 1988. Biogenesis of mitochondria. *Ann. Rev. Cell Biol.* 4: 289–333.

Bagarinao, T., and Vetter, R. D. 1989. Sulfide tolerance and detoxification in shallow-water marine fishes. *Mar. Biol.* 103: 291–302.

Beauchamp, R. O., Bus, J. S., Popp, J. S., Boreiko, C. J., and Andrejelkovish, D. A. 1984. A critical review of the literature on hydrogen sulfide toxicity. *CRC Crit. Rev. Toxicol.* 13: 25–97.

Cary, S. C., Vetter, R. D., and Felbeck, H. 1989. Habitat characterization and nutritional strategies of the endosymbiont-bearing bivalve, *Lucinoma aequizonata*. *Mar. Ecol. Prog. Ser.* 55: 31–45.

Cavanaugh, C. M. 1983. Symbiotic chemoautotrophic bacteria in marine invertebrates from sulfide-rich habitats. *Nature* 302: 58–61.

Cavanaugh, C. M., Gardiner, S. L., Jones, M. L., Jannasch, H. W., and Waterbury, J. B. 1981. Prokaryotic cells in the hydrothermal vent tube-worm *Riftia pachiptila* Jones: Possible chemoautotrophic symbionts. *Science* 213: 340–342.

Conway, N., Capuzzo, J. M., and Fry, B. 1989. The role of endosymbiotic bacteria in the nutrition of *Solemya velum*: Evidence from a stable isotope analysis of endosymbionts and host. *Limnol. Oceanog.* 34: 249–255.

Corliss, J. B., Dymond, J., Gordon, L. I., Edmond, J. M., von Herzon, R. P., Ballard, R. D., Green, K., Williams, D., Bainbridge, A., Crane, K., and van Andel, T. H. 1979. Submarine thermal springs on the Galapagos Rift. *Science* 203: 1073.

Distel, D. L., Lane, D. J., Olsen, G. J., Giovannoni, S. J., Pace, B., Pace, N. R., Stahl, D. A., and Felbeck, H. 1988. Sulfur-oxidizing bacterial endosymbionts: Analysis of phylogeny and specificity by 16S rRNA sequences. *J. Bacteriol.* 170: 2506–2510.

Doeller, J. E., Kraus, D. W., Calacino, J. M., and Wittenberg, J. B. 1988. Gill hemoglobin may deliver sulfide to bacterial symbionts of *Solemya velum* (Bivalvia, Molluska). *Biol. Bull.* 175: 388–396.

Felbeck, H. 1981. Chemoautotrophic potential of the hydrothermal vent tubeworm *Riftia pachyptila* Jones (Vestimentifera). *Science* 213: 336–338.

Felbeck, H. 1983. Sulfide oxidation and carbon fixation in the gutless clam *Solemya reidi*: An animal-bacterial symbiosis. *J. Comp. Physiol.* 152: 3–11.

Fiala-Medioni, A., Felbeck, H., Childress, J. J., Fisher, C. R., and Vetter, R. D. 1990. Lysosomic resorption of bacterial symbionts in deep-sea bivalves. In: Nardon, P., Gianinazzi-Pearson, V., Greneir, A. M., Margulis, L., and Smith, D. C., eds., *Endocytobiology IV*. INRA, Paris.

Fisher, C. R., 1990. Chemoautotrophic and methylotrophic symbioses in marine invertebrates. *CRC Crit. Rev. Aquat. Sci.* 2.

Fisher, C. R., Childress, J. J., and Sanders, N. K. 1988. The role of vestimentiferan hemoglobin in providing an enviroment suitable for chemoautotrophic sulfide-oxidizing endosymbionts. *Symbiosis* 5: 229–246.

Fisher, C. R., Childress, J. J., and Minnich, E. 1989. Autotrophic carbon fixation by the chemoautotrophic symbionts of *Riftia pachyptila. Biol. Bull.* 177: 372–385.

Fry, B., Guest, H., and Hayes, J. M. 1983. Sulfur isotopic composition of deep-sea hydrothermal vent animals. *Nature* 306: 51–52.

Gray, M. W., and Doolittle, W. F. 1982. Has the endosymbiont hypothesis been proven? *Microbiol. Rev.* 46: 1–42.

Gustafson, R. G., and Reid, R. G. B. 1988. Association of bacteria with larvae of the gutless protobranch bivalve *Solemya reidi* (Cryptodonta: Solemyidae). *Mar. Biol.* 97: 389–401.

Hand, S. C., and Somero, G. N. 1983. Energy metabolism of hydrothermal vent animals: Adaptations to a feed-rich and sulfide-rich deep-sea environment. *Biol. Bull.* 165: 167–181.

Hartl, F., Pfanner, N., Nicholson, D. W., and Neupert, W. 1989. Mitochondrial protein import. *Biochim. Biophys. Acta* 988: 1–45.

Haymon, R. M., and Koski, R. A. 1985. Evidence for an ancient hydrothermal vent community: Fossil worm tubes in Cretaceaous sulfide deposits of the Sumail Ophiolite, Oman. *Biol. Soc. Wash. Bull.* 6: 57–65.

Jones, M. L. 1985. On the Vestimentifera, new phylum: six new species, and other taxa, from the hydrothermal vents and elsewhere. In: Jones, M. L., ed. The Hydrothermal Vents of the Eastern Pacific: an Overview. *Bull. Biol. Soc. Wash.* 6: 117–158.

Jones, M. L and Gardner, S. L. 1988. Evidence for a transient digestive tract in vestimentifera. *Proc. Biol. Soc. Wash.* 101: 423–433.

Levings, C. S. III, and Brown, G. G. 1989. Molecular biology of plant mitochondria. *Cell* 56: 171–179.

Lloyd, D., Kristensen, B., and Degn, H. 1981. Oxidative detoxification of hydrogen sulfide detected by mass spectroscopy in the soil amoeba *Acanthamoeba castellanii*. *J. Gen. Microbiol.* 126: 167–170.

Mangum, C. P., Woodin, B. R., Bonaventura, C., Sullivan, B., and Bonaventura, J. 1987. The role of coelomic and vascular hemoglobin in the annelid family Terebellidae. *Comp. Biochem. Physiol.* 51A: 281–294.

Margulis, L., and Bermudes, D. 1985. Symbiosis as a mechanism of evolution: Status of cell symbiosis theory. *Symbiosis* 1: 101–124.

Moon, E., Kao, T. H., and Wu, R. 1988. Rice mitochondrial genome contains a rearranged chloroplast gene cluster. *Mol. Gen. Genet.* 213: 247–253.

Muscatine, L., and McNeil, P. L. 1989. Endosymbiosis in *Hydra* and the evolution of internal defense systems. *Amer. Zool.* 29: 371–386.

Nagley, P., and Devenish, R. J. 1989. Leading organellar proteins along new pathways: The relocation of mitochondrial and chloroplast genes to the nucleus. *Trends in Biochem. Sci.* 14: 31–35.

National Research Council. 1979. *Hydrogen Sulfide.* University Park Press.

Nugent, J. M., and Palmer, J. D. 1988. Location, identity, amount and serial entry of chloroplast DNA sequences in crucifer mitochondrial DNAs. *Curr. Genet.* 14: 501–509.

O'Brien, J., and Vetter, R. D. 1990. Production of thiosulphate during sulphide oxidation by mitochondria of the symbiont-containing bivalve *Solemya reidi*. *J. Exp. Biol.* 149: 133–148.

Powell, M. A., and Arp, A. J. 1989. Hydrogen sulfide oxidation by abundant non-hemoglobin heme compounds in marine invertebrates from sulfide-rich habitats. *J. Exp. Zool.* 249: 121–132.

Powell, M. A., and Somero, G. N. 1983. Blood components prevent sulfide poisoning of respiration of the hydrothermal vent tube worm *Riftia pachyptila*. *Science* 219: 297–299.

Powell, M. A., and Somero, G. N. 1985. Sulfide oxidation occurs in the animal tissue of the gutless clam, *Solemya reidi*. *Biol. Bull.* 169: 164–181.

Powell, M. A., and Somero, G. N. 1986. Hydrogen sulfide oxidation is coupled to oxidative phosphorylation in mitochondria of *Solemya reidi*. *Science* 233: 563–566.

Pritchard, A. E., Venuti, S. E., Ghalambor, M. A., Sable, C. L., and Cummings, D. J. 1989. An unusual region of *Paramecium* mitochondrial DNA containing chloroplast-like genes. *Gene* 78: 121–134.

Rau, G. H. 1981. Hydrothermal vent clam and tube worm $^{13}C/^{12}C$: further evidence of non-photosynthetic food sources. *Science* 213: 338–340

Reid, R. G. B. 1980. Aspects of the biology of the gutless species of *Solemya* (Bivalvia: Protobranchia). *Can. J. Zool.* 58: 386–393.

Reid, R. G. B., and Brand, D. G. 1986. Sulfide oxidizing symbiosis in Lucinaceans: Implications for bivalve evolution. *Veliger*. 29: 3–24.

Sogin, M. L. 1989. Evolution of eukaryotic microorganisms and their small subunit ribosomal RNAs. *Amer. Zool.* 29: 487–499.

Somero, G. N., Childress, J. J., and Anderson, A. E. 1989. Transport, metabolism, and detoxification of hydrogen sulfide in animals from sulfide-rich marine environments. *Critical Reviews in Marine Science* 1: 591–614.

Southward, E. C. 1988. Development of the gut and segmentation of newly settled stages of Ridgeia (Vestimentifera): Implications for relationship between vestimentifera and pogonophora. *J. Mar. Biol. Ass. U.K.* 68: 465–487.

Stein, J., Haygood, M., and Felbeck, H. 1990. Diversity of ribulose 1,5 biphospate carboxylase genes in thiotrophic symbioses. In: Nardon, P., Gianinazzi-Pearson, V., Grenier, A. M., Margulis, L., and Smith, D. C., eds., *Endocytobiology IV*. INRA, Paris.

Stern, D. B., and Palmer, J. D. 1984. Extensive and widespread homologies between mitochondrial DNA and chloroplast DNA in plants. *Proc. Nat. Acad. Sci.* 81: 1946–1950.

van der Land, J., and Norevang, A. 1977. Structure and relationships of Lamellibranchia (Annelida, Vestimentifera). *Kongelige Danske Videnskabernes Selskab, Biologiske Skrifter* 21: 1–102.

Vestweber, D., and Schatz, G. 1989. DNA-protein conjugates can enter mitochondria via the protein import pathway. *Nature* 338: 170–172.

Vetter, R. D. 1985. Elemental sulfur in the gills of three species of clams containing chemoautotrophic symbiotic bacteria: A possible energy storage compound. *Mar. Biol.* 88: 33–42.

Vetter, R. D. 1990. Cultured gill filaments from *Solemya reidi*: A model system for the study of thiotrophic symbioses. In: Nardon, P., Gianinazzi-Pearson, V., Grenier, A. M., Margulis, L., and Smith, D. C., eds., *Endocytobiology IV*. INRA, Paris.

Vetter, R. D., and Bagarinao, T. 1989. Detoxification and exploitation of hydrogen sulphide by marine organisms. In Proceedings of the International Symposium on Hydrogen Sulphide Toxicity, Banff, Alberta.

Vetter, R. D., Wells, M. E., Kurtsman, A. L., and Somero, G. N. 1987. Sulfide detoxification by the hydrothermal vent crab *Bythograea thermydron* and other decapod crustaceans. *Physiol. Zool.* 60: 121–137.

Vetter, R. D., Matrai, P. A., Javor, B., and O'Brien, J. 1989. Reduced sulfur compounds in the marine environment: analysis by HPLC. In: *Biogenic Sulfur in the Environment*. American Chemical Society.

Vierny, C., and Iaccarino, M. 1989. Comparative study of the symbiotic plasmid DNA in free-living bacteria and bacteriods of *Rhizobium leguminosarum*. *FEMS Microbiol. Lett.* 60: 15–20.

Waxman, L. 1975. The structure of annelid and mollusc hemoglobins. *J. Biol. Chem.* 250: 3790–3795.

Wilmot, D. B., and Vetter, R. D. 1989. The bacterial symbiont from the hydrothermal vents tube-worm is a unique sulfide specialist. In Proceedings of Annual Meeting of American Society of Microbiologists.

V Symbiosis and Ecology

17 Fungal Symbioses and Evolutionary Innovations

Bryce Kendrick

Heterotrophic, and therefore dependent on the parallel or prior existence of other organisms, the true fungi (kingdom Eumycota; phyla Zygomycota and Dikaryomycota) are a relatively ancient group that has two life forms: *mycelia* (branching, filamentous, apically extending, exploratory, indefinite, feeding thalli which are osmotrophic and plurivorous) and *spores* (microscopic, sealed, nonmotile propagules (dispersal units) typically produced and released in enormous numbers, and virtually omnipresent in the biosphere). These complementary morphs ensure that fungi are successful opportunistic colonizers of an enormous range of substrates. They grow on substances contained in, exuded, secreted, excreted, or abandoned by adjacent organisms. In view of this long-standing intimacy, and the ultimate dependence of all fungi on carbon autotrophs, it is not surprising that such relationships have in many cases evolved into finely tuned symbioses.

The kind of symbiosis that fascinates many of us most, perhaps because of our inherent idealism, or our belief in the Social Contract or the Gaia Hypothesis, is that in which both or all partners derive some advantage from the association: mutualistic symbiosis. It is fascinating, and to some of us encouraging, that such symbioses have been discovered in so many different corners of the biosphere. Fungi are as deeply involved as any other group of organisms. Most laypersons, and even many scientists, think of the fungi as parasites or saprobes, so they might be surprised to learn that almost one-third of the known fungi are involved in mutualistic symbioses.

Such relationships have been established with many photobionts (cyanobacteria, chlorophytes, bryophytes, pteridophytes, gymnosperms, and angiosperms), and also with some heterotrophic organisms (notably Coleopteran, Dipteran, Homopteran, Hymenopteran, and Isopteran insects). Most important from the point of view of the theme explored here, several of these relationships have given rise to major evolutionary innovations which have conferred on the interdependent organisms the ability to colonize habitats previously unavailable to them.

The most distinctive examples usually cited are the lichens, generally recognized as dual organisms, with external fungal morphology and an indwelling algal component, which can grow in many extremely inhospitable habitats. But it is also widely accepted that a much more important group of organisms, the land plants, arose as a joint venture between fungi and green algae (Jeffrey 1962; Pirozynski and Malloch 1975; Lewis, this volume). Devonian plant fossils contain structures closely resembling the "vesicles" (also called intramatrical spores) produced by modern vesicular-arbuscular mycorrhizal fungi, and over 90 percent of modern plants establish mycorrhizal associations. The fungi evolved an extensive, tubular, branching thallus, adapted to ramifying through substrates such as soil and to scavenging water and low concentrations of available mineral nutrients. The algae, although they photosynthesize efficiently, are accustomed to growing surrounded by water which contains dissolved minerals; they have not developed an extensive root system. Even in plants, fungal hyphae considerably extend the effective volume of soil tapped by the root system.

Atsatt (this volume) has recently proposed an even more radical hypothesis: that land plants arose from the incorporation of a fungal genome into a green algal one, thus coupling "the synthetic power of the chloroplast with the degradative capacities of the fungus." He suggests that absorptive nutrition allowed the evolution of first embryos, then pollen and seeds. He posits the Oomycetes, a group of fungal-like protoctista with cellulosic hyphal walls and diploid somatic nuclei, as possible descendants of the early mycobionts.

It is particularly interesting that, while the results of the symbioses with algal groups have long been recognized as distinct although dual organisms (or triple or quadruple, in the cases of some lichens; see Hawksworth 1988a), the almost equally dual (or multiple) nature of such organisms as the Pinaceae, in which one tree may concurrently have several different mycobionts (Trappe 1977), was not widely recognized until quite recently. Mycologists are still busy reminding plant ecologists that there is an absolutely vital underground component of their community studies which they have almost totally neglected for many years.

Lichens

In lichens, the mycobiont typically provides a physical framework within which the alga exists. The mycobiont therefore mediates all interactions with the outside world, acquiring whatever water and mineral nutrients become available and cushioning the impact of environmental extremes on the photobiont. The photobiont, conditioned to leak much of its photosynthate to the fungus, has a straightforward role. The partnership can withstand extemely hostile physical conditions, but is vulnerable to many atmospheric pollutants; lacking roots and frequently growing on impermeable substrates, it depends on mineral nutrients brought to it through the air. Reduced lichen populations are a well-known result of the atmospheric pollution attendant on industrial development.

It is also well known that the algal components of lichens are usually exploited by their mycobiont in a condition that is often called "controlled parasitism." There seems little doubt that the fungus gains more from the association than does the alga. Yet its presence inside a lichen often permits an alga to exist in many habitats in numbers that would not have been possible without the protection afforded by the fungus. So despite its subordinate position, the alga gains selective advantage by being lichenized, and the symbiosis must be considered mutualistic.

Hawksworth (1988a) has compiled the various kinds of symbioses between dikaryomycotan fungi and "algae" (eukaryotic Chlorophyta and prokaryotic Cyanobacteria). All but 2 percent of the fungi involved in lichens are ascomycetes. They are drawn from 16 of the 46 recognized orders, but only 6 of those orders are exclusively lichenized. In addition, there are some basidiomycetous and conidial lichens. Even in the absence of a fossil record, these facts demonstrate that the lichen association has evolved many, many times. Unlike many other groups, lichens have no common ancestor—only a widely shared process springing from natural affinity, opportunity, or need.

Everyone is familiar with the case in which the fungus establishes the general shape of the organism and the alga dwells within. But only mycologists (sometimes in the guise of lichenologists) appear to be aware of some of the variations on this theme which hint at a greater degree of flexibility in the establishment of such symbioses than has been widely recognized. Hawksworth (1988a,b) points out that there are special cases in which three or even four bionts are associated in a stable relationship; for example, no fewer than 500 lichens contain two taxa of photobionts—one a chlorophyte, one a cyanobacterium—within a single thallus, and many conspecific lichens may incorporate algae of different, even if closely related, taxa. It is intriguing that over 13,000 fungi from 300 genera are known only in the lichenized condition. Yet a much smaller number of algae are involved in these relationships (about 30 genera, of which only 10 account for most lichens), and many of those can also be found free-living, sometimes even alongside the lichen containing their presumed siblings.

In most lichens, the fungus reproduces sexually and the alga does not. But the lichen spectrum encompasses many levels of intimacy and obligation. Most fungi in the Arthopyreniaceae grow independently, but some are patchily associated with a green alga, and some are invariably lichenized. This seems to represent an ongoing evolutionary progression toward a fully integrated symbiosis. In other cases, fungi are always found associated with particular algae, but no specialized dual structures are produced; this is true of the fascinating cryptoendolithic associations documented in Antarctica (Friedmann 1982) and also known from the hot deserts of Namibia. In other cases, the mycobiont encloses the photobiont but does not affect its morphology. In yet others, the algae are modified but are uniformly distributed throughout the thallus. In the typical lichen, as found in exclusively lichenized orders of fungi (e.g., Pertusariales, Teloschistales), the algal cells are restricted to a zone just beneath the upper surface of the thallus. Lichen associations are constantly being formed and dissolved. Some are being transmuted, as when a second fungus attacks the thallus, supplants the original mycobiont, and finally establishes its own lichenized relationship with the original alga.

So the fine-tuning process, and perhaps also the major evolutionary process, still continues. The taxonomy of most lichens is basically fungal. Lichenized fungi usually display truly ascomycetous or basidiomycetous characteristics, and should therefore be placed in those groups rather than being recognized as a separate phylum or division (Margulis and Schwartz 1988).

Lichens are well known and understood as a dual phenomenon. But there are a few other putatively mutualistic fungus-alga associations in which the alga, not the fungus, is the major partner. The best-known example of such a relationship is that between the macroscopic brown algae *Ascophyllum nodosum* and *Pelvetia canaliculata* and the ascomycete *Mycosphaerella ascophylli* (Kohlmeyer and Kohlmeyer 1972; Hawksworth 1988a). Here the fungus grows and forms reproductive structures inside the algae. It might easily be thought that the fungus was parasitizing the algae, but the algae are never found in nature without the fungus. Although this relationship is still not well understood, it is possible that such relationships may be a form of evolutionary novelty in which the fungus confers on the alga some resistance to desiccation, permitting it to grow in habitats that are only sporadically wetted (*Pelvetia* grows well above high-tide level) and thus perhaps initiating another move from marine to terrestrial life.

Mycorrhizae

Mycorrhizal associations represent the second major fungal excursion into mutualism with photobionts. Mycorrhizae can be structurally divided into seven different kinds: ectomycorrhizae, ectendomycorrhizae, arbutoid, monotropoid, vesicular-arbuscular, ericoid, and orchid mycorrhizae. Among these, two dichotomies are observed: (1) an incomplete morphological one between mycobionts that penetrate the cells of the root cortex and develop exchange organs within them (the endomycorrhizal pattern) and mycobionts that surround but do not penetrate the cells of the root cortex (the ectomycorrhizal pattern) and (2) the more important dichotomy between truly mutualistic associations, in which the net carbon flux is from plant to fungus (ecto, ectendo, arbutoid, ericoid, most V-A) in exchange for minerals, and those in which the net flux is from fungus to plant (monotropoid, orchid, some V-A).

About 6000 species of fungi are known to be involved in mutualistic mycorrhizal symbioses. Over 5000 basidiomycetes and several hundred ascomycetes form ectomycorrhizae; perhaps 200 or more ancient, nonseptate, obligately biotrophic fungi of isolated and uncertain taxonomic position form vesicular-arbuscular mycorrhizae. Whereas in the lichens it was the photobionts that lacked specificity, here it is the fungi. This is inevitable, since almost 300,000 plants can establish mycorrhizal relationships. Harley (1989) notes that, among the British flora, all the gymnosperms, 80 percent of the angiosperms, and 70 percent of the pteridophytes are potentially mycorrhizal. Some groups of photobionts, such as the Pinaceae, are obligately ectomycorrhizal—2000 species of conifers, with as many as 5000 species of fungi—and one tree may not only establish associations with a succession of different fungi during its lifetime (Last et al. 1987; Mason et al. 1984), but may also have several or many fungi associated with its roots at any one time (Trappe 1977).

It has often been suggested that mycorrhizal associations arose from the mitigation of earlier parasitic relationships (fungi attacking roots). Yet the majority of mycorrhizal fungi, both dikaryomycotan and nonseptate, have few or no close parasitic relatives. The enigmatic asexual fungi that initiate the ubiquitous vesicular-arbuscular endomycorrhizae are clearly ancient and have no known nonmycorrhizal relatives, so it is virtually impossible to comment on their pre-mycorrhizal existence. The more modern fungi responsible for ectomycorrhizae have many free-living relatives. Knowledge of these enabled Malloch (1987) to suggest that, since plants had always been intimately associated with fungi, the advent of large numbers of highly successful saprobic basidiomycetes (Aphyllophorales, Agaricales, Russulales, Hymenogastrales) and ascomycetes (Pezizales) presented an opportunity for newly evolved plants moving into difficult but available habitats to establish new associations, though of a different kind. Another theory is that some groups of plants, in developing chemical defenses against insects, inadvertently expelled their endomycorrhizal partners, and had to establish another, less intimate kind of root symbiosis with the new fungi (Pirozynski 1983). Once again we are contemplating a diverse assemblage of fungi representing several different orders, and two different subphyla, among which the ectomycorrhiza must have evolved many times. New forms are indeed still evolving. Another hypothesis (Malloch et al. 1980) contrasts the tree diversity of essentially ectomycorrhizal boreal forests (very low) with that of endomycorrhizal tropical rainforests (very high), and suggests that in boreal forests (and in low-diversity tropical forests) many differently adapted ectomycorrhizal fungi make the necessary adjustments to changes in soil microhabitats, even within the root system of individual trees, while in the rainforest it is the multiplicity of tree species which must fit those microhabitats, since there are many fewer endomycorrhizal mycobionts available.

The investment of ectomycorrhizal fungi in the mycorrhizal organ is considerable: far more, and more concentrated, biomass is produced than by vesicular-arbuscular endomycorrhizal fungi. The fungal mantle surrounding the root is used as a nutrient sink by both fungus and plant. This seems to be an adaptation to the stressful conditions and to the poor soils in which many ectomycorrhizal trees, especially the conifers, often grow as pioneer colonizers of waste lands. Kendrick and Berch (1985) review these ideas.

Trappe and Fogel (1977) wrote: "Most woody plants require mycorrhizae to survive, and most herbaceous plants need them to thrive." The apparently logical and inevitable evolution of mycorrhizal relationships may be perceived as a major step in the original colonization and in the ongoing exploitation of the terrestrial habitat by plants and fungi. Lewis' chapter in this volume reviews the routes by which these symbioses may have evolved.

Symbioses with Animals

After the all-pervasive mycorrhizal and lichen symbioses, the relatively stable relationships established by fungi with animals seem little more than evolutionary footnotes. Symbioses with carbon-fixers make sense, but fungi and insects are often in competition for the same food supply. What can tie such carbon heterotrophs together in an obligate relationship? The fungi have cellulolytic enzymes that are almost unknown in the animal kingdom; this makes them desirable exosymbionts for colonial animals that have a large labor force and access to massive but indigestible food sources. The two organisms are physically distinct, so there is no question of a shared name. The morphology of the animals involved has not changed, but their behavior has been profoundly influenced. The fungi appear little modified, and the entire symbiotic process appears to have been serendipitously initiated and elaborated by the animals, though there is no doubt that the fungus derives advantage from the relationship (Kendrick 1985).

There are two entirely separate and independent examples of this symbiosis, involving social insects from different orders. In Central and South America, the Attine ants climb trees and cut a harvest of leaf segments, carrying them flamboyantly to their subterranean nest, where they are further masticated by the ants, then digested by a domesticated cellulase-possessing species of *Leucoagaricus* or *Lepiota*

(Holobasidiomycetes: Agaricales), or occasionally *Xylaria* (Ascomycetes: Sphaeriales) or *Auricularia* (Phragmobasidiomycetes: Auriculariales). The ants provide their garden fungus with a constantly replenished food supply, ideal growing conditions, and freedom from fungal competitors, which are excluded by physical and chemical means (Weber 1972). In return, the mycobiont transmutes the inedible leaves into edible fungal mycelium, which constitutes the sole food of the ant colony. The fungi never form reproductive structures in or near the gardens, and have been identified only in axenic culture. Leaf-cutting ants may be a nuisance, in that they sometimes strip economically important trees or other crops of their leaves, but they have ecological significance. In tropical forests the turnover of organic matter on the ground is usually very rapid, and even trees do not penetrate very deeply into the soil. A large colony of *Atta*, with hundreds of fungal gardens, greatly increases the aeration and the organic content of the soil, and opens it up for subsequent colonization by other organisms. Since there may be one colony of *Atta* per 2 square meters, these ants and their fungi are important factors in soil ecology and nutrition.

In Africa and Asia, mound-building termites of the Macrotermitinae also maintain subterranean gardens, which are usually monocultures of species of *Termitomyces* (Holobasidiomycetes: Agaricales) or occasionally *Xylaria* (Ascomycetes: Sphaeriales). Again, the termites exhibit many behavioral modifications paralleling those in the Attinae (Batra 1979). Other termites do not cultivate fungi, but have cellulolytic microorganisms as gut endosymbionts. It would be interesting to compare the relative efficiencies of these two disparate symbioses.

Other Fungus-Insect Symbioses

The 200 members of the Septobasidiales (Phragmobasidiomycetes) are always found in a mutualistic relationship with scale insects, although the insects can survive without the fungi (Couch 1938; Evans 1988). Some gall midges (Diptera: Cecidomyiidae) have mutualistic relationships with species of the coelomycetous anamorph-genus *Macrophoma* (teleomorph = *Botryosphaeria* [Ascomycetes: Dothideales]), which inhabit what are called "ambrosia galls" (Neger 1909; Bissett and Borkent 1988). The larvae are dependent on the fungal mycelium for food, while the fungus needs the adult midge as vector. The female midge carries the fungus in a pair of specialized pouches called mycangia, and deposits conidia with the eggs. *Macrophoma* probably has other passive means of dispersal, but the intervention of the adult midge as a long-range aerial vector almost certainly improves the fungus' chances of establishing itself in another appropriate substrate. Bark beetles of the Lymexylidae, the Platypodidae, and the Scolytidae (Coleoptera) also carry "ambrosia fungi" in mycangia as they fly from tree to tree. When the female beetle burrows into weakened or newly cut trees to lay eggs, she introduces the fungus, which grows in the wood and sporulates in the tunnels. The larvae subsequently consume the slimy spores as their major food source, since they, like the ants and the termites, cannot digest wood. When adult beetles are ready to leave the tree in search of fresh substrate, they first rock back and forth to fill their mycangia with the slimy spores. Most of the mycobionts are yeasts or yeast-like (*Ascoidea, Dipodascus, Endomyces, Endomycopsis, Hansenula, Saccharomyces, Torula*), ascomycetous anamorphs (*Acremonium, Ambrosiella, Diplodia, Scopulariopsis*), or basidiomycetes or their anamorphs.

Woodwasps of the genus *Sirex* (Hymenoptera: Siricidae) penetrate wood with their long ovipositor and lay eggs in the xylem. Mycangia associated with the ovipositor ensure that some of the fungal spores they contain are deposited with the egg. The specific nature of the contributions made by the fungi is not understood, but wasps lacking fungi cannot breed. The relationship ensures that the fungi, members of the genera *Stereum* and *Amylostereum* (Holobasidiomycetes: Aphyllophorales), are dispersed and inoculated into new substrate.

Anobiid beetles (Coleoptera: Anobiidae) carry yeast-like fungi of the genera *Torula* and *Symbiotaphrina* in pouches called mycetomes, which are located at the beginning of the mid-gut. The fungi apparently provide some vitamins and essential amino acids to the beetles. Adult beetles smear their eggs with fungal material, and the larvae acquire the fungus by eating some of the shell. Beetles without endosymbionts do not develop normally. It is believed that many other beetles also have endosymbiotic yeasts (Buchner 1965).

We (Brundrett and Kendrick 1987) have discovered an apparently mutualisitic relationship between an agaric, *Boletinellus merulioides*, and an aphid, *Meliarhizophagus fraxinifolii* (Homoptera: Aphidae). While parasitizing the roots of *Fraxinus*, the aphid was found enclosed by hollow storage structures (sclerotia) of the fungus. It was suggested that in exchange for housing and protecting the aphid, the fungus received nutrients excreted by the aphid in its honeydew.

Fungi Endosymbiotic in Plant Tissue

Plant tissues, once assumed to be effectively sterile or axenic, are often permeated by fungi, which provoke no signs of disease or abnormality. The best-known examples are grasses, which frequently contain the mycelia of ascomycetes of the order Clavicipitales. These fungi produce toxic alkaloids that have direct effects on herbivores. Although it had long been known that consumption of the ergots of *Claviceps* could be lethal, fresh interest was aroused after cattle, sheep, horses, or deer that had eaten certain apparently healthy grasses were found to suffer from a neurotoxicosis ("staggers"). The grasses were found to contain mycelia of the clavicipitaceous genera *Atkinsonella*, *Balansia*, *Epichloë*, and *Myriogenospora*, or related *Acremonium* anamorphs. Hardy et al. (1986) found that the presence of such fungi in the food plants of the fall armyworm, *Spodoptera frugiperda* (Lepidoptera: Noctuidae), decreased larval growth rate and increased larval mortality. So here we have a probably long-standing and finely tuned mutualism in which the plant derives protection from herbivory while the fungus presumably obtains food. Evolution appears to have proceeded from parasitic organ-specific infections, as in *Claviceps*, to mutualistic systemic infections, as in *Balansia* and *Epichloë*.

When the teleomorphs of the Clavicipitaceae develop, they usually exact a price from their hosts: *Claviceps* supplants some of the ovaries, and the developing stroma of *Epichloë* effectively prevents the development of the plant's inflorescence. Clearly there is a threat here: if the fungus appropriates or short-circuits all the reproductive energy of the plant, it will be in danger of rendering its host extinct, at least locally. The persistence of the symbiosis indicates that such extremes have been avoided. *Cyperus virens*, the host of *Balansia cyperi*, has compensated for its sterility by evolving vivipary: its aborted inflorescences develop into vegetative plantlets, which disperse both plant and fungus. Some host species have reestablished their ability to reproduce sexually. In response, the mycobionts have lost the ability to produce their teleomorph; instead, their mycelia invade the host embryo and are dispersed with the seed—they have become totally integrated into the photobiont. It is thought that the presence of these endosymbiotic fungi has played an important part in establishing the primacy of grasses in many ecosystems (Clay 1986, 1988).

The highest concentrations of symbiotic fungi ever found in any plant have recently been isolated from a species of palm in Brazil. The fungi most commonly isolated are members of the Xylariaceae (Ascomycetes: Sphaeriales) (G. Samuels, personal communication). No mutualistic significance of this observation has yet been determined.

Many of the symbioses described above are undoubtedly ancient, but in view of the morphological, physiological, and genetic flexibility of the fungi it would be surprising if these versatile organisms were not negotiating new mutualistic symbioses at this moment.

References

Batra, L. R., ed. 1979. *Insect Fungus Symbiosis.* Allenheld, Osmun.

Bissett, J., and Borkent, A. 1988. Ambrosia galls: The significance of fungal nutrition in the evolution of the Cecidomyiidae (Diptera). In: Pirozynski, K. A., and Hawksworth, D. L., eds., *Coevolution of Fungi with Plants and Animals.* Academic Press.

Brundrett, M. C., and Kendrick, B. 1987. The relationship between the Ash Bolete (*Boletinellus merulioides*) and an aphid parasitic on ash tree roots. *Symbiosis* 3: 315–320.

Buchner, P. 1965. *Endosymbiosis of Animals with Plant Microorganisms.* Wiley.

Clay, K. 1986. Induced vivipary in the sedge *Cyperus virens* and the transmission of the fungus *Balansia cyperi* (Clavicipitaceae). *Can. J. Bot.* 64: 2984–2988.

Clay, K. 1988. Clavicipitaceous fungal endophytes of grasses: Coevolution and the change from parasitism to mutualism. In: Pirozynski, K. A., and Hawksworth, D. L., eds., *Coevolution of Fungi with Plants and Animals.* Academic Press.

Couch, J. N. 1938. *The Genus Septobasidium.* University of North Carolina Press.

Evans, H. C. 1988. Coevolution of entomogenous fungi and their insect hosts. In: Pirozynski, K. A., and Hawksworthy, D. L., eds., *Coevolution of Fungi with Plants and Animals.* Academic Press.

Friedmann, E. I. 1982. Endolithic microorganisms in the Antarctic cold desert. *Science* 215: 1045–1053.

Hardy, T., Clay, K., and Hammond, A. 1986. Leaf age and related factors affecting endophyte-mediated resistance to fall armyworm (Lepidoptera: Noctuidae) in tall fescue. *Environ. Entomol.* 15: 1083–1089.

Harley, J. L. 1989. The significance of mycorrhiza. *Mycol. Res.* 92: 129–139.

Hawksworth, D. L. 1988a. The variety of fungal-algal symbioses, their evolutionary significance, and the nature of lichens. *Bot. J. Linn. Soc.* 96: 3–20.

Hawksworth, D. L. 1988b. Coevolution of fungi with algae and cyanobacteria in lichen symbioses. In: Pirozynski, K. A., and Hawksworth, D. L., eds., *Coevolution of Fungi with Plants and Animals*. Academic Press.

Jeffrey, C. 1962. The origin and differentiation of archegoniate land plants. *Botaniska Notiser* 115: 446–454.

Kendrick, B. 1985. *The Fifth Kingdom*. Mycologue (Waterloo, Ontario).

Kendrick, B., and Berch, S. M. 1985. Mycorrhizae: Applications in agriculture and forestry. In: Robinson, C., ed., *Comprehensive Biotechnology*, volume 3. Pergamon.

Kohlmeyer, J., and Kohlmeyer, E. 1972. Is *Ascophyllum nodosum* lichenized? *Bot. Mar.* 15: 109–112.

Last, F. T., Dighton, J., and Mason, P. A. 1987. Successions of sheathing mycorrhizal fungi. *Trends in Ecol. and Evol.* 2: 157–161.

Lewis, D. H. 1986. Evolutionary aspects of mutualistic associations between fungi and photosynthetic organisms. In: Rayner, A. D. M., Brasier, C. M., and Moore, D., eds., *Evolutionary Biology of the Fungi*. Cambridge University Press.

Malloch, D. 1987. The evolution of mycorrhizae. *Can. J. Pl. Path.* 9: 398–402.

Malloch, D., Pirozynski, K. A., and Raven, P. H. 1980. Ecological and evolutionary significance of mycorrhizal symbioses in vascular plants. *Proc. Natl. Acad. Sci.* 77: 2113–2118.

Margulis, L., and Schwartz, K. V. 1988. *Five Kingdoms*, second edition. Freeman.

Mason, P. A., Wilson, J., and Last, F. T. 1984. Mycorrhizal fungi of *Betula* spp.: Factors affecting their occurrence. *Proc. Roy. Soc. Edinb.* 85B: 141–151.

Neger, F. W. 1909. Ambrosiapilze II. Die Ambrosia der Holzbohrkafer. *Ber. Deut. Bot. Ges.* 27: 372–389.

Pirozynski, K. A. 1983. Pacific mycogeography: An appraisal. *Aust. J. Bot. Suppl.* 10: 137–159.

Pirozynski, K. A., and Malloch, D. 1975. The origin of land plants: A matter of mycotrophism. *Biosystems* 6: 153–164.

Trappe, J. M. 1977. Selection of fungi for ectomycorrhizal inoculation in nurseries. *Ann. Rev. Phytopath.* 15: 203–222.

Trappe, J. M., and Fogel, R. D. 1977. Ecosystematic functions of mycorrhizae. In *The Below-Ground Ecosystem: A Synthesis of Plant-Associated Processes*. Range Sci. Dep. Ser. No. 26, Colorado State University, Fort Collins.

Weber, N. A. 1972. *Gardening Ants: The Attines*. Memoir 92, American Philosophical Society.

18 The Web of Life: Development over 3.8 Billion Years of Trophic Relationships

Peter W. Price

"Continuing sequences of assemblies are the veins of evolution," wrote Rubenstein (1989, p. 132), but the scale at which he developed this theme was orders of magnitude smaller than the scale on which I shall develop the same idea. I argue that whole organisms have assembled into symbiotic associations of increasing biotic complexity, providing macroevolutionary advances beyond those explicable by gradualism. A symbiosis may begin as a parasitism and evolve into a mutualistic association, becoming so obligate that the original units act almost as one organism. Communities of these units form the basis for major adaptive radiations on this planet.

There is no adequate gradualistic explanation for the wonders of biotic nature and the rapidity with which terrestrial life radiated to fill this earth. The questions go well beyond how the elephant got its trunk. How did large tree-like plants evolve so rapidly within perhaps 50 million years after the colonization of land by photosynthesizing organisms? How did large herbivores evolve? These are the questions I will address, albeit in a very crude manner.

The contribution this paper makes is not to symbiosis as a creative force, because this has been established (see, e.g., Margulis 1981). Rather, I emphasize that the origin of symbionts is likely to derive from parasite-host relationships and to evolve into mutualistic symbioses, providing the basis for extensive macroevolution and adaptive radiation. I will use the terms *mutualism* and *mutualistic symbiosis* interchangeably, although the latter is more accurate (Smith and Douglas 1987).

Mutualists From Parasites

The general argument for the acquisition of beneficial symbionts from parasites is as follows. When two species interact, there are many possible relationships: competition (– , –), to amensalism (○, –), predation and parasitism (+ , –), commensalism (+ , ○), and mutualism (+ , +) (figure 1) . Among all these relationships, the general tendency will be for natural selection to reduce negative effects and favor positive effects (Price 1984). Thus, parasitism may evolve toward neutralism, commensalism, and even mutualistic symbiosis. This does not mean, however, that all organisms ultimately should rejoice in an "orgy of mutual benefaction" (May 1981, p. 95); mutualistic associations are commonly antagonistic to a third species (Addicott 1981).

How, then, do these mutualistic associations develop? How are mutualists acquired? In my opinion there are two important routes for the acquisition of mutualists from parasitic organisms: a parasite of a host may evolve directly into a mutualist with that host, or a parasite of a host may enable another species to exploit that host by associating mutualistically with the parasite.

Because they are finely evolved for living intimately with other organisms, parasites have a predisposition, or preadaptation, for becoming symbiotic mutualists (Price 1980). Therefore, I would argue the point—not pressed by Margulis (1981) and others—that the evolution of the eukaryotes, the acquisition of organelles, and the evolution of biotic complexity have depended on the initial invasion of parasitic organisms, an invasion which became mutualistic secondarily. Of course, commensals could also evolve into mutualists (figure 1), although commensalism appears to be much more sporadic in nature than parasitism and thus provides a poorer source of relationships from which to develop novel associations.

An excellent model for this process has been developed by K. W. Jeon and co-workers (Jeon and Lorch 1967; Jeon 1972; Jeon and Jeon 1976; Jeon and Ahn 1978; Jeon 1987). A strain of *Amoeba proteus* became infected by a strain of parasitic bacteria in 1966. These parasitic bacteria originally had many deleterious effects (Jeon 1972; Jeon, this volume), but in only 200 generations (18 months) the *Amoeba* population had become largely dependent upon the bacterium. The mutualistic association had become established, and in 10 years 100 percent of cells formed clones in the presence of the bacterium. In

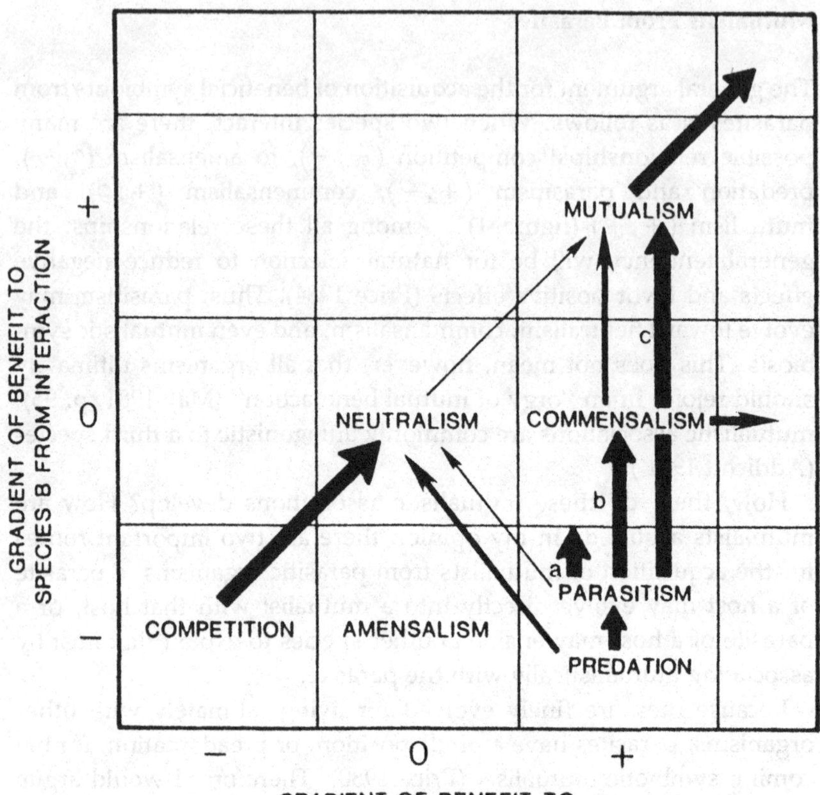

Figure 1
Predicted evolutionary pathways followed in two-species-interactive systems in response to natural selection for minimization of negative effects. Thick solid arrows indicate the most probable evolutionary routes, and thin arrows show other possible pathways. the possibilities for parasites to reduce impact on the host (a), to become commensals (b), and to become mutualists (c) are illustrated. From Price 1984.

evolutionary time, 10 years is almost instantaneous. This example illustrates the remarkably rapid shift from parasite to mutualist, a process which probably has been repeated millions of times in the evolution of life.

Atsatt (1988 and this volume) argued, as did Pirozynski and Malloch (1975), that land plants are "inside-out" lichens in which a semiaquatic green alga became the thallus and an aquatic fungus the endosymbiont essential in mineral nutrition for the new association. In Atsatt's words, "it began with fungal parasitism, may have evolved into a mutualism, and culminated in the acquisition of the fungal genome by the plant host" (1988, p. 17). What is more, the association between alga and fungus is proposed to be mediated by a virus-like element that makes possible a connection of cytoplasm between parasite and host across the cellulose cell wall. Atsatt's (1988, p. 21) argument for "a series of escalating parasitic events" in the evolution of land plants captures the essence of the theme I develop in more general terms here.

Another case of the shift from parasite to mutualist involves the development of mycorrhizal fungi, which are mutualists on the majority of terrestrial plants (Malloch et al. 1980).

Harley writes that "mycorrhizal fungi have evolved from aggressive parasites by progressive selection of nonlethal varieties so that the symbiotic stage becomes indefinitely prolonged" (1972, pp. 28–29). Some species of fungi are still mycorrhizal on some plant species and parasitic on others. Fungus and plant mutualisms also include endobiotic fungi which produce toxins effective against the plant's insect and mammalian herbivores; again these endobionts were derived from parasitic fungi (Carroll 1988; Clay 1988). Thus, the evolution of large plants is intimately associated with the utilization of parasitic species, ultimately as mutualists. And, of course, large animals frequently depend on large plants as a food source, or at least as a substrate to support the microcosms in their guts. The potential for parasites ultimately to benefit their hosts is obviously widespread.

The second important route to acquisition of a mutualist is for the parasite to mediate the interaction between its host and another antagonist of the host. This case is developed extensively in Price et al. 1986 and in Price et al. 1988.

One example is that of the blue-stain fungal pathogens in the genus *Ceratocystis*, which have long been parasites of coniferous trees. Many species of bark beetle in the genus *Dendroctonus* are mutualistic with these fungi (Whitney 1982). The fungi are vectored by mites, which are phoretic on the beetles (Bridges and Moser, 1983), to live trees; they invade the tissues of their host and suppress its resistance by resinosis. As a result, the beetles breed successfully in the cambium of the tree, and their progeny transmit the fungus again. In fact, *Dendroctonus* is the preeminent genus of parasites among bark beetles because they attack healthy trees, whereas most species of bark beetles attack only dead or dying trees. By association with the fungal parasite, *Dendroctonus* has very successfully moved from saprophyte to parasite.

Another case concerns ichneumonid wasps whose larvae are endoparasitic on other insects, although the origin of the mediating parasite is less clear. The larvae can survive in the host only in the presence of a virus transmitted by the female wasp and injected with the egg (Vinson et al. 1979; Edson et al. 1981). The virus suppresses the immune response of the host, so the wasp egg survives and is not encapsulated and killed by host hemocytes. It is very likely that the acquisition of this viral mutualist occurred early in the development of endoparasitism in these wasps and was significant in the adaptive radiation of the group, resulting in tens of thousands of species.

Thus, when we look at the intervention by parasites in feeding by organisms on such refractory material as plants, with their cellulose cell walls, lignified xylem, and tough bark, it becomes evident that food webs based on plants are strongly affected (figure 2). In many cases, metazoans, which do not possess cellulases, utilize cellulose in plants by associating with bacteria, protoctists, or fungi, which are able to synthesize cellulases. Although the microorganisms were probably parasitic first on the plant or on the herbivore, major radiations have resulted from this mutualistic association: about 2000 species of termites, over 10,000 species of wood-boring beetles, and about 200 species of Artiodactyla (deer, antelope, camels, etc.). The last group contains many of the large herbivores conspicuous on the grasslands of Africa, which rely heavily on archaeobacterial methanogens for decomposition of plant food.

Nematodes parasitic on beetles can survive only when bacteria are present which produce antibiotics preventing the decay of host tissue. One example of this type is provide by *Neoaplectana* and the bacterium *Xenorhabdus* (Maggenti 1981; Paul et al. 1981, Nealson, this volume).

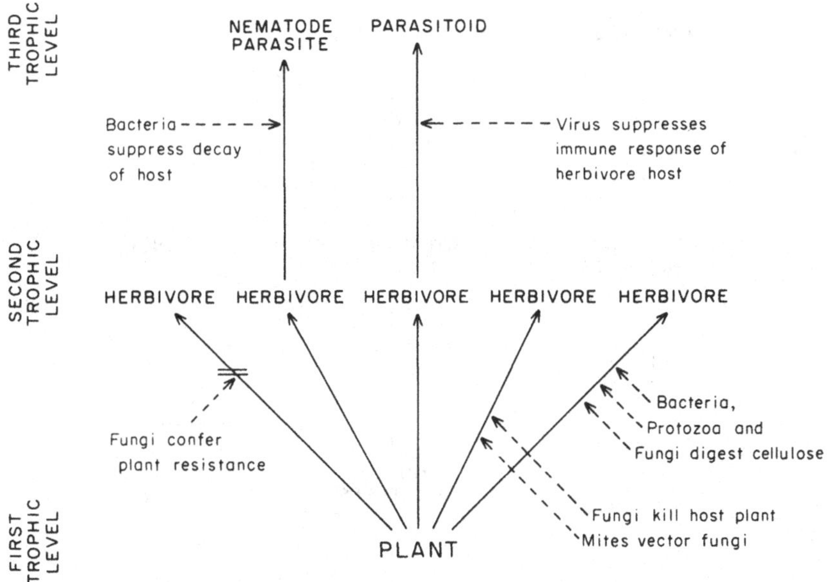

Figure 2
A summary of some important links between herbivores and plants, and parasites of herbivores and their hosts, that are facilitated by associated parasitic/mutualistic species. Specific cases given in the text illustrate the common microbial intervention between interacting macroscopic organisms.

I find the evidence overwhelming that parasites have provided important building blocks in the evolution of biotic complexes and still play vital roles in trophic relationships. The majority of feeding relationships between plants and herbivores and their parasites are probably mediated by symbionts, accounting for over half the species on Earth (table 1). Once herbivores and their associates accomplished the difficult task of converting plants into animal protoplasm, it was simple for large predators to evolve, probably without the need for symbionts to mediate the conversion of food.

Nesting in the Evolution of Life

Thus, when we gain a microscopic perspective of relationships between organisms, we obtain the strong impression that the evolution of the five kingdoms of organisms, as proposed by Whittaker (1969), has depended on a nesting of derived groups within the capabilities of the earlier groups (figure 3; see also Price 1988). Through many

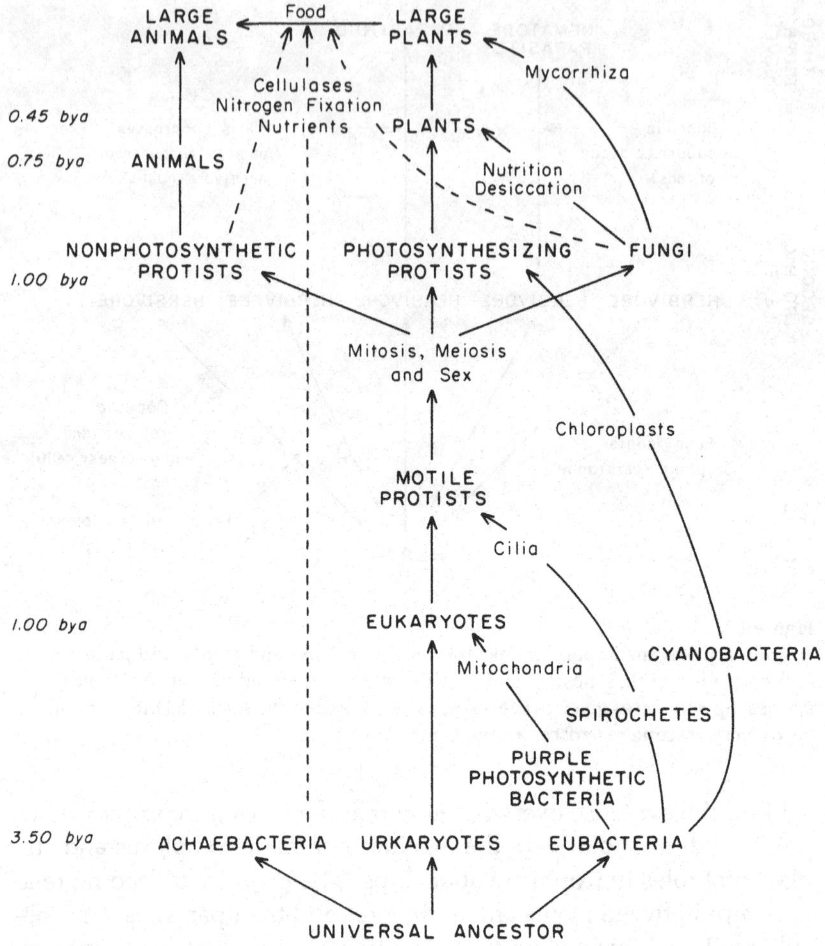

Figure 3
Major steps in the evolution of complex design of organisms and complex interactions resulting in webs of interacting organisms discussed in the text. Capitalized names are taxa of organisms or groups of taxa. Solid lines with arrows indicate links between evolutionary steps, and the contributions made by the taxa in the evolution of biotic complexity are given in capital and lower-case lettering. Dashed lines indicate the utilization of members of early kingdoms to enable the utilization of plants by large animals. For example, the methanogenic bacteia, important as mutualists in the ruminant gut, are members of the Archaeobacteria. This figure is simplified purely to illustrate some of the linkages in the evolution of biotic complexity, and is based on references cited in the text.

Table 1
Some examples of major adaptive radiations probably involving parasites evolving into mutualisms, with estimates on the extent of radiation. Note that the land plants, the invertebrate herbivores, and the parasitic wasps add up to 54 percent of the Earth's species.

	Extent of radiation*			
Radiating macroorganisms	Associated microorganisms	Number of species	Percent of species on earth	Source
Land plants	Aquatic algae and fungi	300,000	21	Pirozynski and Malloch 1975 Atsatt 1988
Land plants	Land plants and fungi as mycorrhiza	280,000	20	Malloch et al. 1980; Harley 1972
Invertebrate herbivores	Insects and microorganisms	360,000	26	Buchner 1965; Martin 1987; Price 1984; Whitcomb and Hackett 1989
Parasitic wasps	Endoparasitic wasps and viruses	100,000	7	Vinson et al. 1979; Edson et al. 1981
Vertebrate herbivores	Artiodactyla and bacteria	200	<<1	Hungate 1975

*These estimates are based on Southwood 1978, Price 1980, and Strong et al. 1984, with some modification, and must be regarded as first approximations.

instances of parasitism resulting in mutualism, the complexity of life has increased. The niche space occupied by derived groups depends heavily upon the broader capabilities of the older kingdoms, which allow broader niche occupation (table 1).

The present ecology of many species essentially recapitulates the phylogeny of life on earth (Price 1988). The capabilities of derived groups depend on associations with the earlier taxa. Close associations have frequently passed from parasitic to mutualistic because the parasites are adapted for living in close association with the host.

With a clear perspective of time, we see that microbial life has had billions of years (perhaps 3.8 billion years; see Woese 1984) to radiate into every conceivable quarter of the Earth, to solve every conceivable biochemical problem, and to surmount every obstacle of biotic interaction. By associating with these masters of intrigue and cunning, larger forms of life have thrived, and close association is the speciality of parasites.

What we see in a Noah's Ark perspective of biotic diversity—large plants and large animals—is actually large microcosms, many of them assembled through parasitic association. Parasites and their hosts have formed many of the building blocks for macroscopic life on Earth.

Acknowledgment

Partially supported by grant BSR-8705302 from the National Science Foundation.

References

Addicott, J. F. 1981. Stability properties of two-species models of mutualism: Simulation studies. *Oecologia* 49: 42–49.

Atsatt, P. R. 1988. Are vascular plants "inside-out" lichens? *Ecology* 69: 17–23.

Bridges, J. R., and Moser, J. C. 1983. Role of two phoretic mites in transmission of bluestain fungus, *Ceratocystis minor*. *Ecological Entomology* 8: 9–12.

Buchner, P. 1965. *Endosymbiosis of Animals with Plant Microorganisms*. Wiley.

Carroll, G. 1988. Fungal endophytes in stems and leaves: From latent pathogen to mutualistic symbiont. *Ecology* 69: 2–9.

Clay, K. 1988. Fungal endophytes of grasses: A defensive mutualism between plants and fungi. *Ecology* 69: 10–16.

Edson, K. M., Vinson, S. B., Stoltz, D.B., and Summers, M. D. 1981. Virus in a parasitoid wasp: Suppression of the cellular immune response in a parasitoid's host. *Science* 211: 582–583.

Harley, J. L. 1972. *The Biology of Mycorrhiza,* third edition. Hill, London.

Hungate, R. E. 1975. The rumen microbial ecosystem. *Annual Review of Ecology and Systematics* 6: 39–66.

Jeon, K. W. 1972. Development of cellular dependence on infective organisms: micrurgical studies in amoebas. *Science* 176: 1122–1123.

Jeon, K. W. 1987. Change of cellular "pathogens" into required cell components. In: Lee, J. J., and Fredrick, J. F., eds., *Endocytobiology III*. New York Academy of Sciences.

Jeon, K. W., and Ahn, T. I. 1978. Temperature sensitivity: A cell character determined by obligate endosymbionts in amoebas. *Science* 202: 635–637.

Jeon, K. W., and Jeon, M. S. 1976. Endosymbiosis in amoebae: Recently established endosymbionts have become required cytoplasmic components. *Journal of Cell Physiology* 89: 337–344.

Jeon, K. W., and Lorch, I. J. 1967. Unusual intra-cellular bacterial infection in large, free-living amoebae. *Experimental Cell Research* 48: 236–240.

Maggenti, A. 1981. *General Nematology*. Springer.

Malloch, D. W., Pirozynski, K. A., and Raven, P. H. 1980. Ecological and evolutionary significance of mycorrhizal symbioses on vascular plants (a review). *Proceedings of the National Academy of Sciences* 77: 2113–2118.

Margulis, L. 1981. *Symbiosis in Cell Evolution: Life and Its Environment on the Early Earth*. Freeman.

Martin, M. M. 1987. *Invertebrate-Microbial Interactions: Ingested Fungal Enzymes in Arthropod Biology*. Cornell University Press.

May, R. M. 1981. Models for two interacting populations. In: May, R.M., ed., *Theoretical Ecology: Principles and Applications,* second edition. Sinauer.

Paul, V. J., Frautschy, S., Fenical, W., and Nealson, K. H. 1981. Antibiotics in microbial ecology: Isolation and structure assignment of several new antibacterial compounds from the insect-symbiotic bacteria *Xenorhabdus* spp. *Journal of Chemical Ecology* 7: 589–597.

Pirozynski, K. A., and Malloch, D. W. 1975. The origin of land plants: A matter of mycotrophism. *BioSystems* 6: 153–164.

Price, P. W. 1980. *Evolutionary Biology of Parasites*. Princeton University Press.

Price, P. W. 1984. *Insect Ecology,* second edition. Wiley.

Price, P. W. 1988. An overview of organismal interactions in ecosystems in evolutionary and ecological time. *Agriculture Ecosystems and Environment* 24: 369–377.

Price, P. W., Westoby, M., Rice, B., Asatt, P. R., Fritz, R. S., Thompson, J. N., and Mobley, K. 1986. Parasite mediation in ecological interactions. *Annual Review of Ecology and Systematics* 17: 487–505.

Price, P. W., M. Westoby, and B. Rice. 1988. Parasite-mediated competition: Some predictions and tests. *American Naturalist* 131: 544–555.

Rubenstein, E. 1989. Stages of evolution and their messengers. *Scientific American* 260(6): 132.

Smith, D. C., and Douglas, A. E. 1987. *The Biology of Symbiosis*. Edward Arnold, London.

Southwood, T. R. E. 1978. The components of diversity. In: Mound, L. A., and Waloff, N., eds., *Diversity of Insect Faunas*. Royal Entomological Society, London.

Strong, D. R., Lawton, J. H., and, Southwood, R. 1984. *Insects on Plants: Community Patterns and Mechanisms*. Harvard University Press.

Vinson, S. B., Edson, K. M., and Stoltz, D. B. 1979. Effect of a virus associated with the reproductive system fo the parasitoid wasp. *Campoletis sonorensis*, on host weight gain. *Journal of Invertebrate Pathology* 34: 133–137.

Whitcomb, R. F., and Hackett, K. J. 1989. Why are there so many species of mollicutes? An essay on prokaryote diversity. In: Knutson, L., and Stoner, A. K., eds., *Biotic Diversity and Germplasm Preservation, Global Imperatives*. Kluwer.

Whitney, H. S. 1982. Relationships between bark beetles and symbiotic organisms. In: Mitton, J. B., and Sturgeon, K. B., eds., *Bark Beetles in North American Conifers*. University of Texas Press.

Whittaker, R. H. 1969. New concepts on kingdoms of organisms. *Science* 163: 150–160.

Woese, C. R. 1984. *The Origin of Life*. Carolina Biological Supply, Burlington.

19

Bacteria and Bacteria-like Objects in Endomycorrhizal Fungi (Glomaceae)

Silvano Scannerini and Paola Bonfante-Fasolo

As Burnett (1987) has pointed out, "the limits of Fungi are not yet adequately defined, the origin and the affinities of the group conventionally classified as fungi are not known and their characteristic modes of speciation remain to be adequately assessed." Nonetheless, it seems plausible to suggest that fungi, on the strength of their nutritional modes, biological cycles, and genetics, can be described as a kingdom (Margulis and Schwartz 1988; Webster 1988) embracing the zygo-, asco-, and basidiomycetes (Margulis and Schwartz 1988; Lewis 1987; Cavalier-Smith 1987).

The question of the origin and evolution of the fungi is far from being solved. Despite the considerable progress toward an understanding of the general biological features of the more ancient (zygomycetes) and more recent (asco- and basidiomycetes) fungi, no agreement has been reached on the question of mono- or polyphyletic descent, though the critical revision of Cavalier-Smith (1988) indicates that the former is more probable within the limits of the classification proposed above.

Mutualistic and pathogenetic symbiotic phenomena are common in the fungal kingdom (including the presence of plasmids and cryptic viruses; see Esser and Meinhardt 1986). Evidence suggests the primary importance of mycotrophism in the evolutionary innovation of photosynthetic organisms (Pirozynski and Malloch 1975; Pirozynski 1981; Lewis 1987; Lewis, this volume; Scannerini and Bonfante-Fasolo 1990). However, no demonstration has yet been made of the innovative effects of mutualistic symbioses on fungi. The aim of the present paper is therefore to discuss the possible role of vesicular-arbuscular mycorrhizae fungal evolution, since together with lichens they represent the most widespread form of symbiosis involving fungi.

Lichens and Mycorrhizae as Source of Novelty in Fungi?

Lichens are often considered the product of innovative events resulting from a symbiotic way of life (Honneger, this volume): they act as unique, integrated systems (Lewis 1987) in which the mycobiont controls the phycobiont population. The association induces the production of characteristic new metabolites in the fungus. Some ascolichens reproduce asexually by symbiotic diaspores. Thus lichens can be regarded as "new organisms" evolved from long-term symbiotic associations. Lichens, traditionally, are classified as whole organisms according to Linnaean rules. This information suggests lichens resulted from innovative processes involving the basidiomycetes and ascomycetes, but they cannot be easily seen as responsible for further evolutionary innovation within fungi. For example, the new metabolic pathways expressed by the lichen-forming fungus during the symbiosis are not expressed when the fungus is in pure culture (Margulis and Schwartz 1988). Moreover, at the cellular level, special features resulting in a differentiated cortex with a water-repellent surface and the differentiation of three-dimensional fungal tissues are quite similar to those found not only in ectomycorrhizae (Scannerini and Bonfante-Fasolo 1983) but also in carpophores of saprotrophic asco- and basidiomycetes (McLaughlin 1982); this indicates that these features are not exclusive to these mutualistic associations.

Vesicular-arbuscular mycorrhizae (VAM) have never been regarded as a source of innovation for the fungal kingdom, despite the amount of literature on their importance in plant evolution (Scannerini and Bonfante-Fasolo 1990). On the contrary, they would deserve closer attention, since the fungal partner belongs to a restricted group of Zygomycetes—the Glomaceae—according to Pirozynski and Dalpé (1989).

Some common features can be recognized in the Glomaceae:

• They are obligate symbionts and agents of VAM in Angiospermophyta, Coniferophyta, and Ginkgophyta and probably of mycothalli in the gametophytes and sporophytes of Filicinophyta and Bryophyta (Bonfante-Fasolo 1984).

• Their spores germinate *in vitro,* but do not proceed beyond this stage in the absence of a host root (Burggraaf and Beringer 1989).

• They display a complex, host-regulated morphogenesis (Bonfante-Fasolo 1984, 1987; Scannerini 1985a,b) with a differentiation into spores, extraradical mycelia, and both inter- and intracellular intraradical mycelia.

From an evolutionary standpoint, there are at least three reasons why the Glomaceae are interesting:

• They are among the most ancient fungi, as is shown by fossil evidence (Pirozynski and Malloch 1975), and they colonized both the stems and the roots of fossil plants (Pirozynski and Dalpé 1989). Fungi may be derived monophyletically from the Glomaceae (Pirozynski and Dalpé 1989).

• Responsible for the present vesicular-arbuscular mycorrhizae, they are among the most widespread fungi (Lewis 1987).

• Vesicular-arbuscular fungi display endocytobionts similar to those of other evolutionarily significant biological systems (Jeon 1983 and this volume; Nardon 1988 and this volume; Tiivel 1987) in which endocytobionts are responsible for the survival or the improved fitness of their hosts.

To elucidate the role of VAM fungi on the evolution of fungi, closer attention should be paid to evidence for endocytobiosis in the Glomaceae.

The Glomaceae and Their Endocytobionts

The presence of endocytobionts in VAM fungi was casually discovered during ultrastructural investigation of the development of the infection unit in mycorrhizae and mycothalli. Although some papers deal exclusively with the endocytobionts (Protsenko 1975; Macdonald and Chandler 1981; Macdonald et al. 1982; Macdonald 1983), the available information remains scanty.

Conventional techniques of transmission electron microscopy (TEM) have shown that a relatively large number of VAM species and strains contain two types of endocytobionts (table 1):

• rod-like bacteria with cell walls, a nucleoid, dense inclusions, and ribosomes. These bacteria are enclosed within the vacuoles of normal hyphae (figure 1) of *Glomus fasciculatum* (Bonfante-Fasolo 1987), *Glomus caledonium* (Macdonald and Chandler 1981), and an unidentified VAM fungus (Scannerini et al. 1975).

Table 1
Vesicular arbuscular fungi (VA) harboring bacteria and/or bacteria-like organelles (BLOs).

VA fungus	Endocytobiont	Fungal Phase	References
Acaulospora laevis	BLOs	spore	Mosse 1970
Gerdemann & Trappe	BLOs	spore	MacDonald et al. 1982
Gigaspora heterogama (Nicol. & Gerd.) Gerdemann & Trappe	BLOs	spore	MacDonald et al. 1982
Gigaspora margarita	BLOs	spore	MacDonald et al. 1982
Becker et al.		*in vitro* germinated spore	Bonfante-Fasolo (unpublished results)
Glomus caledonium	Bacteria	spore	MacDonald and Chandler 1981
(Nicol. & Gerd.)	BLOs	spore, intramatrical mycelia	MacDonald and Chandler 1981
Nicolson & Gerdemann	BLOs	intramatrical mycelia, arbuscules	Bonfante-Fasolo and Fontana 1985
	BLOs		
Glomus fasciculatum Gerdemann & Trappe (Trappe sensu Gerdemann)	BLOs	intraradical mycelia, arbuscules	Bonfante-Fasolo and Scannerini 1977 Bonfante-Fasolo and Fontana 1985 Bonfante-Fasolo 1984
Glomus macrocarpum Tul. & Tul.	BLOs	spore, extramatrical mycelia, intraradical mycelia, arbuscules	Bonfante-Fasolo and Fontana 1985
Glomus mosseae	BLOs	intraradical mycelia, arbuscules,	MacDonald et al. 1982
(Nicol. & Gerd.)		spore	Bonfante-Fasolo (unpublished results)
Gerdemann & Trappe		*in vitro* germinated spore	
Glomus versiforme (Kartsen Berch) =	BLOs	extramatrical and intramatrical mycelia, arbuscules	Bonfante-Fasolo and Fontana 1985
Glomus epigaeum Daniels & Trappe		spore	Bonfante-Fasolo (unpublished results)
Glomus sp strain E3	BLOs	*in vitro* germinated spore	Bonfante-Fasolo (unpublished results)
"white reticulate fungus"	BLOs	spore	MacDonald et al. 1982
Unidentified endomycorrhiza	Bacteria	intraradical mycelia, arbuscules	Scannerini et al. 1975
	Bacteria	intercellular hyphae	Bonfante-Fasolo 1978
	BLOs	arbuscules	Protsenko 1975
	BLOs	arbuscule-like structure	Ligrone 1988
	BLOs	arbuscule-like structure	Ligrone and Lopes 1989

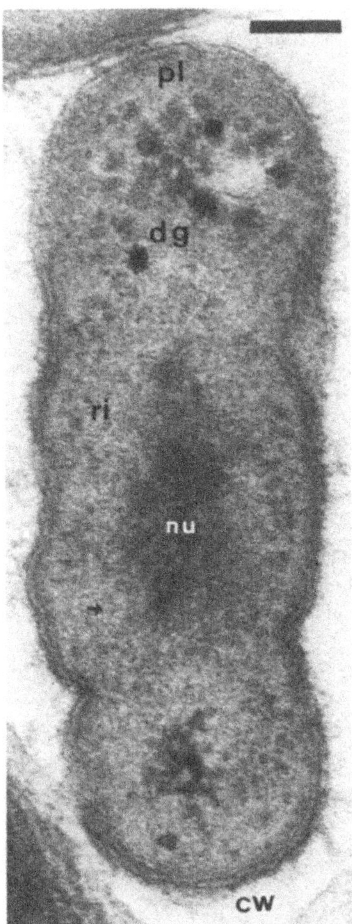

Figure 1
Intravacuolar bacterium in the intercellular mycelium of an unidentified fungus colonizing *Vitis vinifera* roots. The cell wall (cw) shows loose fibrillar material. Electron-dense granules (dg) are found in the protoplasm. The central electron-dense area could correspond to the nucleoid (nu). pl: plasma membrane.

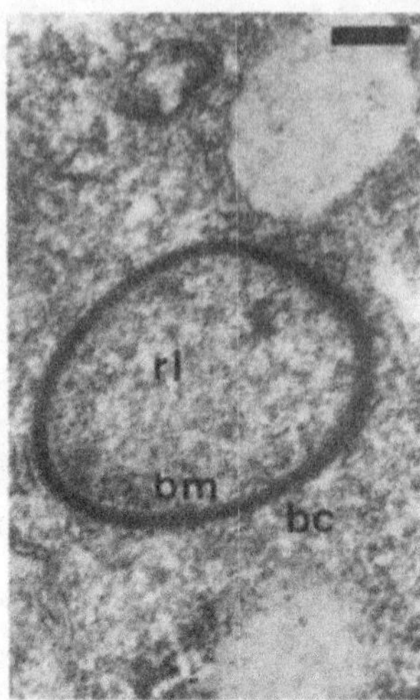

Figure 2
Bacteria-like organelle in the arbuscule of *Glomus fasciculatum* colonizing *Vitis vinifera* roots. The BLO cell coat is electron dense (bc). Only scattered ribosome-like granules (rl) can be observed in the protoplasm. bm: unit membrane.

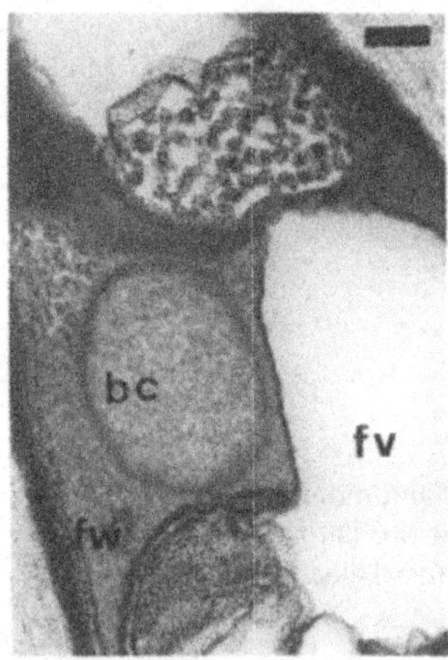

Figure 3
Bacteria-like organelle in the arbuscule of *Glomus fasciculatum* colonizing *Vitis vinifera* roots. The section was subjected to the PATAg reaction for polysaccharide localization. The BLO cell coat (bc) is weakly reactive to the silver reaction; the fungal cell wall (fw) is strongly labeled. fv: fungal vacuole.

- bacteria-like organelles (BLOs), or organisms lacking walls. These are found in the fungal cytoplasm (figures 2–7).[1]

The plants that host these mycorrhizal fungi are by no means similar taxonomically. They can be found among Anthocerophyta, Hepatophyta, Pteridophyta, Ginkgophyta, Dicots, and Monocots (table 2).

All BLOs are very similar in structure. More or less coccoid, they measure less than 1 μm in diameter and are bounded by both an electron-dense coat and a unit membrane. They contain ribosome-like granules scattered inside an amorphous protoplasm, in which electron-dense areas are found by following the usual TEM preparation treatments. Their morphology recalls that of bacteria (figures 2–7). Post-embedding reactions with wheat germ agglutinin (WGA), the lectin specific for N-acetylglucosamine residues, demonstrated a strong gold labeling on the chitinic cell walls of *Glomus versiforme* (Bonfante-Fasolo et al. 1989), while the cell coat of the BLOs living inside the fungus was not labeled (unpublished results). This result, demonstrating significant differences between the fungal wall and the BLO coat, suggests that the N-acetylglucosamine residues, which might occur in the peptide glycans of a bacterial cell coat, are not present in BLOs or do not react with lectin.

The position of BLOs inside the hyphae appears to be unspecific: although Macdonald and Chandler (1981) maintain that their preferential location is alongside the nucleus, BLOs may occupy a position closer to mitochondria, fat deposits, glycogen, vacuoles, and fungal walls (figures 2–7). They have been observed in nature, in mycorrhizae obtained by synthesis, and in mycelia obtained by germinating spores. All phases of the life cycle of BLO-containing Glomaceae, excluding those of arbuscle degeneration, contain endocytobionts.

The functional significance of both these types of endocytobionts remains unknown. No BLOs or bacteria living in vacuoles have been cultured; Smith (1979) has suggested that microorganisms not bounded by host membranes are so thoroughly integrated into the host that they probably cannot be grown *in vitro*.

In view of the physiological differences in bacteria, ultrastructural evidence alone does not suffice to clarify the intracellular bacterial species. Their presence in extensively vacuolized hyphae containing clumps of myelinic membranes does not exclude the possibility that they may be hyperparasites. Lastly, the scanty documentation available refers neither to degenerative forms of these bacteria nor to host

Table 2
Plants with VA mycorrhizae harboring bacteria and/or bacteria-like organelles (BLOs).

Plants	Fungi	References
Anthocerophyta		
Phaeoceros laevis (L.) Prosk	unidentified	Ligrone 1988
Hepatophyta		
Conocephalum conicum (L.) Dum	unidentified	Ligrone and Lopes 1989
Pteridophyta		
Lastrea phegopteris (L.) Bory	unidentified	Bonfante-Fasolo and Zappi 1985
Ginkgophyta		
Ginkgo biloba L.	*Glomus fasciculatum*	Bonfante-Fasolo and Fontana 1985
	Glomus macrocarpum	
	Glomus caledonium	
Angiospermophyta		
Pisum sativum L.	unidentified	Protsenko 1975
Vitis vinifera L.	unidentified	Bonfante-Fasolo 1978
	Glomus fasciculatum	Bonfante-Fasolo 1984
Trifolium parviflorum Ehrk.	*Glomus caledonium*	MacDonald and Chandler 1981
Allium cepa L.	*Glomus caledonium*	MacDonald et al. 1982
Allium porrum L.	*Glomus versiforme*	Bonfante-Fasolo (unpublished results)
Ornithogalum umbellatum L.	unidentified	Scannerini et al. 1975
	Glomus fasciculatum	Bonfante-Fasolo and Scannerini 1977

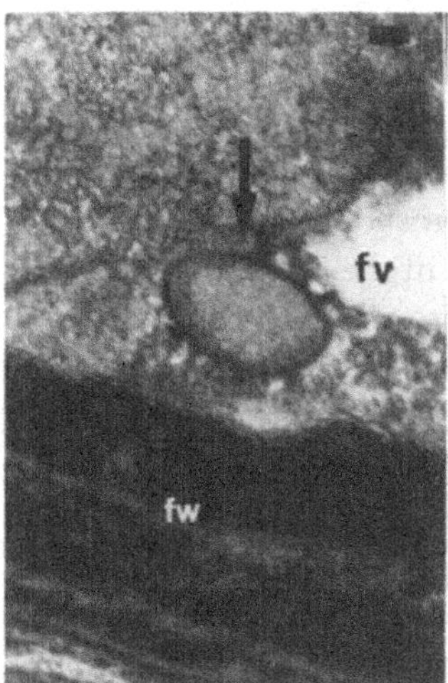

Figure 4
Bacteria-like organelle (arrow) in the young spore of *Gigaspora magarita*. fw: wall of the spore.

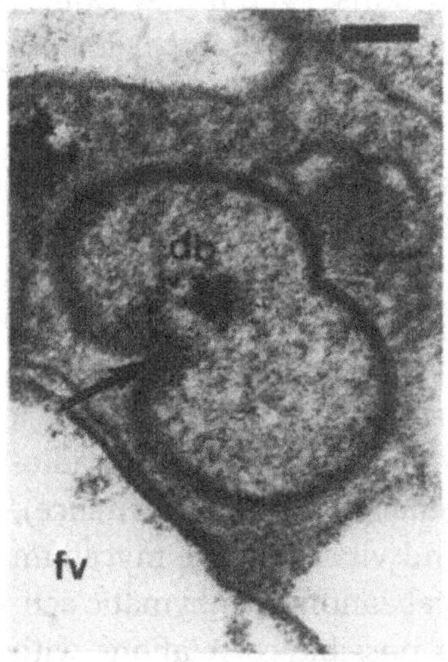

Figure 5
Bacteria-like organelle in the intercellular mycelium of an unidentified VA fungus colonizing *Lastrea phegopteris* roots. The BLO seems to show an asymmmetrical initial division (arrow). db: electron-dense area.

mechanisms controlling the number of endocytobionts. This lends strength to the supposition of casual superinfection.

BLO structures are comparable with that of cytoplasmic xenosomes (Soldo 1983) and of the type III endocytobionts transferred via the ovaries in *Sitophilus* (Nardon 1988). Their presence in all stages of the life cycle of their host strains, and above all in the mycelia of axenically germinated spores, indicates that they are a stable, heritable component of the fungus. As Macdonald (1983) has pointed out, interpretation of these structures is critical. Indeed, it cannot be said for certain that they are bacterial until they are cultured *in vitro*. Experimental data concerning their possible sensitivity to antibiotics or culture conditions promoting or inhibiting their development have not yet been gathered, since the mycelium containing them cannot be grown *in vitro*. The information available with respect to BLOs, therefore, remains purely descriptive.

Possibilities and Prospects Offered by a Study of the BLOs

Mycelia have been obtained in axenic conditions, even if they cannot be subcultured (Becard and Piché 1989). Extensive improvements have been made in ultrastructural cytochemistry and cytofluorimetry (Bonfante-Fasolo and Perotto 1988; Berta et al. 1990a). The presence of BLOs in aseptically germinated spores makes it possible to work with growth factors, hormones and antibiotics under controlled experimental conditions, and highly specific cytochemical reactions may be used to differentiate components of BLOs.

BLOs can thus be screened with enough precision to elucidate their significance as bacteria-like organisms or cytoplasmic organelles (by a direct search of their DNA with EM cytochemical and autoradiographic techniques and the factors influencing their maintenance), their possible role in spore germination and vitality of the mycelium (by using vital dyes and probes for the localization of enzymatic activities; see Schubert et al. 1987), and their possible correlations with the infectivity and efficiency of VAM fungi (by utilizing mathematical techniques that allow the evaluation of growth and morphogenesis of mycorrhizal plants as a function of infection with fungal spores; see Berta et al. 1990b).

Studies of BLOs could provide new information on the meaning of endocytobiosis in VAMs, and on their role in the establishment of the mycorrhizal symbiosis (which is surely the greatest novelty in

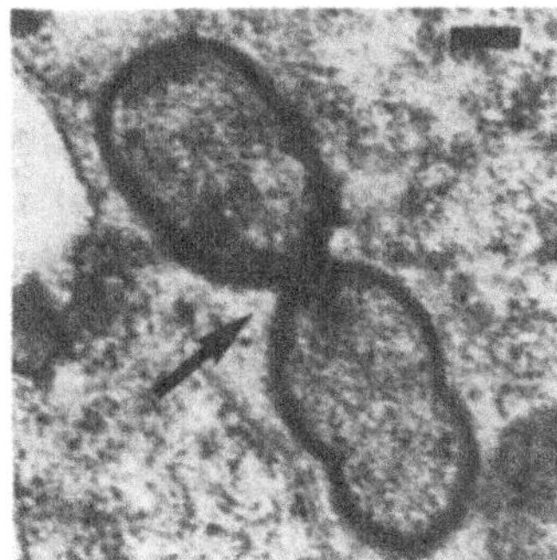

Figure 6
Bacteria-like organelles in the arbuscule of *Glomus fasciculatum* colonizing *Ginkgo biloba* roots. The picture probably shows the end of a division. The two BLOs seems to be still attached to each other via a small piece of coat (arrow).

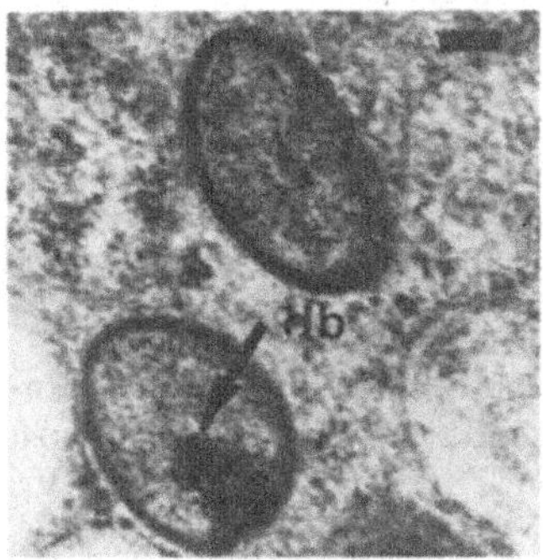

Figure 7
Bacteria-like organelles in the arbuscule of *Glomus fasciculatum* colonizing *Ginkgo biloba* roots. The BLOs are adjacent. They show an electron-dense cell coat, and electron-dense areas (db) similar to the bacterial nucleoids (cf. figure 1).

zygomycetes, and which is correlated with the evolutionary success of Glomaceae). However, research on BLOs requires the development of isolation and *in vitro* cultivation techniques; only when this goal is reached will we be able to reliably assess the hypothesis that a fungal ancestor acquired endocytobionts subsequently passed on to Glomaceae zygomycetes. Endocytobionts such as BLOs in the Glomaceae represent an important opportunity, which has been so far neglected, for investigation of symbioses capable of generating novelty in fungi.

Note

1. The figures in this chapter are transmission electron micrographs of bacteria and bacteria-like organelles in vesicular-arbuscular fungi. All the samples were fixed in glutaraldehyde-OsO_4, and counterstained with uranyl acetate–Pb citrate, with the exception of figure 3. The bar corresponds to 0.1 μm.

References

Becard, G., and Piché, Y. 1989. New aspects on the acquisition of biotrophic status by a vesicular-arbuscular mycorrhizal fungus *Gigaspora margarita. New Phytol*. 112: 77–83.

Berta, G., Sgorbati, S., Trotta, A., Fusconi, A., and Scannerini, S. 1990a. Correlations between host-endophyte interactions and structural changes in host cell chromatin in a VA mycorrhiza. In: Nardon, P., Gianinazzi-Pearson, V., Grenier, A. M., Margulis, L., and Smith, D. C., eds., *Endocytobiology IV*. INRA, Paris.

Berta, G., Fusconi, A., Trotta, A., and Scannerini, S. 1990b. Morphogenetic modifications induced by the mycorrhizal fungus *Glomus* E_3 on the root system of *Allium porrum. New Phytol*. 114: 207–215.

Bonfante-Fasolo, P. 1978. Some ultrastructural features of vesicular-arbuscular mycerrhiza in the grapevine. *Vitis*. 1: 386–395.

Bonfante-Fasolo, P. 1984. Anatomy and morphology of VA Mycorrhizae. In: Powell, C. L., and Bagiaraj, D. J., eds., *VA Mycorrhiza*. CRC Press.

Bonfante-Fasolo P. 1987. Vesicular-arbuscular mycorrhizae: Fungus plant interactions at the cellular level. *Symbiosis* 3: 249–268.

Bonfante-Fasolo, P., and Fontana, A. 1985. VAM Fungi in *Ginkgo biloba* roots. Their interactions at cellular level. *Symbiosis*. 1: 53–67.

Bonfante-Fasolo, P., and Perotto, S. 1988. Ericoid mycorrhiza: New insights from ultrastructure allied to cytochemistry. In: Ghiara, G., ed., *Cell Interactions and Differentiation*. University of Naples.

Bonfante-Fasolo, P., and Scannerini, S. 1977. A cytological study of the vesicular-arbuscular mycorrhize in *Ornithogalum umbellarom* L. *Allionia* 22: 5–21.

Bonfante-Fasolo, P., and Zappi, C. 1985. Interazioni cellular tra funghi VAM e pteridofite in radici infertate naturalmente. *Giorn. Bot. Ital.* 119 (suppl. 2): 93–94.

Bonfante-Fasolo, P., Faccio, A., Perotto, S., and Schubert, A. 1989. Correlation between chitin distribution and cell wall morphology in the mycorrhizal fungus *Glomus versiforme. Mycological Research* 94: 157–165.

Burggraaf, A. J. P., and Beringer, J. E. 1989. Absence of nuclear DNA synthesis in vesicular-arbuscular mycorrhizal fungi during *in vitro* development. *New Phytol.* 111: 25–33.

Burnett, J. H. 1987. Aspects of the macro- and micro-evolution of the Fungi. In: Rayner, A. D. M., Brasier, C. M., and Moore, D., eds., *Evolutionary Biology of the Fungi.* Cambridge University Press.

Cavalier-Smith, T. 1987. The origin of fungi and pseudofungi. In: Rayner, A. D. M., Brasier, C. M., and Moore, D. H., eds., *Evolutionary Biology of Fungi.* Cambridge University Press.

Cavalier-Smith, T. 1988. Eukaryote cell evolution. In: Greuter, W., and Zimmer, B., eds., *Proceedings of XIV International Botanical Congress.* Koeltz, Königstein.

Esser, K., and Meinhardt, F. 1986. Ectomycorrhizal fungi: State of the art, application and perspectives for research under consideration of molecular biology. *Symbiosis* 2: 125–138.

Jeon, K. W. 1983. Integration of bacterial endosymbionts in amoebae. *International Review of Cytology (Intracellular Symbiosis)* Suppl. 14: 29–47.

Lewis, D. H. 1987. Evolutionary aspects of mutualistic associations between fungi and photosynthetic organisms. In: Rayner, A. D. M., Brasier, C. M., and Moore, D. H., eds., *Evolutionary Biology of Fungi.* Cambridge University Press.

Ligrone, R. 1988. Ultrastructure of a fungal endophyte in *Phaeoceros laevis* (L.) Prosk (Anthocerophyta). *Botanical Gazette* 149: 92–100.

Ligrone, R., and Lopes, C. 1989. Cytology and development of a mycorrhiza-like infection in the gametophyte of *Conocephalum conicum* (L.) Dum. (Marchantiales, Hepatophyta). *New Phytol.* 111: 423–433.

Macdonald, R. M. 1983. Bacterium-like organelles in VA mycorrhizal fungi. In: Schenk, H. E. A., and Schwemmler, W., eds., *Endocytobiology II: Intracellular Space as Oligogenetic System.* Walter de Gruyter.

Macdonald, R. M., and Chandler, M. R. 1981. Bacterium-like organelles in the vesicular-arbuscular mycorrhizal fungus *Glomus caledonius*. *New Phytol*. 89: 241–246.

Macdonald, R. M., Chandler, M. R., and Mosse, B. 1982. The occurence of bacterium-like organelles in vesicular-arbuscular mycorrhizal fungi. *New Phytol*. 90: 659–663.

Margulis, L., and Schwartz, K. W. 1988. *Five Kingdoms: An Illustrated Guide to the Phyla of Life on Earth*, second edition. Freeman.

McLaughlin, D. 1982. Ultrastructure and cytochemistry of basidia and basidiospore development. In: Wells, K., and Wells, E. K., eds., *Basidium and Basidiocarp*. Springer-Verlag.

Mosse, B. 1970. Honey coloured, sessile *Endogone* spores. Changes in fine structure during spore development. *Arch. Mikrobiol*. 74: 129–145.

Nardon, P. 1988. Cell-to-cell interactions in insect endocytobiosis. In: Scannerini, S., Smith, D. C., Bonfante-Fasolo, P., and Gianinazzi-Pearson, V., eds., *Cell-to-Cell Signals in Plant, Animal and Microbial Symbiosis*. Springer-Verlag.

Pirozynski, K. A. 1981. Interactions between fungi and plants through the ages. *Can. J. Bot*. 59: 1824–1827.

Pirozynski, K. A., and Dalpé, Y. 1989. Geological history of the Glomaceae with particular reference to mycorrhizal symbiosis. *Symbiosis* 7: 1–36.

Pirozynski, K A., and Malloch, D. W. 1975. The origin of land plants: A matter of mycotrophism. *BioSystems* 6: 153–164.

Protsenko, M. A. 1975. Microorganisms in the hyphae of mycorrhiza forming fungus. *Microbiologiya* 44: 1121–1124.

Scannerini, S. 1985a. Mycorrhizal Symbiosis. 1. The structures. *Riv. Biol*. 78: 430–439.

Scannerini, S. 1985b. Mycorrhizal Symbiosis. 2. The process. *Riv. Biol*. 78: 546–553.

Scannerini, S., and Bonfante-Fasolo, P. 1983. Comparative ultrastructural analysis of mycorrhizal associations. *Can. J. Bot*. 61: 917–943.

Scannerini, S., and Bonfante-Fasolo, P. 1990. Plants and mycorrhizal fungi: Coevolution or not coevolution? In: Nardon, P., Gianinazzi-Pearson, V., Grenier, A. M., Margulis, L., and Smith, D. C., eds., *Endocytobiology IV*. INRA, Paris.

Scannerini, S., Bonfante-Fasolo, P., and Fontana, A. 1975. An ultrastructural model for the host-symbiont interactions in the endotrophic mycorrhiza of

Ornithogalum umbellatum L. In: Sanders, P. E., Mosse, B., and Tinker, P. B., eds., *Endomycorrhizas*. Academic Press.

Schubert, A., Marzachí, C., Mazzitelli, M., Cravero, M. C., and Bonfante-Fasolo, P. 1987. Development of total and viable extraradical mycelium in the vesicular-arbuscular mycorrhizal fungus *Glomus clarum* Nicol. and Schenk. *New Phytol.* 107: 183–190.

Smith , D. C. 1979. From extracellular to intracellular: The establishment of a symbiosis. *Proc. Roy. Soc. Lond. B* 204: 67–286.

Soldo, A. T. 1983. The biology of xenosome, an intracellular symbiont. *International Review of Cytology (Intracellular Symbiosis)* Suppl. 14: 29–47.

Tiivel, T. 1987. Leafhopper endocytobiosis. *Endocyt. C. Res.* 4: 25–38.

Webster, J. 1988. Botany and mycology. In: Greuter, W., and Zimmer, B., eds., *Proceedings of XIV International Botanical Congress*. Koeltz, Königstein.

20 Mutualistic Symbioses in the Origin and Evolution of Land Plants

David H. Lewis

Mutualism may be regarded (even defined) as the outcome of genomic recombinations in which the resultant phenotypes of the interacting organisms are fitter than each would have been when not interacting. The genomic recombinations that had the greatest evolutionary potential in primeval *aquatic* environments were those that led to the eukaryotic protoctists. These probably remain the most diverse of nature's experimental symbioses.

The thesis to be developed here is that the initial exploitation of the *terrestrial* environment by "land plants" also depended on mutualistic symbioses. In this case, the organisms involved were "algae" and "fungi." The use of quotation marks is essential here because it has become clear that "fungi," "algae," and "land plants" are very collective nouns which each encompass a range of polyphyletic organisms. It is therefore no surprise that the terms "lichen" (which refers to symbiotic associations between fungi and algae) and "mycorrhiza" (which refers to symbiotic associations between fungi and land plants) also refer to a miscellaneous range of associations. Generalizations about them should therefore be made with considerable circumspection. The same applies to the associations between fungi and the photosynthetic parts of land plants that have been recently recognized as mutualistic (Clay 1988) and for which the term "mycophylla" has been proposed (Lewis 1987). Table 1 summarizes the mutualisms between photobionts and mycobionts currently recognized. This paper will primarily concern mutualistic mycorrhizas of vascular plants. Coevolutionary aspects of lichens, mycophylla, and bryophyte-fungal associations likely to be mutualistic have been reviewed recently by Hawksworth (1988a), Clay (1988), and Boullard (1988).

Table 1
Mutualistic symbioses between photobionts and mycobionts.

Photobionts	Symbioses
Whole organism associated with mycobiont[a]	
Cyanobacteria	Lichens
Algae (various phyla)	Lichens
Part of organism associated with mycobiont	
Photosynthetic tissue	Mycophylla
Nonphotosynthetic tissue	Some mycorrhizas[b]

a. "Mycophycobioses" (associations between the ascomycete *Mycosphaerella ascophylli* and the phaeophytes *Ascophyllum nodusum* and *Pelvetia canaliculata*) are not included in the table, since there is no evidence that the associations are mutualistic. Unlike in lichens where the fungi are exhabitants, *M. ascophylli* is an inhabitant (Kohlmeyer and Kohlmeyer 1972; Hawksworth 1988b).
b. The qualification "some" is necessary because not all associations described as mycorrhizal are mutualistic (Lewis 1987).

Phylogenetic Origin of Relevant Taxa

To develop the hypothesis for the symbiotic origin of land plants, as proposed by Church (1921) (see also Corner 1964) and as considered in more detail by Jeffrey (1962) and by Pirozynski and Malloch (1975), requires a brief summary of relevant phylogenies of the interacting taxa.

It is evident that land plants (bryophytes and tracheophytes) arose from green algae, and it is generally agreed that the particular ancestral group of green algae were similar to extant Charophyta. Raven (1987) has reviewed in detail the morphological, ultrastructural, biochemical, and biophysical evidence in support of the latter assertion, and earlier (Raven 1977) he proposed a scenario for the evolutionary development of the supracellular transport processes upon which terrestrial plant life depends. (See also Atsatt 1988 and Atsatt's chapter in this volume.)

There is a view that fungi—in the restricted sense of zygomycetes, ascomycetes, basidiomycetes, and their allies (including their asexual forms), all of which lack motile stages—are derived from organisms similar to extant chytridiomycetes. Cavalier-Smith (1987) regards this last group as the most primitive of the Kingdom Fungi, its zygomycetes being derived from an *Allomyces*-like ancestor. He goes on to derive ascomycetes, the Endomycota segregated from this group, and basidomycetes from the entomophthoralean zygomycetes along parallel rather than sequential lines. It is somewhat arbitrary whether chytrids are regarded as "advanced" protoctists from which fungi (*sensu stricto*) developed (Margulis 1981) or as "primitive" fungi from which nonciliated forms arose (Cavalier-Smith 1987). The important point is that zygomycetes had aquatic ancestors which potentially could interact with charophytan green algae.

Morphological Interactions between Photobionts and Mycobionts in Relation to the Initial Colonization of Land

Extant photobionts, including both algae and tracheophytes, often respond to infection with fungi by gross morphological disturbances such as galls or witches' brooms (Wood 1963, chapter 8). Fungus-like protoctists, including chytrids, oomycetes, and slime molds, can also induce morphological change. Whereas the outcomes of these interactions are now antagonistic (i.e., the reproductive fitness of the photobionts is reduced), it is not difficult to imagine situations where morphological change could be advantageous (e.g., increase in area of photosynthetic surface) and so be favored by natural selection. Pirozynski (1988 and in this volume) has discussed and extended the postulate that major morphological features of vascular plants (e.g., flowers, fruits, and organs of food storage and vegetative propagation) are products of symbioses involving *Agrobacterium*-like horizontal gene transfer from fungi and gall-inducing insects to plants (see also Atsatt 1988 and Atsatt's chapter in this volume). Such transfer, although it would stabilize and canalize what Pirozynski terms "quantum evolutionary innovations," is not necessary for phenotypic change *per se*, as the examples mentioned at the start of this section confirm.

With specific regard to mutualistic symbioses involving photobionts and fungi, figure 1 summarizes the four kinds of morphological outcomes that are theoretically possible. The evolutionary success (or failure) of the individual kinds of interaction is dependent on the capacity (or lack of capacity) of interacting unicellular or filamentous symbionts to respond morphologically to each other by forming a three-dimensional vegetative tissue.

Where neither symbiont could respond significantly [−/−:A], the result was an evolutionary dead end of the kind possibly represented

		Capacity of Photobiont to form a 3-D Tissue	
		—	+
Capacity of Mycobiont to form a 3-D Tissue	—	A	C
	+	B	D

A : -/- ? Geosiphon

B : -/+ Lichens

C : +/- (a) VA Mycorrhizas
(b) Ericoid Mycorrhizas

D : +/+ Ectomycorrhizas

Figure 1
Potential morphological interactions between photobionts and mycobionts.

today by the symbiosis that involves *Geosiphon pyriforme*. This symbiosis is essentially an association between a zygomycetous fungus (similar to the species of *Glomus* that form vesicular-arbuscular mycorrhizas) and a bryophyte, but it incorporates *Nostoc* (if available) to form the characteristic bladders (Mollenhauer 1988).

Where the fungal component responded by forming a tissue and the photobiont remained undifferentiated [−/+:B], the lichen habit became established. This is currently the situation for about half of the extant ascomycetous fungal species; with the possible exception of some homoisomerous species, the form of the dual organism is determined by developmental processes induced in the mycobiont by the photobiont—processes to which little experimental investigation has been addressed (but see Honegger, this volume).

Where the mycobiont remained undifferentiated but the photobiont responded by forming tissues [+/−:C(a)], the resulting mutualisms were the early land plants, either bryophytes or vascular plants with vesicular-arbuscular (VA) mycorrhizas. (Discussion of interactions C(b) and D of figure 1 is temporarily deferred since they are not relevant to the *initial* colonization of land but only to that of particular habitats which developed much later in geological time.) It should be noted that colonizations of land by lichens, by nonvascular plants, and by vascular plants are not mutually exclusive but depend on the time of origin of the different symbioses.

Until recently, it was thought that ascomycetes (and therefore lichens) had a Mesozoic origin (Pirozynski 1976). However, if the interpretation of particular Silurian fossils as ascomycetous spores is correct (Sherwood-Pike and Gray 1985), then they must be of much greater antiquity. Evidence from the geographical distribution of lichens in relation to plate-tectonic events, studies of obligately lichenicolous fungi, and investigations of ascus structures of host lichens also support a pre-Mesozoic origin (Hawksworth 1988a). Whatever the precise time of their origin, the nutritional mode adopted by the fungi has been very successful. Nevertheless, their global ecological impact has been restricted by the poikilohydric nature of terrestrial lichens and the limited photosynthetic capacity of the approximately forty genera of photobionts now involved.

Land plants arose during the Silurian or even the Ordovician. The later date is firmly based on macro-fossils (*Cooksonia*) (Edwards and Fanning 1985); the earlier date depends on interpretation of microfossils (Gray 1985). Whatever the precise date, one particular facet of the process was the overcoming of the restricted photosynthetic capacity of algae just mentioned. It is hypothesized that these limits to photosynthesis in the proto-land-plants (i.e., charophytan algae) were removed by fungus-induced morphological changes. These morphological changes provided the raw material on which natural selection acted along the lines proposed by Raven (1977). It is probable that many interactive evolutionary experiments occurred, involving a range of algae and a spectrum of fungi along the chytridiomycete-zygomycete-ascomycete continuum. The outcomes were hepatic bryophytes with essentially two-dimensional structures, nonvascular erect organisms such as the Devonian *Aglaophyton* (Edwards 1986), and vascular plants.

Fungi would have had to infect both the photosynthetic and the nonphotosynthetic parts of the heterotrichous alga. In these ways, colonization of land depended on both mycophyllous and mycorrhizal fungi (table 1). The evidence for the former is limited; however, fungal infection of the nonphotosynthetic, rhizoidal parts of the proto-land-plants would have been a *sine qua non*, since without roots acquisition of water and mineral salts from soil would have been very inefficient. That such vesicular-arbuscular (VA) mycorrhizal infection of early land plants existed is borne out by the demonstration of mycobionts in their horizontal but rootless axes (Pirozynski and Malloch 1975). As was discussed in Lewis 1987, such VA mycorrhizas (which persist to the present day) would have enabled fungi to forage for phosphorus in particular. This element is present in soil solutions in very low concentrations—two or three orders of magnitude less than other major required elements for plant growth, such as potassium, calcium, magnesium, nitrogen, and sulfur (Epstein 1972)—and is very immobile (Nye and Tinker 1977).

Infection by fungi would also have induced lignification as a defensive reaction, a biochemical response exploited during the Silurian in the evolution of vascular and supporting tissues (Raven 1977, 1984; Atsatt, this volume).

Interactions between Land Plants and Fungi in Relation to the Colonization of Particular Habitats

The VA mycorrhizas that were necessary for the initial colonization of land would have had to exploit essentially mineral soils, i.e., ones with low organic content. Such soils remain dominated by VA mycorrhizal plants. However, the ectomycorrhizal and ericoid mycorrhizal types tend to dominate the more organic soils that now prevail at higher altitudes and latitudes (Read 1984). These soils are represented by the mull-moder-mor-peat series, which can be recognized from the early Carboniferous on (Wright 1985). It has become clear that an important attribute that enables plants with these two kinds of mycorrhiza to colonize such soils effectively is the ability of the fungi involved to obtain nitrogen from organic sources. In the case of the ericoid mycorrhizas, this can be derived from organic nitrogen complexed with phenolic material such as that found in peat (Read, Leake, and Langdale 1989). Other attributes of the fungi include a ready ability to acquire other elements, such as phosphorus, and tolerance of low pH and of high concentrations of ions and compounds toxic to nonmycorrhizal plants.

The less extreme forms of these soils are now dominated by ectomycorrhizal trees in which the fungi are mainly basidiomycetous, although zygomycetous and ascomycetous forms exist. In these associations, both partners form tissues, i.e., they represent type +/+:D of figure 1. Here, however, the photobionts are themselves the products of the earlier-formed kind of symbiosis discussed in the previous section.

Table 2
Phylogenetic relationships of the principal ectomycorrhizal angiospermous families to the Hamamelidaceae.

Family	Percentage of species ectomycorrhizal[a]	Sporne's index[b]	Chapman's rank[c]	Chapman's evolutionary distance[c]
Myrtaceae	92	45	15.0	0
Rosaceae	26	43	15.0	0
Euphorbiaceae	11	37	15.0	0
Dipterocarpaceae	100	38	39.5	6
Leguminosae	19	48	39.5	6
Fagaceae	100	43	102.0	14
Betulaceae	100	38	102.0	14
Salicaceae	96	44	102.0	14

a. from Newman and Reddell 1987
b. from Sporne 1980
c. from Chapman 1987

The ectomycorrhizal habit is scattered among gymnospermous and angiospermous families in an apparently random manner. No testable hypothesis has yet been formulated to explain this distribution. (See Trappe 1987 for a series of questions that must be addressed.) Some conclusions do, however, emerge when data on distribution of ectomycorrhiza among angiospermous families (Newman and Reddell 1987) are compared with the phylogenetic positions of these families.

Sporne (1980) arranged nearly 300 families of angiosperms by means of an index derived from the distribution of thirty characters. As part of a reanalysis of Sporne's compilation, Chapman (1987) has ranked families in terms of "evolutionary distance" from the primitive Hamamelidaceae. According to Newman and Reddell (1987), five families of angiosperms have nearly 100% of their species ectomycorrhizal and another three have percentages between 10 and 25. Table 2 shows that, whereas the range of values for Sporne's index for these eight families is very restricted (37–48 from the observed range for all families of 23–87), "evolutionary distance" from the Hamamelidaceae is clustered into three discrete values within its overall range (0–60). Furthermore, there is at least one fully ectomycorrhizal family within each cluster. This suggests that the ectomycorrhizal habit has arisen independently at several stages during the evolution of the angiosperms, with major bursts of ectomycorrhizal speciation within three main phylogenetic lines (Hamamelidae—Fagaceae and Betulaceae; Dilleniidae—Dipterocarpaceae and Salicaceae; Rosidae—Myrtaceae). Furthermore, within each of these distinct lines of evolution, the ectomycorrhizal habit is most extensively exhibited in relatively primitive families (i.e., those with "evolutionary distances" less than 30% of the maximum value found). The other three (also relatively primitive) families with a significant number of ectomycorrhizal species listed in table 2 are large ones, and closer analysis of the data available may well reveal concentration of the ectomycorrhizal habit in particular taxonomic divisions of the families. For example, Newbery et al. (1988) have examined the Caesalpinoideae of the Leguminosae in particular.

Ectomycorrhizas are also characteristic of the gymnospermous Pinaceae, conifers which did not become a significant part of the Earth's flora until the Mesozoic (Malloch, Pirozynski, and Raven 1980). The late Mesozoic (Cretaceous) was, until very recently, also considered to be the time during which the angiosperms arose; however, an earlier (Carboniferous) origin has now been suggested (Martin, Gierl, and Saedler 1989). By this Period, basidiomycetes had also evolved (Stubblefield and Taylor 1988), and I have speculated (Lewis 1987) on the significance of concentration of atmospheric oxygen in the development of ligninolytic activity by this group of fungi, an activity that had to be suppressed in ectomycorrhizal species.

Whereas it is thus possible for there to have been in excess of 300 million years of coevolution between ectomycorrhizal fungi and plants, it is more probable that the associations originated less than 150 million years ago. This is certainly so for the ericoid type [+/−:C(b)] of figure 1. Here, as for the ectomycorrhizal symbiosis, the photobionts are products of the primeval VA mycorrhizal symbiosis and, as in that initial association, the fungi do not form a three-dimensional vegetative structure. Whereas VA mycorrhizas are now almost the norm for angiosperms, and ectomycorrhizas are found among many families in addition to the major ones listed in table 2 (Harley and Smith 1983; Newman and Reddell 1987; Trappe 1987), the ericoid type is confined to three angiospermous families within the Ericales, all of which are evolutionarily more advanced than the principal ectomycorrhizal families. The combined evidence from the fossil record and the geographical distribution indicates a Cretaceous origin, perhaps with the Epacridaceae older than the Ericaceae (Raven and Axelrod 1974; K. A. Pirozynski, personal communication). However, indices of evolutionary advancement suggest that the Ericaceae is more primitive than the Epacridaceae, which, in turn, is less advanced than the Empetraceae (Chapman 1987). Very few species of fungi are involved. Koch's postulates have been confirmed for *Hymenoscyphus* (= *Pezizella*) *ericae* and a small number of other ascomycetes (Read 1983; Dalpé 1986, 1989). The resulting symbioses produce vegetation which dominates heathlands characteristic of high altitudes and latitudes (Read 1984) by virtue of physiological and biochemical attributes discussed by Read et al. (1989).

Conclusions

It is evident from the above that three distinct kinds of mycorrhiza enabled algae and then bryophytes and tracheophytes to colonize land and, later, particular types of soil. These soils have themselves evolved in response to both biotic and abiotic factors. Other terrestrial surfaces such as rocks and bark, and also some soils, have become dominated by lichens. In these ways, it is abundantly clear that mutualistic symbioses have been powerful sources of evolutionary innovation which have increased ecological ranges. The early land plants were essentially algal-fungal associations in which the morphology of the alga dominated (cf. the "mycophycosymbioses" between littoral brown algae and *Mycosphaerella ascophylli* [Kohlmeyer and Kohlmeyer 1972]). The gross morphology of lichens is determined by influences of their photobionts on their fungi. A closer analysis of the relationship between habitats and morphologies of lichens may well reveal patterns correlated with particular physiological and biochemical features, as is the case for mycorrhizas. Among these, the ectomycorrhizas (and their variants, ectendo- and arbutoid) and the ericoid mycorrhizas represent secondary symbioses superimposed on the evolved land plant. They are the terrestrial equivalent of the inter-eukaryote symbioses found among aquatic protoctists. The degree of physiological dependence of the fungi on their co-symbionts is correlated with the time of evolution of the association. VA mycorrhizal fungi have no free-living existence in nature and have not yet been cultured in the laboratory. Many ectomycorrhizal fungi are difficuit to culture or are slow-growing. Frequently, their growth (or germination of spores) is stimulated by root exudates (Melin 1963; Fries 1987). Most of these fungi probably have a limited capacity to live asymbiotically. At the other extreme, ericoid mycorrhizal fungi can be found free-living and are readily cultured in the laboratory.

There are, of course, ecologically successful free-living terrestrial fungi, algae, and land plants. The selective forces and mechanisms responsible for nonmycorrhizal plants are discussed in Lewis 1987, Tester et al. 1987, and Trappe 1987.

References

Atsatt, P. R. 1988. Are vascular plants "inside-out" lichens? *Ecology* 69: 17–23.

Boullard, B. 1988. Observations on the coevolution of fungi with hepatics. In *Coevolution of Fungi with Plants and Animals*, ed. Pirozynski, K. A., and Hawksworth, D. L. Academic Press.

Cavalier-Smith, T. 1987. The origin of fungi and pseudofungi. In *Evolutionary Biology of the Fungi*, ed. Rayner, A. D. M., Brasier, C. M., and Moore, D. Cambridge University Press.

Chapman, J. L. 1987. Sporne's advancement index revisited. *New Phytologist* 106: 319–332.

Church, A. H. 1921. The lichen as transmigrant. *Journal of Botany* 59: 7–13, 40–46.

Clay, K. 1988. Clavicipitaceous fungal endophytes of grasses: Coevolution and the change from parasitism to mutualism. In *Coevolution of Fungi with Plants and Animals*, ed. Pirozynski, K. A., and Hawksworth, D. L. Academic Press.

Corner, E. J. H. 1964. *The Life of Plants*. Weidenfeld and Nicolson.

Dalpé, Y. 1986. Axenic synthesis of ericoid mycorrhiza in *Vaccinium angustifolium* Ait. by *Oidiodendron* species. *New Phytologist* 103: 391–396.

Dalpé, Y. 1989. Ericoid mycorrhizal fungi in the Myxotrichaceae and Gymnoascaceae. *New Phytologist* 113: 523–527.

Edwards, D. F. 1986. *Aglaophyton major*, a non-vascular land-plant from the Devonian Rhynie Chert. *Botanical Journal of the Linnean Society* 93: 177–204.

Edwards, D., and Fanning, U. 1985. Evolution and environment in the late Silurian-early Devonian: The rise of the pteridophytes. *Philosophical Transactions of the Royal Society, series B* 309: 147–165.

Epstein, E. 1972. *Mineral Nutrition of Plants: Principles and Perspectives*. Wiley.

Fries, N. 1987. Ecological and evolutionary aspects of spore germination in the higher basidiomycetes. *Transactions of the British Mycological Society* 88: 1–7.

Gray, J. 1985. The microfossil record of early land plants: Advances in understanding of early terrestrialisation, 1970–1984. *Philosophical Transactions of the Royal Society, series B* 309: 167–195.

Harley, J. L., and Smith, S. E. 1983. *Mycorrhizal Symbiosis*. Academic Press.

Hawksworth, D. L. 1988a. Coevolution of fungi with algae and cyanobacteria in lichen symbioses. In *Coevolution of Fungi with Plants and Animals*, ed. Pirozynski, K. A., and Hawksworth, D. A. Academic Press.

Hawksworth, D. L. 1988b. The variety of fungal-algal symbioses, their evolutionary significance, and the nature of lichens. *Botanical Journal of the Linnean Society* 96: 3–20.

Jeffrey, C. 1962. The origin and differentiation of archegoniate land plants. *Botaniska Notiser* 115: 446–454.

Kohlmeyer, J., and Kohlmeyer, E. 1972. Is *Ascophyllum nodosum* lichenized? *Botanica Marina 15: 109–112.*

Lewis, D. H. 1987. Evolutionary aspects of mutualistic associations between fungi and photosynthetic organisms. In *Evolutionary Biology of the Fungi*, ed. Rayner, A. D. M., Brasier, C. M., and Moore, D. Cambridge University Press.

Malloch, D. W., Pirozynski, K. A., and Raven, P. H. 1980. Ecological and evolutionary significance of mycorrhizal symbioses in vascular plants (a review). *Proceedings of the National Academy of Sciences* 77: 2113–2118.

Margulis, L. 1981. *Symbiosis in Cell Evolution*. Freeman.

Martin, W., Gierl, A., and Saedler, H. 1989. Molecular evidence for pre-Cretaceous angiosperm origins. *Nature* 339: 46–48.

Melin, G. 1963. Some effects of forest tree roots on mycorrhizal basidiomycetes. In *Symbiotic Associations*, ed. Nutman, P. S., and Moss, B. Cambridge University Press.

Mollenhauer, D. 1988. Weitere Untersuchungen an *Geosiphon pyriforme*—Einer Lebensgemeinschaft von Pilz und Blauälge. *Natur und Museum* 118: 289–320.

Newbery, D. M., Alexander, I. J., Thomas, D. W., and Gartlan, J. S. 1988. Ectomycorrhizal rain-forest legumes and soil phosphorus in Korup National Park, Cameroon. *New Phytologist* 109: 433–450.

Newman, E. I., and Reddell, P. 1987. The distribution of mycorrhizas among families of vascular plants. *New Phytologist* 106: 745–751.

Nye, P. H., and Tinker, P. B. 1977. *Solute Movement in the Soil-Root System*. Blackwell.

Pirozynski, K. A. 1976. Fossil fungi. *Annual Review of Phytopathology* 14: 237–246.

Pirozynski, K. A. 1988. Coevolution by horizontal gene transfer: A speculation on the role of fungi. In *Coevolution of Fungi with Plants and Animals*, ed. Pirozynski, K. A., and Hawksworth, D. A. Academic Press.

Pirozynski, K. A., and Hawksworth, D. L. (eds.) 1988. *Coevolution of Fungi with Plants and Animals*. Academic Press.

Pirozynski, K. A., and Malloch, D. W. 1975. The origin of land plants: A matter of mycotrophism. *BioSystems* 6: 153–164.

Raven, J. A. 1977. The evolution of vascular plants in relation to supracellular transport processes. *Advances in Botanical Research* 5: 153–219.

Raven, J. A. 1984. Physiological correlates of the morphology of early vascular plants. *Botanical Journal of the Linnean Society* 88: 105–126.

Raven, J. A. 1987. Biochemistry, biophysics and physiology of chlorophyll *b*-containing algae: Implications for taxonomy and phylogeny. *Progress in Phycological Research* 5: 1–121.

Raven, P. H., and Axelrod, D. I. 1974. Angiosperm biogeography and past continental movements. *Annals of the Missouri Botanical Garden* 61: 539–673.

Read, D. J. 1983. The biology of mycorrhiza in the Ericales. *Canadian Journal of Botany* 61: 985–1004.

Read, D. J. 1984. The structure and function of the vegetative mycelium of mycorrhizal roots. In *The Ecology and Physiology of the Fungal Mycelium*, ed. Jennings, D. H., and Rayner, A. D. M. Cambridge University Press.

Read, D. J., Leake, J. R., and Langdale, A. N. 1989. The nitrogen nutrition of mycorrhizal fungi and their host plants. In *Nitrogen, Phosphorus and Sulphur Utilization by Fungi*, ed. Boddy, L., Marchant, R., and Read, D. J. Cambridge University Press.

Sherwood-Pike, M. A., and Gray, J. 1985. Silurian fungal remains: Probable records of the class Ascomycetes. *Lethaia* 18: 1–20.

Sporne, K. R. 1980. A re-investigation of character correlations among dicotyledons. *New Phytologist* 85: 419–449.

Stubblefield S. P., and Taylor, T. N. 1988. Recent advances in palaeomycology. *New Phytologist* 108: 3–25.

Tester, M., Smith, S. E., and Smith, F. A. 1987. The phenomenon of "nonmycorrhizal" plants. *Canadian Journal of Botany* 65: 429–431.

Trappe, J. M. 1987. Phylogenetic and ecologic aspects of mycotrophy in the angiosperms from an evolutionary standpoint. In *Ecophysiology of VA Mycorrhizal Plants,* ed. Safir, G. R. CRC Press.

Wood, R. K. S. 1963. *Physiological Plant Pathology*. Blackwell.

Wright, V. P. 1985. The precursor environment for vascular plant colonization. *Philosophical Transactions of the Royal Society, Series B* 309: 143–145.

21 Fungi and the Origin of Land Plants

Peter R. Atsatt

Algae, having a photosynthetic mode of nutrition, differ fundamentally from fungi, which have a degradative and absorptive mechanism of nutrient acquisition. Land plants, and particularly the seed plants, differ from green algae and fungi in combining highly integrated forms of both modes of nutrition, so that in many ways they are effectively photosynthetic fungi. The complex architecture of the vascular plants is far more than the sum of what is achieved by the synthesis of chloroplastic products; it is equally achieved and maintained by specialized degradative processes. Protoplast disassembly and death is a tissue-specific or cell-specific event in plants, an event that occurs repeatedly throughout an individual's life and for which the primary control mechanisms are not known (Leshem et al. 1986). The digestion of cell walls is fundamental to the differentiation of specialized tissues, such as abscission zones, perforated cell plates, fusing vessel elements, germinating seeds, and the senescing and softening tissues of fruits. Maturing pollen grains and embryos are also nourished by digestive processes, and both gametophytes and sporophytes exhibit invasive growth and absorptive feeding. The occurrence of haustorial cells and specialized degradative processes in an evolutionary lineage derived from the green algae has several possible explanations, including convergent evolution, the expression of ancient genes conserved in both fungi and green algae, or a more recent acquisition of fungal genes, either via a single symbiotic event (Atsatt 1988) or via stepwise horizontal gene transfer (Lamboy 1984; Pirozynski 1988). Though it is currently impossible to distinguish between these alternative hypotheses, it may nevertheless be instructive to examine the evolution of plants from a perspective that emphasizes their fungus-like qualities.

As a heuristic exercise, I will consider the idea that plants are fundamentally parasitic, and that they began their evolution when juvenile sporophytes became biotrophic parasites on the genetically and phenotypically distinct gametophyte. After the development of vascular tissue and large independent sporophytes, male and female gametophytes in turn became parasitic on the sporophyte. Mothers could then provide their dispersing offspring (seeds) with energy-rich reserves that increased sporophyte survival in marginal habitats. The packaging of male gametes in fungus-like propagules capable of entering sporophytic tissue permitted the closure of embryos within carpels and intensified competition among large numbers of insect-delivered male gametophytes—perhaps the two most significant factors in the success of the angiosperms (Mulcahy 1979). Here I explore the idea that digestive and absorptive nutrition were key innovations that permitted this symbiosis between generations, and the idea that the outcomes—the embryo, pollen,and the seed—are linked together by the common thread of parasitism.

Evolution of the Embryo

The embryo is conspicuously absent from most discussions that list necessities for existence on land. The requirements for colonization of land are often described as resistance to desiccation, support of aerial parts, and ability to obtain water from the soil. These represent solutions to problems associated with increasing height above the moisture-laden soil; of great significance, they are perhaps secondary in importance relative to the development of an embryo. Algae lack embryos. Thus, the appearance of the embryo was a seminal event in the origin of land plants. It may have solved a fundamental problem associated with becoming sedentary and anchored in young, nutrient poor soils. Initially, habitats suitable for growth and reproduction were probably rare, and offspring, no longer automatically dispersed by water currents, may have accumulated near their parents, where competition for limited resources undoubtedly became intense. Under these competitive conditions, an embryo-producing plant had great selective advantage; offspring capable of acquiring nutrients from their parents increased their individual fitness, and parental genotypes that invested in their offspring gained greater control over their inclusive fitness (Westoby and Rice 1982; Queller 1983; Mazer 1987).

A central issue underlying the evolution of plant embryos involves the genetic conflict of interest that exists among siblings and between parents and their offspring, i.e., between gametophytes and sporophytes (see Mazer 1987 and references therein). Although there is some overlap in the loci transcribed by gametophytes and sporophytes (Mulcahy 1971), these genomes nevertheless produce distinctive phenotypes with different fitness requirements, and thus their nutritional integration would have required an aggressive sporophyte countered by a defensive gametophyte. Selection would have favored parasitic sporophytes (selfish embryos) capable of eliciting the release of maternal nutrients at the expense of their siblings; conversely, the maternal gametophyte had to evolve countermeasures that allowed it to allocate resources in a way that increased its fitness at the expense of the inclusive fitness of selfish offspring. Westoby and Rice (1982) suggest that the angiosperm endosperm (with two doses of the maternal genome) and integuments (which act to seal off haustorially aggressive endosperms) provide mothers with increased control over how maternal investment is distributed to embryos.

We do not know whether the embryo originated as a unique event or whether it had a polyphyletic origin, as Sluiman (1985) suggests. Graham (1985) recognizes four successive stages in plant embryo evolution; oogamous sexual reproduction, egg fertilization on the parental gametophyte, retention and development of the zygote on the gametophyte, and the establishment of a nutritional and developmental relationship (mutualism) between the dependent sporophyte and the parental gametophyte. The last event was the key innovation; it required physiological mechanisms that permitted the genetically distinct zygote expressing the sporophyte phenotype to parasitize the enveloping gametophyte. Graham (1985) contends that resolution of the mystery of land-plant evolution depends on understanding how this close developmental and nutritional relationship between generations originated. It is my thesis that this mutualistic innovation may have been engendered by the incorporation of a fungal genome specialized for extracellular digestion and absorption.

Bridging the Generation Gap

A key problem in the evolution of the embryo was the cytoplasmic connection between the gametophyte and the integrated sporophyte generation. Apparently the earliest solution to this problem was the evolution of a distinctive wall-membrane apparatus consisting of wall ingrowths and amplified plasma membrane. Cells that develop these specialized transport properties are called *transfer cells*. The wall-membrane ingrowths develop only on certain areas of the cell wall, apparently in response to stimuli emanating from neighboring compartments (Pate and Gunning 1972). Such ingrowths occur at many sites, ensuring short-distance transport, but are particularly well developed at the junctions between sporophyte and gametophyte generations, where cytoplasmic contact is almost invariably lacking (Gunning and Pate 1974). The earliest evolutionary occurrence of this specialized wall-membrane apparatus appears to coincide with the earliest occurrence of an embryo-like zygote in the green alga *Coleochaete* (Graham 1985).

How and why did the transfer wall-membrane apparatus evolve? Its presence in *Coleochaete* suggests an origin that predates the evolution of the embryo. Perhaps the transfer cell evolved in response to fungal parasites as an early evolutionary solution to the problem of nutrient acquisition on land. Lewis (1986) has emphasized that because phosphate is present in the soil at concentrations two to three orders of magnitude lower than other macro-nutrients and is extremely immobile, plants (and their mycorrhizal symbionts) need to grow toward phosphate rather than wait for it to diffuse to them. The mycorrhizal gametophytes of extant Bryophytes and Psilophytes suggest that early terrestrial gametophytes were probably also mycorrhizal. Transfer cells, a cell type that often develops at the interface with fungal cells in extant mycorrhizal associations (Ashford and Allaway 1982; Robertson and Robertson 1982; Duddridge and Read 1984; Toth and Miller 1984; Allaway et al. 1985), may have been involved in early fungal symbiosis. Wall ingrowths also occur in mycorrhizal fungal cells adjacent to plant transfer cells. This fungal membrane amplification results from the proliferation of incomplete septa (Ashford and Allaway 1982; Allaway et al. 1985; Massicotte et al. 1986). Both morphologically and functionally, the nutritional junction between plants and their fungal symbionts is quite similar to the nutritional union between the gametophyte and the parasitic sporophyte. Did selection for the ability to form transfer cells with mycorrhizal fungi preadapt some gametophyte lineages to incorporate a second mutualist, the diploid sporophyte?

Equally important, what kinds of biochemical similarities exist between the amplified plasma membranes of transfer cells and the plasma membranes of fungal haustorial cells? In vesicular-arbuscular mycorrhizae, the arbuscules, highly branched hyphal intrusions, have high ATPase activity localized along the host plasma membrane surrounding the arbuscule branches, which implies a specialized modification of the invaginated plant membrane (Marx et al. 1982; Smith and Gianinazzi-Pearson 1988). Similarly, in the moss haustorium, the wall-membrane invaginations at the sporophyte-gametophyte junction are sites of intense localized enzyme activity (Maier and Maier 1972), apparently reflecting their specialized transport properties (Caussin et al. 1983). The haustoria of biotrophic fungi, such as the powdery mildews and rusts, apparently control nutrient efflux from hosts by inducing modification of membrane function at localized areas of the host cell (Woods et al. 1988). Does the invaginated wall-membrane apparatus of transfer cells represent an inverted haustorial "extension" with locally specialized plasma membranes that function similarly to those of fungal haustoria? The properties of the plasma membrane ATPase from plants and fungi (yeast, *Neurospora*) are remarkably similar (Sze 1985). An important question is whether the coding sequences for this plant enzyme are more similar to those of fungi or to those of green algae ancestral to the charophycean lineage.

Invasive Protoplasts

Within the embryophyte lineage, the origin of intrusively growing fungus-like cells such as laticifers, fibers, sclerids, and tyloses appears to be coincident with the evolution of vascular tissue (Atsatt 1988). A similar innovation apparently closely connected to the development of vascular tissue was the origin of invasive protoplasts that resulted in the fusion of fully differentiated cells.

The union of fully differentiated cells is a fundamental characteristic of the fungi. Hyphal fusion and clamp cell fusion involve localized wall lysis and enlargement of a single fusion pore (Todd and Aylmore 1984). Septal dissolution is also a regular event in many fungi, where cross-walls are apparently synthesized so as to resist or yield to wall lytic enzymes (Wessels and Sietsma 1981). The vascular plants similarly exhibit a range of cell interconnections that includes the complete dissolution of intervening walls in compound laticifers and vessel elements, the enlargement of primary plasmodesmata during the formation of sieve plate pores, and the formation of secondary plasmodesmata.

Secondary plasmodesmata are cytoplasmic strands that dissolve and penetrate fully formed walls. This type of localized enzymatic dissolution of cell walls is unknown in the green algae. The plasmodesmata of green algae are found only between offspring cells, and are not known to develop secondarily between adjacent cells derived from different parent cells (Marchant 1976). In vascular plants, secondary plasmodesmata develop between sieve tube elements and sister companion cells during the initiation of lateral roots, between laticifers and adjacent cells, between fused cells of similar or different genetic origin (such as graft unions), and between parasitic plants and their hosts (Jones 1976). The endoplasmic reticulum appears to be involved in these localized dissolution processes, which may be mediated by membrane-associated hydrolases (Fincher and Stone 1981).

Invasive protoplasts also appear to be involved in the feeding of embryos. In the forest tree *Nothapodytes foetida* (Icacinaceae), the endosperm haustoria produce short branches that at first grow intercellularly but then form plasmodesmata-like penetrations of adjacent cells, and the two protoplasts unite. The nucleus of the invaded cell degenerates while the nucleus of the haustorial cell becomes greatly enlarged and distorted and protrudes into each newly created extension of the haustorium. The haustorium thus continues its growth by replacing the nuclei of adjacent cells with its own modified nucleus (Swamy and Ganapathy 1957). Similarly, the invasion of one cell by another can apparently occur in the absence of haustorial extensions. Studies by Zang and colleagues (1980, 1984) suggest that the feeding of wheat embryos by surrounding nucellus cells involves considerably more than wholesale digestion. The breakdown of nucellar cells begins with the extrusion of nuclear material into vesicles that remain within the ER cavity. The endomembrane system then degenerates, and mass extrusion of disassembled protoplasm occurs through enlarged plasmodesmata that constitute wall ruptures. Intercellular movement of protoplasm occurs within nucellar tissues and between the nucellus and the embryo sac. In this process, nucleolus-like bodies as well as nuclear and protoplasmic fragments apparently enter the antipodal cells of the developing embryo.

Haustorial Cells of Seed Plants

Haustoria are intrusively growing cell protrusions that bring recipient cells into intimate association with donor cells and increase the surface area available for nutrient absorption. Rudimentary haustorial cells are present in the primitive vascular plants, but their real rise to prominence appears to coincide with the origin of seed plants. Seeds presumably evolved when precociously developing female gametophytes were able to parasitize the enveloping sporophyte, and when male gametophytes were able to reach these deeply buried eggs. Both innovations were achieved by fungus-like cells capable of invading plant tissue and absorbing nutrients from distant sources. The digestive and absorptive haustorial cell represents the essence of the parasitic fungi, and here I would like to stress its fundamental role in the evolution of the male and female gametophytes of seed plants, as well as the parasitic seed plants.

Male Gametophytes

The pollen tube of seed plants constituted a major innovation, a shift from swimming sperm to nuclei transported through hyphae-like tubes. Analogous systems are rare among the green algae, but there is strong precedence for this mechanism in the fungi, where sexual reproduction involves the fusion of tubes that bring together genetically distinct nuclei. The gnetophyte *Welwitschia* provides an interesting comparison, because the eggs of this unusual plant send out tubular processes that grow upward and unite with descending pollen tubes; nuclear union occurs within the united tubes (Bold et al. 1987). The evolution of a diminutive parasitic male gametophyte capable of invading genetically dissimilar sporophytes was a revolutionary event that permitted the origin of seeds and set the stage for the rise of flowering plants (Mulcahy 1979).

Pollen tubes have several characteristics which suggest that their morphological similarity to hyphae is more than evolutionary convergence. Friedman (1987a,b) has described in detail the pervasive, highly branched haustorial system of the male gametophyte of *Ginkgo biloba*. Like fungi, this gametophyte branches subapically as well as apically, and also exhibits a yeast-like developmental dimorphism, shifting from diffuse spherical growth to tubular growth in response to nutrient changes. For descriptions of similar haustorial male gametophytes in both gymnosperms and angiosperms, see Friedman 1987a. Cytologically, both pollen tubes and fungal hyphae have a distinctive zone of mitochondria just behind the tip, and both cytoskeleton systems respond to antagonistic cytochalasins by rapidly forming swollen tips (Steer and Steer 1989). Calcium ions similarly modulate tip growth in both systems (Jackson and Heath 1989). There are also common physiological and biochemical features between pollen-pistil interactions and fungus-host interactions. These include a recognition reaction involving glycoproteins, an active response requiring enzyme synthesis, and the production of biostatic phytoalexin-like compounds. In both pollen-pistil and fungus-host interactions, callose is a characteristic product of incompatibility (Hodgkin et al. 1988).

Female Gametophytes

In most angiosperms, the entire surface of the female gametophyte (embryo sac) serves an absorptive function, demolishing adjacent layers of the parental sporophyte. In some species, more active growth occurs at the ends of the embryo sac, where extensions protrude out of the ovule and digest their way into the placenta (Maheshwari 1950; Raghavan 1976). In some genera of the parasitic Loranthaceae, several female gametophytes simultaneously grow up through the stylar tissue, sometimes striking the base of the stigma and then recurving downward (Maheshwari and Kapil 1966). The development of tubular haustorial growth by female gametophytes is not entirely unexpected, in view of the pronounced expression of this growth form in male gametophytes.

Within the female gametophytes of flowering plants, double fertilization produces two zygotes; one develops into the embryo and the other into the endosperm. Traditional descriptions of the endosperm as a "tissue" that nourishes the embryo seem to be less than accurate. Since the endosperm results from the fusion of a sperm nucleus with a maternal (usually diploid) nucleus, the resulting zygote would more correctly be characterized as a "sister sporophyte" of the embryo. This specialized "sister sporophyte" often produces extensively branched haustorial processes that digest their way through the gametophyte and into surrounding maternal sporophyte tissue. Although these ramifying extensions of the endosperm are by no means limited to plant families containing parasitic members, their expression appears to be particularly aggressive within parasitic seed plants. In members of the parasitic Santalales the endosperm haustoria become long enough to reach the base of the ovule and establish contact with the vascular supply of the placenta, or in other cases may even extend out into the flower pedicel (Maheshwari 1950; Raghavan 1976).

The embryo of vascular plants is also a digestive, haustorial organism. The elongating suspensor cells may push the embryo itself into new food sources, or they may function as haustorial extensions of the stationary embryo. For example, the suspensor haustoria of *Gnetum* embryos aggressively parasitize the endosperm and sometimes continue their foraging into the maternal sporophyte (Vasil 1959). In angiosperm plant families with completely suppressed endosperm, suspensor haustoria show a much greater range of growth and form (Raghavan 1976).

Parasitic Seed Plants

The brief consideration of the parasitic gametophytes and sporophytes of seed plants above provides a reasonable basis for understanding the multiple origins of parasitism within this phylogenetic group (Atsatt 1973). When we fully appreciate the degree to which the embryonic sporophytes of seed plants parasitize the gametophytes and sporophytes that envelop them, we can see that nothing fundamentally new occurred when germinating sporophytes began parasitizing more distantly related sporophytes. The achlorophyllous parasitic plants and those that live hyphae-like within photosynthetic tissue attest to the "completeness" with which plants have become obligate biotrophic parasites. If all embryophytes are potentially parasitic, what we need to ask is how and why certain groups (the seed plants and the parasitic seed plants) have more fully exploited this potential—a daunting question not yet fully answered for the fungi (Heath 1986).

Concluding Remarks

The embryo fundamentally distinguishes plants from the green algae, and among the embryophytes those forms producing pollen and seeds have largely inherited the terrestrial Earth. Few would argue with the notion that the origins of the embryo, of pollen, and of seeds were cornerstone events in the history of plant evolution. A fundamental connection between these entities is that each is dependent upon the digestive and absorptive mode of nutrition that characterizes the kingdom Fungi. Indeed, the ability of plants to selectively degrade and recycle extends well beyond the mutualistic union of two generations, and may be the core feature that has raised the green algal chloroplast to such lofty evolutionary heights.

Perhaps the parasitic qualities of the embryophytes did not require the symbiotic input of fungal genes. The ability to secrete wall-degrading enzymes in a controlled and localized manner may simply be a refinement of the widespread production of such enzymes by heterotrophs that predate the origin of eukaryotic cells (Glenn 1976). Neither is the invasive and absorptive haustorium a phylogenetically unique innovation, since it is shared by plants, fungi, oomycetes, and a number of other heterotrophic protists. Parasitism in general has been regularly achieved within diverse and independent eukaryotic lineages, so the importance of convergent processes cannot be overestimated. But even these considerations and the fact that plants have clear phylogenetic affinities with the charophycean green algae (Graham 1985) should not dissuade against the possibility that they also have fungal ancestors. The brown algae may provide an analogous example. Recent ribosomal DNA-sequence comparisons have shown a close phylogenetic relationship between the phloem-containing brown algae and the fungus-like Oomycetes (Bhattacharya and Druehl 1988). The advanced brown algae have many characteristics that parallel those found in vascular plants (Prichard and Bradt 1984), and this makes the discovery of fungus-like ancestors in this lineage particularly interesting.

The ideas I have formulated about the parasitic nature of embryo-producing plants do not provide evidence for the specific hypothesis that plant nuclei contain a fungal genome. Rather, my purpose is to develop a point of view that provides a reasonable basis for questioning and seriously investigating the genetic affinities of plant traits that appear to be shared by fungi. Near the top of my list would be wall-degrading enzymes, such as β-1,3 glucanases and chitinases and the vacuolar and plasma membrane ATPases. Among secondary compounds I would focus on the alkaloids, and among hormones on the gibberellins. Hormones are also instrumental in the mobilization of resources into seeds and fruits (Stephenson 1981 and references therein), and the gibberellins are particularly important in the acquisition of nutrients by embryos (Pharis and King 1985). Plants produce at least 26 different gibberellins that are structurally identical to those of the ascomycete *Gibberella fujikuroi*, and the recent discovery of GA3 in very low levels in *Neurospora crassa* may suggest a wide distribution of gibberellins in fungal ascomycetes (Takahashi et al. 1986). The vacuoles of yeast and plants are very similar (Boller and Wiemken 1986), and comparison of the vacuolar proton ATPase from *Neurospora*, yeast, and corn, including immunological cross-reaction for the 70-kD subunit, has revealed a conspicuous functional and structural similarity (Bowman et al. 1986). Indeed, recent sequence analysis of plant and fungal proton ATPase subunits showed the carrot cDNA sequence to be over 70% homologous with exons of a *Neurospora* 69-kDa genomic clone (Zimniak et al. 1988).

In the absence of competing hypotheses, the "green algal paradigm" of land-plant ancestry severely limits the kinds of phylogenetic comparisons that might otherwise be attempted. For example, similar vacuolar-type ATPases are present in all eukaryotes (Nelson 1987), and the many reports of gibberellic acid-like substances in algae make it apparently easy to explain similarities, or seemingly unnecessary to look outside the algae for genetic sources. If the parasitic nature of the embryophytes does derive from the symbiotic introduction of fungal genes, this event should have predated the embryo; thus, members of the charophycean green algae may contain the predicted genome. Critical tests of a fungal symbiosis model will require comparing specific gene sequences of plants with those of fungi (ascomycetes) and green algae within and ancestral to the charophycean lineage.

Acknowledgments

Financial support of this research by the Whitehall Foundation is greatly acknowledged. The comments of reviewers were very helpful.

References

Allaway, W. G., Carpenter, J. L., and Ashford, A. E. 1985. Amplification of inter-symbiont surface by root epidermal transfer cells in the Pisonia mycorrhizae. *Protoplasma* 128: 227–231.

Ashford, A. E., and Allaway, W. G. 1982. A sheathing mycorrhiza on *Pisonia grandis* R. Br. (Nyctaginaceae) with development of transfer cells rather than a Hartig net. *New Phytologist* 90: 511–519.

Atsatt, P. R. 1973. Parasitic flowering plants: How and why did they evolve? *Am. Nat.* 107: 502–510.

Atsatt, P. R. 1988. Are vascular plants inside-out lichens? *Ecology* 69: 17–23.

Bhattacharya, D., and Druehl, L. D. 1988. Phylogenetic comparison of the small-subunit ribosomal DNA sequence of *Costaria costata* (Phaeophyta) with those of other algae, vascular plants and Oomycetes. *Phycol.* 24: 539–543.

Bold, H. C., Alexopoulos, C. J., and Delevoryas, T. 1987. *Morphology of Plants and Fungi.* Harper and Row.

Boller, T., and Wiemken, A. 1986. Dynamics of vacuolar compartmentation. *Ann. Rev. Plant Physiol.* 37: 137–164.

Bowman, E. J., Mandala, S., Taiz, L., and Bowman, B. J. 1986. Structural studies of the vacuolar membrane ATPase from *Neurospora crassa* and comparison with the tonoplast membrane ATPase from *Zea mays*. *Proc. Natl. Acad. Sci.* 83: 48–52.

Caussin, C., Fleurat-Lessard, P., and Bonnemain, J. L. 1983. Absorption of some amino acids by sporophytes isolated from *Polytrichum formosum* and ultrastructural characteristics of the haustorium transfer cells. *Ann. Bot.* 51: 167–173.

Duddridge, J. A., and Read, J. A. 1984. Modification of the host-fungus interface in mycorrhizas synthesized between *Suillus bovinus* (Fr.) O. Kuntz and *Pinus sylvestris* L. *New Phytologist* 96: 583–588.

Fincher, G. B., and Stone, B. A. 1981. Metabolism of noncellulosic polysaccharides. In: Tanner, W., and Loewus, F. A., eds., *Encyclopedia of Plant Physiology*, new series, volume 13B, *Plant Carbohydrates II. Extracellular Carbohydrates*. Springer-Verlag.

Friedman, W. E. 1987a. Growth and development of the male gametophyte of *Ginkgo biloba* within the ovule (*in vivo*). *Amer. J. Bot.* 74: 1797–1815.

Friedman, W. E. 1987b. Morphogenesis and experimental aspects of growth and development of the male gametophyte of *Ginkgo biloba* in vitro. *Amer. J. Bot.* 74: 1816–1830.

Glenn, A. R. 1976. Production of extracellular proteins by bacteria. *Ann. Rev. Microbiol.* 30: 41–62.

Graham, L. E. 1985. The origin of the life cycle of land plants. *Amer. Sci.* 73: 178–186.

Gunning, B. E. S., and Pate, J. S. 1974. Transfer cells. In: Robards, A. W., ed., *Dynamic Aspects of Plant Ultrastructure*. McGraw-Hill.

Heath, M. 1986. Evolution of parasitism in the fungi. In: Rayner, A.D.M., Brasier, C.M., and Moore, D., eds., *Evolutionary Biology of the Fungi*. Cambridge University Press.

Hodgkin, T., Lyon, G. D., and Dickinson, H. G. 1988. Recognition in flowering plants: A comparison of the *Brassica* self-incompatibility system and plant pathogen interactions. *New Phytologist* 110: 557–569.

Jackson, S. L., and Heath, I. B. 1989. Effects of exogenous calcium ions on tip growth, intracellular Ca^{2+} concentration, and actin arrays in hyphae of the fungus *Saprolegnia ferax*. *Experimental Mycology* 13: 1–12.

Jones, M. G. K. 1976. The origin and development of plasmodesmata. In: Gunning, B. E. S., and Robards, A. W., eds., *Intercellular Communication in Plants: Studies on Plasmodesmata*. Springer-Verlag.

Lamboy, W. F. 1984. Evolution of flowering plants by fungus to host horizontal gene transfer. *Evolutionary Theory* 7: 45–51.

Leshem, Y. Y., Halevy, A., and Chaim, F. 1986. *Processes and Control of Plant Senescence*. Elsevier.

Lewis, D. H. 1986. Evolutionary aspects of mutualistic associations between fungi and photosynthetic organisms. In: Rayner, A. D. M., Brasier, C. M., and Moore, D., eds., *Evolutionary Biology of the Fungi*. Cambridge University Press.

Maheshwari, P. 1950. *An Introduction to the Embryology of Angiosperms*. McGraw-Hill.

Maheshwari, P., and Kapil, R. N. 1966. Some Indian contributions to the embryology of angiosperms. *Phytomorphology* 16: 239–291.

Maier, K., and Maier, U. 1972. Localization of beta-glycerophosphatase and Mg^{++}-activated adenosine triposphatase in a moss haustorium, and the relation of these enzymes to the cell wall labyrinth. *Protoplasma* 75: 91–112.

Marchant, H. J. 1976. Plasmodesmata in algae and fungi. In: Gunning, B. E. S., and Robards, A. W., eds., *Intercellular Communication in Plants: Studies on Plasmodesmata*. Springer-Verlag.

Marx, C., Dexheimer, J., Gianinazzi-Pearson, V., and Gianinazzi, S. 1982. Enzymatic studies on the metabolism of vesicular-arbuscular mycorrhizas. IV. Ultracytoenzymological evidence (ATPase) for active transfer processes in the host-arbuscule interface. *New Phytologist* 90: 37–43.

Massicotte, H. B., Peterson, R. L., Ackerley, C. A., and Piche, Y. 1986. Structure and ontogeny of *Alnus crispa-Alpova diplophloeus* ectomycorrhizae. *Can. J. Bot.* 64: 177–192.

Mazer, S. J. 1987. Maternal investment and male reproductive success in angiosperms: Parent-offspring conflict or sexual selection? *Biol. Soc. Linn. Soc.* 30: 115–133.

Mulcahy, D. L. 1971. A correlation between gametophytic and sporophytic characteristics in *Zea mays* L. *Science* 171: 1155–1156.

Mulcahy, D. L. 1979. The rise of the angiosperms: A genecological factor. *Science* 206: 20–23.

Nelson, N. 1987. Structure, functions and evolution of proton-ATPases. *Plant Physiol.* 86: 1–3.

Pate, J. S., and Gunning, B. E. W. 1972. Transfer cells. *Ann. Rev. Plant Physiol.* 23: 173–196.

Pharis, R. P., and King, R. W. 1985. Gibberellins and reproductive development in seed plants. *Ann. Rev. Plant. Physiol.* 36: 517–568.

Pirozynski, K. A. 1988. Coevolution by horizontal gene transfer: A speculation on the role of fungi. In: Pirozynski, K. A., and Hawksworth, D. L., eds., *Coevolution of Fungi with Plants and Animals*. Academic Press.

Pritchard, H. N. and Bradt, P. T. 1984. *Biology of Nonvascular Plants*. Mosby.

Queller, D. C. 1983. Kin selection and conflict in seed maturation. *J. Theor. Biol.* 100: 153–172.

Raghavan, V. 1976. *Experimental Embryogenesis in Vascular Plants*. Academic Press.

Robertson, D. C., and Robertson, J. A. 1982. Ultrastructure of *Pterospora andromedea* Nuttall and *Sarcodes sanquinea* Torrey mycorrhizas. *New Phytologist* 92: 539–551.

Sluiman, H. J. 1985. A cladistic evaluation of the lower and higher green plants (Viridiplantae). *Plant Syst. Evol.* 149: 217–232.

Smith, S. E., and Gianinazzi-Pearson, V. 1988. Physiological interactions between symbionts in vesicular-arbuscular mycorrhizal plants. *Ann. Rev. Plant Physiol.* 39: 221–244.

Steer, M. W., and Steer, J. M. 1989. Tansley review no. 16: Pollen tube tip growth. *New Phytologist* 111: 323–358.

Stephenson, A. G. 1981. Flower and fruit abortion: Proximate causes and ultimate functions. *Ann. Rev. Ecol. Syst.* 12: 253–279.

Swamy, B. G. L., and Ganapathy, P. M. 1957. A new type of endosperm haustorium in *Nothapodytes foetida*. *Phytomorphology* 7: 331–336.

Sze, H. 1985. H^+-translocating ATPases: Advances using membrane vesicles. *Ann. Rev. Plant Physiol.* 36: 175–208.

Takahashi, N., Yamaguchi, I., and Yamane, H. 1986. Gibberellins. In: Takahashi, N., ed., *Chemistry of Plant Hormones*. CRC Press.

Todd, N. K., and Aylmore, R. C. 1984. Cytology of hyphal interactions and reactions in *Schizophyllum commune*. In: Moore, D., Casselton, L. A., Wood,

D. A., and Frankland, J. D., eds., *Developmental Biology of Higher Fungi*. Cambridge University Press.

Toth, R., and Miller, R. M. 1984. Dynamics of arbuscule development and degeneration in a *Zea mays* mycorrhiza. *Amer. J. Bot.* 71: 449–460.

Vasil, V. 1959. Morphology and embryology of *Gnetum ula. Brongn. Phytomorphology* 9: 167–215.

Wessels, J. G. H., and Sietsma, J. H. 1981. Fungal cell walls: A survey. In: Tanner, W., and Loewus, F. A., eds., *Plant Carbohydrates II. Extracellular Carbohydrates*. Springer-Verlag.

Westoby, M., and Rice, B. 1982. Evolution of the seed plants and inclusive fitness of plant tissues. *Evolution* 36: 713–724.

Woods, A. M., Didehvar, F., Gay, J. L., and Mansfield, J. W. 1988. Modification of the host plasmalemma in haustorial infections of *Lactuca sativa* by *Bremia lactucae. Physiol. Mol. Plant Path.* 33: 299–310.

Zang, W. C., and Yan, W. M. 1984. Transport of disassembled protoplasm from degenerated nucellus into embryo sac and its role in feeding the proliferating antipodals in wheat. *Acta Bot. Sinica* 26: 11–18.

Zang, W. C., Yan, W. M., and Wu, S. H. 1980. Intercellular migration of protoplasm and its relations to the development of embryo sac in nucellus of wheat. *Acta Bot. Sinica* 22: 32–36.

Zimniak, L., Dittrich, P., Gogarten, J. P., Kibak, H., and Taiz, L. 1988. The DNA sequence of the 69-kDa subunit of the carrot vacuolar H^+-ATPase. *J. Biol. Chem.* 263: 9102–9112.

VI

Symbiosis and Morphogenesis

22

Fungal Evolution: Symbiosis and Morphogenesis

Rosmarie Honegger

Fungal evolution manifests itself at anatomical and morphological levels mainly in sporocarps and sporangia (Poelt 1986); trophic structures (hyphae, mycelia, budding cells) remain relatively simple in most groups. This holds true even for ecologically obligate biotrophic fungi that live in parasitic or mutualistic symbioses with multicellular, photo-autotrophic hosts (e.g., seaweeds, bryophytes, or vascular plants). However, the invasion of relatively large host structures, such as roots or leaves, and the establishment of a biotrophic relationship necessitated the formation of particular appressorial, haustorial, and/or sheathing structures at the interface between the fungus and the host cell. The monophyletic group of zygomycetous fungi that form vesicular-arbuscular mycorrhizae (VAM) illustrate this. Today about 80 species are distinguished, representing only about 0.12 percent of all fungi, but associating with an estimated number of more than 175,000 plant species (bryophytes, pteridophytes, gymnosperms, and angiosperms). VAM are almost ubiquitous in terrestrial ecosystems that support plant life. Apparently, the extant species differ very little in morphology from the oldest fossil samples, recorded in Devonian Rhyniales and Lycopsids (e.g., *Asteroxylon*; Kidston and Lang 1921). Although it is the only mode of nutrition in VAM fungi, the symbiotic way of life has triggered little innovation at the morphological level in these taxa ever since the development of this highly successful mode of interaction between fungus and host cell (Scannerini and Bonfante-Fasolo, this volume). However, mycorrhizal fungi must have had an immense impact on the evolution of their hosts, and it is likely that the successful colonization of terrestrial ecosystems by plants became possible only with the aid of mycorrhizal symbionts (Pirozynski and Malloch 1975; Lewis 1987; Lewis, this volume).

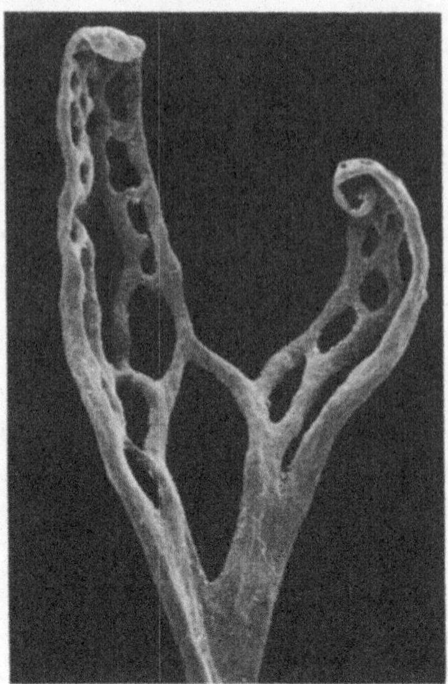

Figure 1
Ramalina menziesii ("lace lichen"), a three-dimensional, internally stratified structure with a species-specific external morphology that harbors the photobiont cell population: the response of a lichen-forming fungus to spatial limitations on the substratum where its photobiont cells normally live. × 27.

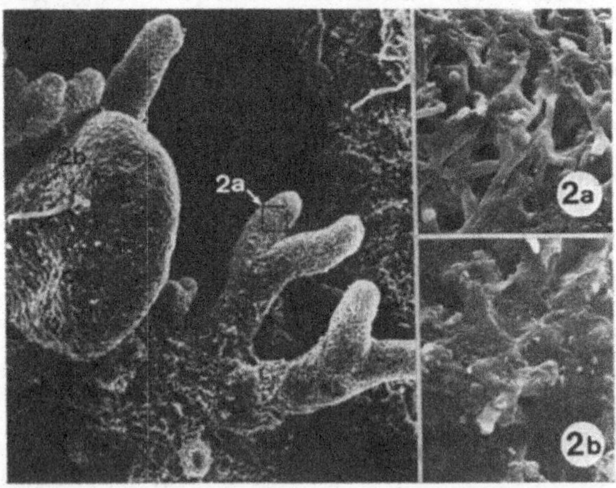

Figure 2
In the cephalodiate, fruticose *Stereocaulon ramulosum* (from Hawaiian lava), the chlorophycean phycobionts (*Pseudochlorella* sp.) trigger the formation of slim, ecorticate branches, whereas sac-like cephalodia (ce) with conglutinate peripheral cortex are formed around populations of the nitrogen-fixing cyanobacteria (e.g. *Scytonema* sp.). (a) Detail of the surface of an ecorticate branch with loosely interwoven hyphae (gas exchange facilitated); ×700. (b) Detail of the surface of a cephalodium with conglutinate cortex (regulation of the O_2 level inside the cephalodium facilitated); ×700.

Lichen Mycobionts

A different situation is encountered in the case of lichen-forming fungi, a polyphyletic group composed mainly of ascomycetes (approximately 13,250 species, or 46 percent of all ascomycetes), a few basidiomycetes (approximately 50 species, or 0.3 percent of all basidiomycetes), and some conidial fungi (approximately 200 species, or 1.2 percent of conidial fungi); together, these species represent about 21 percent of all fungi (Hawksworth 1988a). Of the 46 orders of ascomycetes accepted by Eriksson and Hawksworth (1986), six consist entirely of, and an additional ten include at least some, lichenized taxa. Most lichen mycobionts live as ecologically obligate biotrophs in symbiosis either with unicellular or (rarely) filamentous algal or with cyanobacterial photobionts. Despite the dependence of lichen mycobionts on their photobionts, the majority of lichens can be interpreted as mutualistic symbioses, since the biological fitness of the photobionts, as concluded from their biomass and their ecological breadth, is distinctly increased in the lichenized state (Law and Lewis 1983; Smith and Douglas 1987). A large number of lichen mycobionts are physiologically facultative biotrophs and thus can be cultured in the aposymbiotic state (Bubrick 1988). Isolated mycobiont species in culture form thallus-like colonies that differ markedly in morphology from the symbiotic phenotype.

Apart from their ability to form very peculiar structures, the lichen-forming ascomycetes seem to be typical representatives of their class, as can be concluded from the relatively meager data on cell-wall structure and composition (Boissiere 1987; Honegger and Bartnicki-Garcia 1991), polyol pattern (Honegger 1990a), and cytology. The enigmatic concentric bodies—tiny cell organelles of unknown biogenesis and function—that occur in almost all lichenized ascomycetes have also been found in a range of nonlichenized, parasitic, and saprotrophic ascomycetes (Honegger 1991a).

Lichen Photobionts

So far, phycobionts of approximately 22 genera of chlorophytes (seven orders, three classes; Honegger 1990b), one xanthophycean and one phaeophycean genus, and cyanobionts of 15 genera of cyanobacteria (four orders; all together about 100 species) have been described as lichen photobionts (Tschermak-Woess 1988). About 10

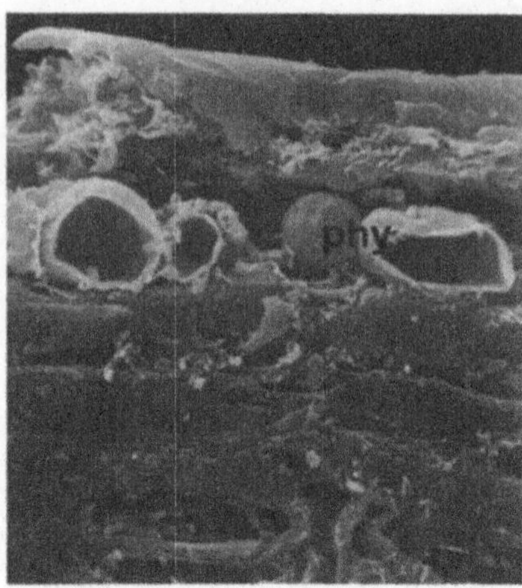

Figure 3
Arthothelium ilicinum: cross-section through part of the crustose, nonstratified endophloeodal thallus in the bark of *Ilex aquifolium* (phy: phycobiont cells); × 1260.

percent of lichen mycobionts associate with cyanobacterial photobionts, which are donors of carbohydrates and fixed nitrogen. More than 85 percent of lichen thalli contain green algal phycobionts, and 3–4 percent of lichens (the so-called cephalodiate species) contain both green algal and cyanobacterial photobionts (James and Henssen 1976; Hertel and Rambold 1988). In some of these triple symbioses, the cyanobionts of mature cephalodia reveal distinctly higher heterocyst frequency and nitrogenase activity than those in young cephalodia or in the aposymbiotic state (Englund 1977). Cephalodial cyanobionts provide fixed nitrogen as ammonia to the symbiotic system and thus play an important role in the nitrogen metabolism of these triple symbioses. Besides the cephalodiate species, in which the mycobiont incorporates the cyanobiont in either internal or superficial (external) gall-like cephalodia (figure 2), an unknown number of cyanotrophic lichens with green algal photobionts form either facultative or obligate associative symbioses with nitrogen-fixing cyanobacteria (Poelt and Mayrhofer 1988).

Some lichen photobionts occur in the aposymbiotic state; others have so far been found exclusively within lichen thalli (Tschermak-Woess 1988; Honegger 1990b). The photobiont has been determined at the species level in fewer than 2 percent of all lichen species (as calculated from table 1 of Tschermak-Woess 1988). For only a few

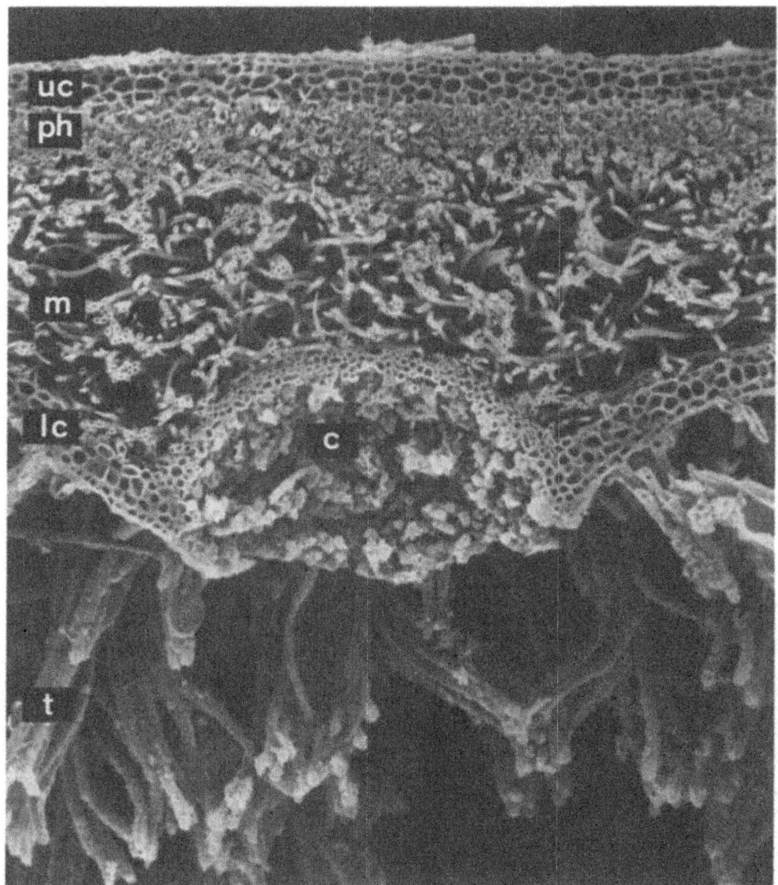

Figure 4
Sticta sp.: cross-section through part of the foliose, internally stratified thallus. The hygrophilous upper (uc) and lower cortex (lc) are built up by isodiametrical, conglutinate fungal cells. Some of the peripheral cells of the lower cortex elongate to form a hygrophilous tomentum (t). The loosely interwoven hyphae of the gas-filled medullary layer (m) have water-repellent wall surfaces. A cyphella (aeration pore, c) is built up by groups of medullary hyphae with globose ends and with a highly water-repellent cell-wall surface layer. The photobiont cells are kept in a uniformly thick layer (ph) underneath the upper cortex. Water and salts are passively taken up by the cortical layers and passively translocated to the photobiont cells within the outermost part of the fungal cell wall underneath a water-repellent surface layer. × 335.

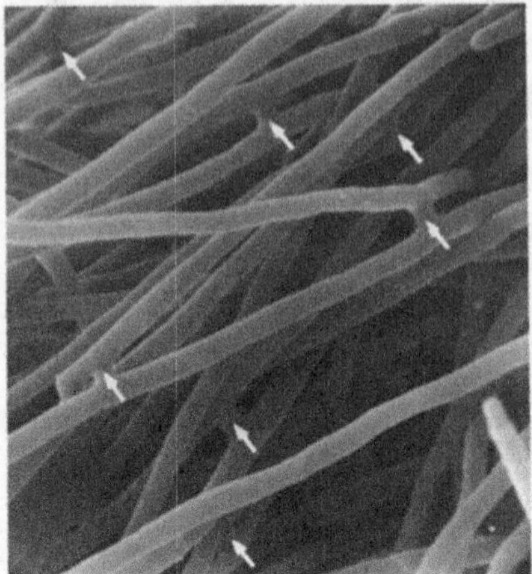

Figure 5
Anastomoses (arrows) between neighboring hyphae in *Peltigera canina*; × 630.

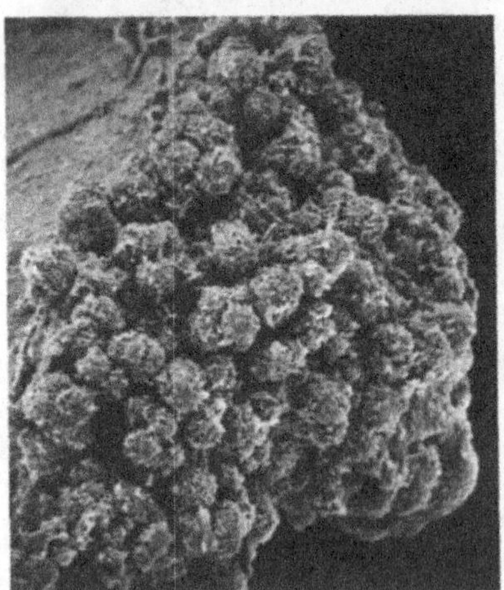

Figure 6
Soredia (noncorticate, powdery granules in which fungal hyphae enclose dividing photobiont cells) in a marginal soralium (site of soredium production) of *Cetraria cetrarioides*; × 125.[1]

lichen mycobionts are experimental data available on the range of compatible algal or cyanobacterial photobionts. Many lichen mycobionts are moderately specific (accept several species of one genus as photobionts), but highly selective with regard to their photobionts. Partner selectivity in lichens has been inferred from observations that some lichen phycobionts are seldom found in the aposymbiotic state, and most of the common taxa of unicellular green algae are not acceptable partners for lichen-forming fungi (Galun and Bubrick 1984; Galun 1988). Some 40 percent of all lichen mycobionts, as estimated by Honegger (1991b), associate with pleurastralean phycobionts of the genus *Trebouxia* (*sensu* Gärtner 1985, including *Pseudotrebouxia* Archibald). Only two of the 22 species of this genus are nonsymbiotic; the others are lichen phycobionts seldom found as aposymbionts (Gärtner 1985; Friedl and Gärtner 1988; Tschermak-Woess 1988) . A long cohabitation of these phycobionts with their fungal hosts is deduced from the observation that most of the morphologically complex lichen species (which may be presumed to be the most evolved ones) contain trebouxioid phycobionts.

The Lichen Symbiosis

Mycorrhizal fungi, as inhabitants of subterranean, photosynthetically inactive, multicellular host structures, provide water and nutrients to the association, while the plant competes for space and secures optimal illumination and gas exchange in its photosynthetically active parts. The fungal partner of the lichen association also usually provides water and nutrients, and in addition, as the quantitatively predominant exhabitant of a population of minute unicellular photobionts cells, has to secure gas exchange and adequate (or even optimal) illumination of its photosynthetically active partner and compete for space. These factors have triggered the formation of morphologically and anatomically complex three-dimensional growing structures that are unique in the fungal kingdom in about 40 percent of lichen-forming fungi (figure 3): squamulose, foliose, or fruticose, internally stratified thalli in which the population of photobiont cells, controlled by the mycobiont, is restricted to a particular layer (figures 1, 4, 19, 20). Lichens are ecologically most successful in terrestrial habitats where plants are at their geographical limits (e.g., arctic-alpine ecosystems). This is due to the limited photosynthetic capacities of lichen photobionts (Lewis, this volume) and to the lack of biopolymers

Figure 7
The soredia of *Lobaria pulmonaria* develop within the soralium into corticate propagules; × 120.[1]

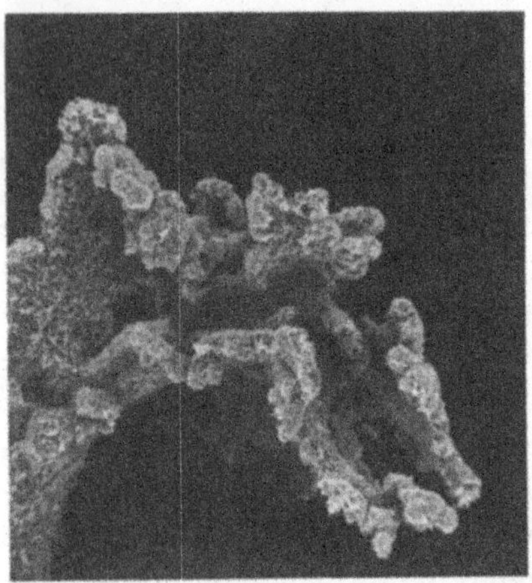

Figure 8
Blastidia (symbiotic diaspores produced by "budding" of the thallus along the margins) in *Cladonia caespiticia*; × 54.[1]

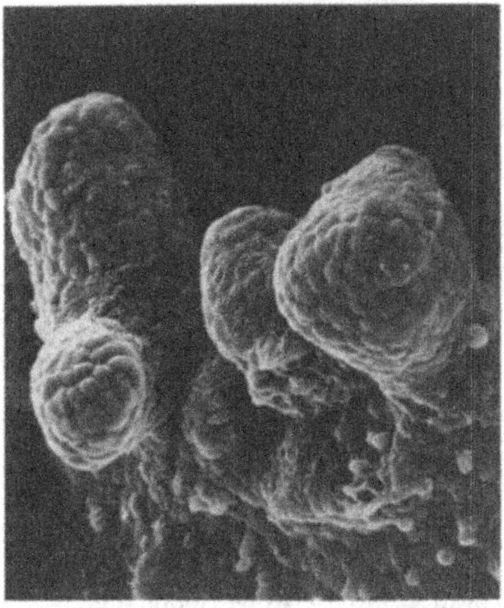

Figure 9
Clavate isidia (corticate, apically constricted protuberances) at the thallus margin of *Sticta fuliginosa*; × 500.[1]

comparable to lignin, which allow the formation of expansive supporting structures. Because of the lack of degradation-resistant polymers, no fossil record of lichen thalli is available.

Association with trebouxioid phycobionts or other photobionts that are rarely found in the aposymbiotic state necessitates that the mycobionts develop particular reproductive strategies. A large number of sexually reproducing lichen mycobionts begin their development as parasites in already-existing thalli, in soredia, or in crustose primordial stages (pre-thalli) of other species, and acquire photobiont cells "by theft" (Poelt and Doppelbauer 1956; Friedl 1987: Ott 1987a–c; Hawksworth 1988b). The crustose pre-thallus stage of development (figures 15, 16) precedes the development of the morphologically complex final thalli, the internally stratified symbiotic phenotypes. Crustose stages can be formed with ultimately incompatible algae in axenic resyntheses (Ahmadjian and Jacobs 1981) and in *in situ* studies (Friedl 1987; Ott 1987c). This phase may be an important feature of the survival strategy employed until compatible photobiont cells become available. Most complex taxa of lichenized fungi have elegantly overcome the problems related to the acquisition of compatible photobiont cells by producing symbiotic diaspores, such as soredia (figures 6, 7), isidia (figures 9, 10), phyllidia (figure 11), or blastidia (figure 8; for summaries see Poelt 1986; Jahns 1988b; Honegger 1991a).

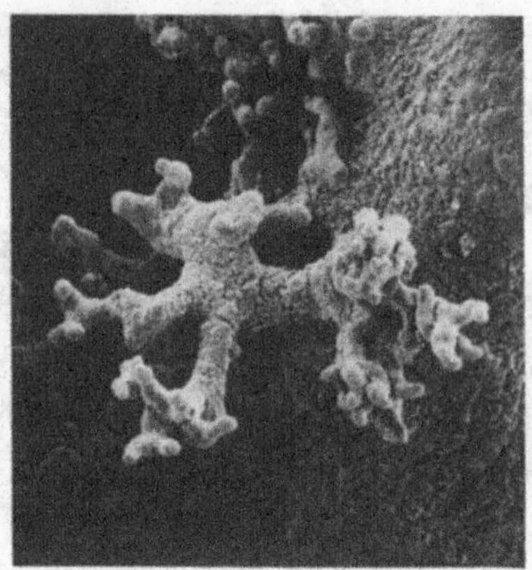

Figure 10
Wind abrasion of a pustule of *Lasallia pustulata* induces intense growth along the wound margins (a) and subsequent formation of coralloid isidia (b); × 60.[2]

The sexual reproductive cycle is often largely reduced in such species. Numerous morphologically and chemically similar species pairs exist, one of which has sexual reproduction, lacks diaspores and shows a relatively limited geographical distribution, and the other of which disperses over relatively large areas exclusively by means of diaspores (Poelt 1970). Wounding by physical factors (e.g., wind abrasion; figure 10) or biological agents (e.g., snail grazing; figure 12) may stimulate the formation of symbiotic propagules in a wide range of species, some of which do not normally form symbiotic diaspores (figure 12).

Lichen thalli represent consortia with unknown numbers of participants. Several fungal genotypes, originating either from different sexual or asexual spores or from different symbiotic diaspores, and several algal genotypes may be incorporated in a single thallus (figure 18; see Jahns 1988a). Experimental evidence has been obtained that hyphae of morphologically distinct lichen species may join to form an intermediate phenotype (Jahns 1988a,b). These chimeras were obtained from species of reindeer lichen, most of which disperse by thallus fragmentation. Similar fusions between hyphae originating from different thallus fragments probably occur regularly in nature; it is not surprising that such lichens are taxonomically difficult. The species concept in lichens requires investigation, and genetic investigations are urgently needed.

Nothing is known about the fate of the different fungal genotypes within individual lichen thalli. Anastomoses between neighboring

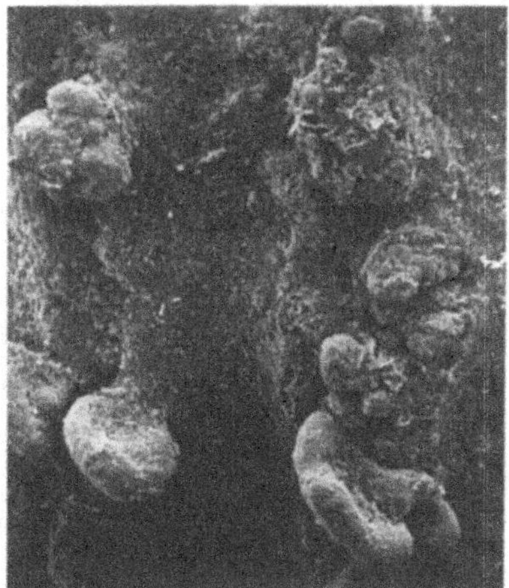

Figure 11
Formation of phyllidia (dorsiventrally organized, internally stratified and basally constricted protuberances) along drought stress cracks in *Peltigera praetextata*.[2]

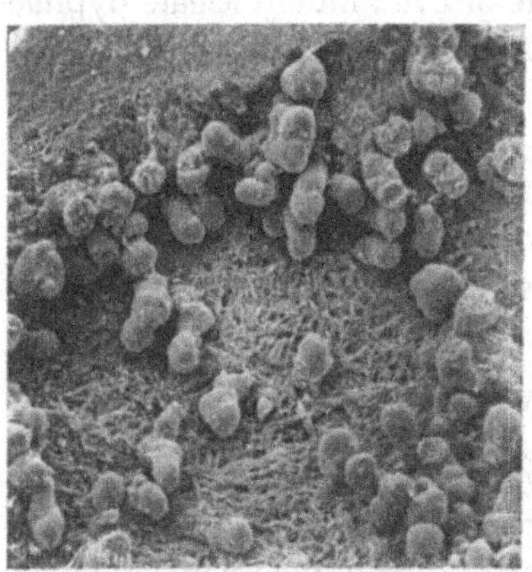

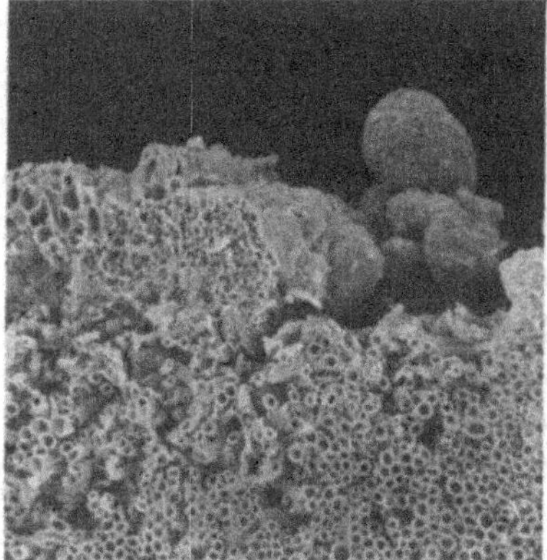

Figure 12
Formation of isidia-like structures in the non-isidiate *Peltigera canina* after wounding by a grazing snail. Left: × 72; right: × 280.[2]

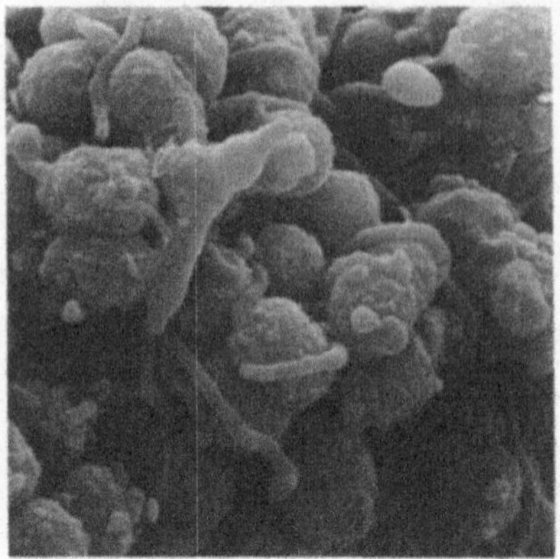

Figure 13
Fungal hyphae overgrowing mixed populations of green algal cells on the bark surface; × 1370.[3]

hyphae are commonly found. In some taxa, such as the Peltigeraceae, anastomoses (figure 5) and binucleate or even multinucleate hyphae are common in the growing thallus (Moreau and Moreau 1919).

Some investigators compare lichen thalli with the stromata of non-lichenized, saprotrophic or parasitic taxa of ascomycetes or conidial fungi (Hawksworth 1988b; Jahns 1988a). The masses of hyphae that build up stromata in these fungi, however, are not involved in the mobilization of nutrients, as are those in lichens. Instead, they support perithecia or conidiomata, within which sexual or asexual spores are produced. Stratified lichen thalli, on the other hand, are the product of a long evolutionary process in nutritionally specialized biotrophic fungi that have elegantly overcome the particular problems related to cohabitation with their very small photobionts by developing symbiotic phenotypes in which they control the relatively favorable physiological conditions of their microbial inhabitants.

Mycobiont-Photobiont Relationships

Many different types of mycobiont-photobiont relationships occur in lichens. In those with green algal photobionts, correlations have been found between the type of mycobiont-phycobiont interaction, the structure and composition of the algal cell wall, and the morphology of the thallus (Tschermak 1941; Plessl 1963; Honegger 1985, 1986a,

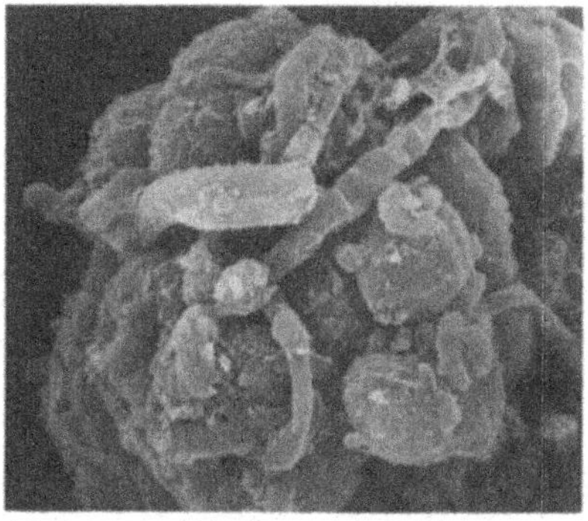

Figure 14
Pre-thallus stage of development: groups of green algal cells are getting more densely enclosed by mycobiont hyphae. Both mycobiont and photobiont cells may be genetically heterogenous. × 1475.[3]

1988). Among the physical relationships, simple wall-to-wall appositions, finger-like or slightly branched intracellular haustoria, and different types of intraparietal haustoria have been recorded. Intracellular haustoria occur exclusively in relatively primitive, nonstratified crustose lichen species, whereas intraparietal haustoria (type 3 according to Honegger 1986a) are restricted to more complex foliose and some fruticose lichen taxa that have trebouxioid phycobionts (Tschermak 1941; Plessl 1963; Honegger 1985, 1986a, 1988). Mycobiont-cyanobiont interactions include intragelatinous protrusions (in foliose taxa), wall-to-wall appositions, and intracellular haustoria in simpler systems with nonstratified thalli (Honegger 1988, 1991b). In lichen species with internally stratified thallus and trebouxioid phycobionts, the haustorial complex fulfills a triple function: (1) It is the site of carbohydrate mobilization (possible modes are summarized in Honegger 1985, 1988, and 1991a; see also Smith and Douglas 1987 and Lines et al. 1989); (2) water, minerals, and fungal metabolites which passively move toward the algal cells within the apoplastic space of the fungal cell wall reach the symbionts at the haustorial interface (the water-repellent cell-wall surface layer of the aerial hyphae in the algal layer is spread over the surface of the algal cell walls and thus plays an important role in translocation; see Honegger 1984, 1985, 1986b, 1988); (3) positioning of the algal cells within the algal layer is performed by growth processes within the haustorial complex of the algae cells (Honegger 1985, 1986a).

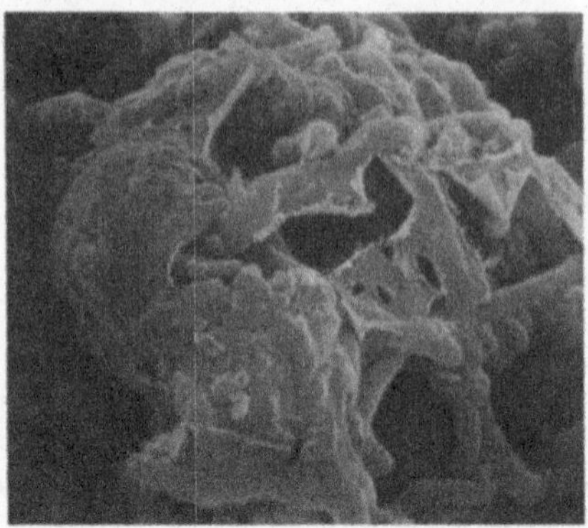

Figure 15
Onset of thallus formation: secretion of mucilaginous material by peripheral hyphae. × 2400.[3]

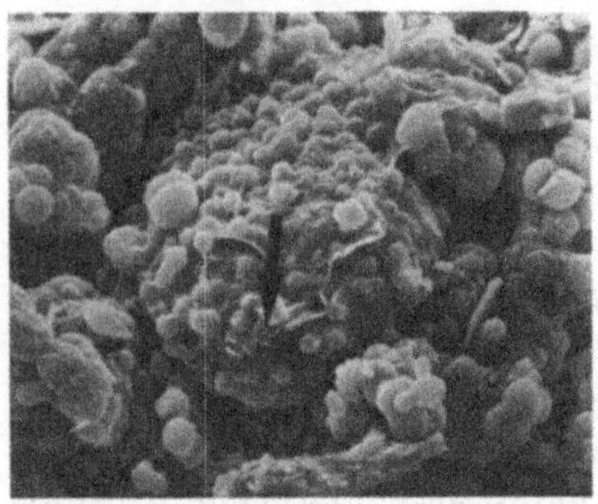

Figure 16
Start of cortex formation and determination of the non-growing basal and growing apical (arrow) poles; × 510.[3]

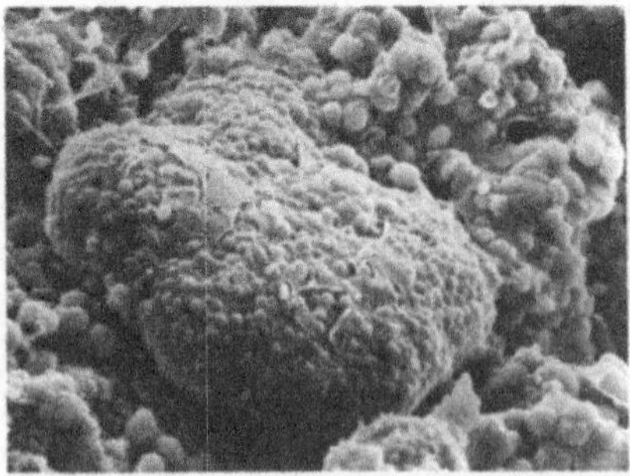

Figure 17
Juvenile thallus with polarized marginal growth; × 310.[3]

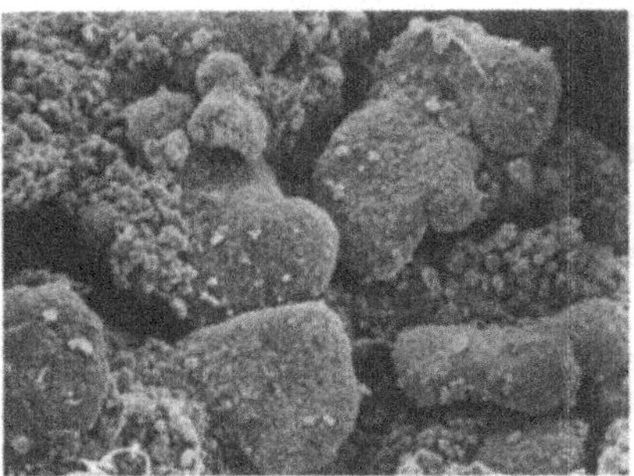

Figure 18
Several juvenile thalli may develop in close vicinity. Later on they may fuse and participate in the formation of a single, genetically heterogenous, rosette-like thallus. × 85.[3]

Morphogenesis in Lichen Thalli with Internal Stratification

In all internally stratified lichen thalli, the mycobiont differentiates into hygrophilic, pseudoparenchymatous conglutinate zones and loosely interwoven plectenchyma that are composed of aerial hyphae with water-repellent cell-wall surfaces (figures 4, 20). In the majority of species, the conglutinate zone forms a multifunctional peripheral cortical layer that covers or even encloses the medulla, a gas-filled internal space built up by aerial hyphae. The photobiont cells are positioned by the haustorial complexes in a uniformly thick layer in the uppermost part of the medulla, an optimal situation with regard to gas exchange, illumination, and the availability of water and minerals (figures 4, 20).

Both conglutinate pseudoparenchymata and aerial hyphae with water-repellent cell-wall surfaces are common in nonlichenized ascomycetes and *fungi imperfecti;* the former structures are present mainly in ascomata and conidiomata, the latter especially in conidiophores. The same structures are found in the thallus-like colonies of aposymbiotically cultured lichen mycobionts; however, they occur in a completely different arrangement from that in the symbiotic phenotype. Aposymbiotic lichen mycobionts grow in a uniformly centrifugal pattern and thus produce globose (in liquid media) or semiglobose colonies (on agar) with cartilaginous, conglutinate central parts and filamentous cells at their periphery.

Only the aerial hyphae of aposymbiotically cultured mycobionts are capable of contacting photobiont cells (Ahmadjian and Jacobs

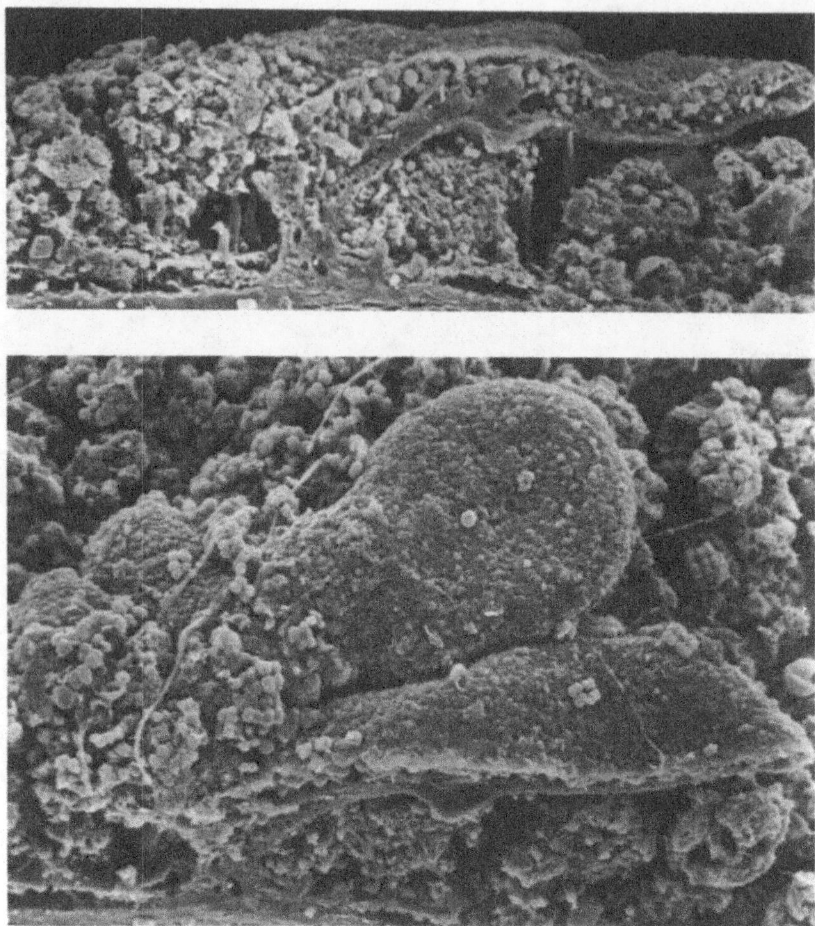

Figure 19
Cross section and laminal view of a juvenile, internally stratified thallus (same specimen) developing from a pre-thallus. Upper and lower cortex ar surrounding the algal layer. × 320.[3]

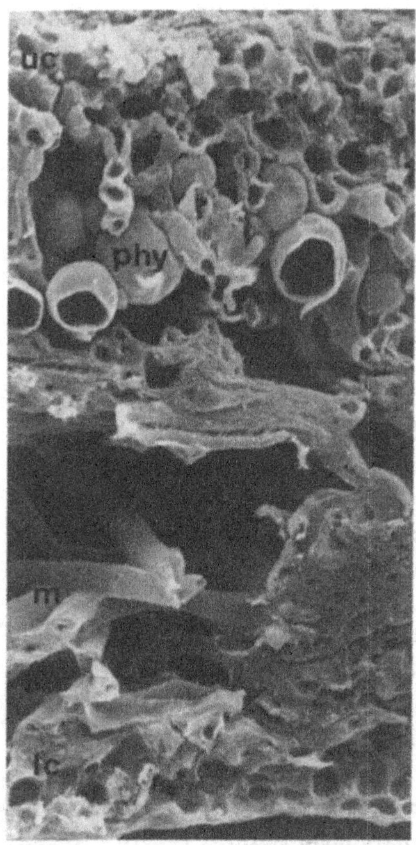

Figure 20
Cross-section through an adult thallus. uc: upper cortical layer; phy: phycobiont cells (Trebouxia sp.); m: medullary layer; lc: lower cortex. × 1230.[3]

1981, 1982, 1985) and forming a crust-like, undifferentiated prethallus (figure 14); this occurs with compatible photobionts and with some incompatible photobionts (Ahmadjian and Jacobs 1981). In a suitable microhabitat, the compatible photobiont triggers, in an unknown manner, the expression of the symbiotic fungal phenotype (figures 15–17). There is strong evidence, although experimental data are lacking, that some mycobionts among the cephalodiate species produce completely different morphotypes in symbiosis with either a compatible green algal or a compatible cyanobacterial photobiont (figure 2; Dughi 1936, 1937, 1944, 1945; James and Henssen 1976). Uncertainties arise through the possible genetic heterogeneity of the fungal material involved in the formation of individual thalli. No resynthesis experiments with single-spore isolates of the mycobiont have so far been carried out with cephalodia-bearing species.

Whereas the aposymbiotic fungus grows uniformly in a centrifugal manner, the symbiotic one reveals polarized growing (figures 16–19). Distinct marginal/apical or intercalary growing zones, or a combination of the two, can be observed (Honegger 1991a). In these growing zones, the highest metabolic activity of the mycobiont and the photobiont (production of secondary metabolites, photosynthesis) and the highest rates of cell division of the photobiont occur (Greenhalgh and Anglesea 1979; Hill 1985, 1989; Larson 1983; Greenhalgh and Whitfield 1987; Honegger 1987). Vertical/radial growth is largely limited to turnover of particular cells in the photobiont layer and cortex and in a slight spatial increase of the medullary layer (figures 19, 20).

Symbiosis and Fungal Evolution

The numerical significance and the enormous ecological success of the lichenized fungi in the most diverse terrestrial ecosystems impressively illustrate the innovative force of the symbiotic way of life in the fungal kingdom. In no other group of fungi are such complex symbiotic phenotypes produced as in lichenized ascomycetes with internally stratified thalli.

Acknowledgments

My sincere thanks are due to Verena Kutasi for doing all the darkroom work, and to D. H. Lewis and René Fester for improving the English style.

Notes

1. Figures 6–12 show examples of symbiotic diaspores (containing mycobiont and photobiont cells).

2. Figures 10–12 show symbiotic diaspores formed as a response to wounding.

3. Figures 13–19 show thallus ontogeny in *Xanthoria parietina*.

References

Ahmadjian, V., and Jacobs, J. B. 1981. Relationship between fungus and alga in the lichen *Cladonia cristatella* Tuck. *Nature* 389: 169–172.

Ahmadjian, V., and Jacobs, J. B. 1982. Artificial reestablishment of lichens. III. Synthetic development of *Usnea strigosa*. *J. Hattori Bot. Lab.* 52: 393–399.

Ahmadjian, V., and Jacobs, J. B. 1985. Artificial reestablishment of lichens. IV. Comparison between natural and synthetic thalli of *Usnea strigosa*. *Lichenologist* 17: 149–165.

Boissiere, J.-C. 1987. Ultrastructural relationship between the composition and the structure of the cell wall of the mycobiont of two lichens. *Bibl. Lichenol.* 25: 117–132.

Bubrick, P. 1988. Methods for cultivating lichens and isolated bionts. In: Galun, M., ed., *Handbook of Lichenology*. CRC Press.

Dughi, R. 1936. Etude comparée du *Dendriscocaulon bolacinum* Nyl. et de la céphalodie fruticuleuse du *Ricasolia amplissima* Scop. Leight. *Bull. Soc. Bot. France* 83: 671–693.

Dughi, R. 1937. Une céphalodie libre lichenogène: le Dendriscocaulon bolacinum Nyl. *Bull. Soc. Bot. France* 84: 430–437.

Dughi, R. 1944. Sur les relations, la position systématique et l'extension du genre *Dendriscocaulon*. *Ann. Fac. Sci. Marseille* 16: 147–157.

Dughi, R. 1945. Une nouvelle céphalodie fruticuleuse de *Ricasolia*. Le *Dendriscocaulon Lesdainei*. *Ann. Fac. Sci. Marseille* 16: 239–242.

Englund, B. 1977. The physiology of the lichen *Peltigera aphthosa*, with special reference to the blue-green phycobiont *Nostoc* sp. *Physiol. Plant.* 41: 298–304.

Eriksson, O., and Hawksworth, D. L. 1986. Outline of the Ascomycetes—1986. *Systema ascom.* 5: 185–324.

Friedl, T. 1987. Thallus development and phycobionts of the parasitic lichen *Diploschistes muscorum*. *Lichenologist* 19: 183–191.

Friedl, T., and Gärtner, G. 1988. *Trebouxia* (Pleurastrales, Chlorophyta) as a phycobiont in the lichen genus *Diploschistes*. *Arch. Protistenkd.* 135: 147–158.

Galun, M. 1988. Lichenization. In: Galun, M., ed., *Handbook of Lichenology*. CRC Press.

Galun, M., and Bubrick, P. 1984. Physiological interactions between the partners of the lichen symbiosis. In: Linskens, H. F., and Heslop-Harrison, J., eds., *Cellular Interactions. Encyclopedia of Plant Physiology*. Springer-Verlag.

Gärtner, G. 1985. Die Gattung *Trebouxia* Puymaly (Chlorellales, Chlorophyceae). *Arch. Hydrobiol*. Suppl. 71: 495–548.

Greenhalgh, G. N., and Anglesea, D. 1979. The distribution of algal cells in lichen thalli. *Lichenologist* 11: 283–292.

Greenhalgh, G. N., and Whitfield, A. 1987. Thallus tip structure and matrix development in *Bryoria fuscescens*. *Lichenologist* 19: 295–306.

Hawksworth, D. L. 1988a. The fungal partner. In: Galun, M., ed., *Handbook of Lichenology*. CRC Press.

Hawksworth, D. L. 1988b. The variety of fungal-algal symbioses, their evolutionary significance, and the nature of lichens. *Bot. J. Linn. Soc*. 96: 3–20.

Hertel, H., and Rambold, G. 1988. Cephalodiate Arten der Gattung *Lecidea* sensu lato (*Ascomycetes lichenisati*). *Plant Syst. Evol*. 158: 289–312.

Hill, D. J. 1985. Changes in photobiont dimensions and numbers during codevelopment of lichen symbionts. In: Brown, D. H., ed., *Lichen Physiology and Cell Biology*. Plenum.

Hill, D. J. 1989. The control of the cell cycle in microbial symbionts. *New Phytol*. 112: 175–184.

Honegger, R. 1984. Cytological aspects of the mycobiont-phycobiont relationship in lichens. Haustorial types, phycobiont cell wall types, and the ultrastructure of the cell wall surface layers in some cultured and symbiotic myco- and phycobionts. *Lichenologist* 16: 111–127.

Honegger, R. 1985. Fine structure of different types of symbiotic relationships in lichens. In: Brown, D. H., *Lichen Physiology and Cell Biology*. Plenum.

Honegger, R. 1986a. Ultrastructural studies in lichens. I. Haustorial types and their frequencies in a range of lichens with trebouxioid phycobionts. *New Phytol*. 103: 785–795.

Honegger, R. 1986b. Ultrastructural studies in lichens. II. Mycobiont and photobiont cell wall surface layers and adhering crystalline lichen products in four Parmeliaceae. *New Phytol*. 103: 797–808.

Honegger, R. 1987. Questions about pattern formation in the algal layer of lichens with stratified heteromerous thalli. *Bibl. Lichenol*. 25: 59–71.

Honegger, R. 1988. The functional morphology of cell-to-cell interactions in lichens. In: Scannerini, S., Smith, D. C., Bonfante-Fasolo, P., and Gianinazzi-

Pearson, V., eds., *Cell to Cell Signals in Plant, Animal and Microbial Symbiosis.* Springer/NATO ASI series.

Honegger, R. 1990a. Acyclic polyols in cultured lichen mycobionts. In: Reisinger, A., and Bresinsky, A., eds., Fourth International Mycological Congress. Abstracts. University of Regensburg.

Honegger, R. 1990b. Surface interactions in lichens. In: Wiessner, W., Robinson, D. G., and Starr, R. C., eds., *Experimental Phycology* 1. Springer-Verlag.

Honegger, R. 1991a. Developmental biology of ascomycetous lichens. In: Read, N. D., and Moore, D., eds., *Developmental Biology of Ascomycetes.* Academic Press.

Honegger, R. 1991b.Lichens. Structural features. In: Reisser, W., ed., *Algal Symbioses*. Blackie.

Honegger, R., and Bartnicki-Garcia, S. 1991. Cell wall structure and composition of cultured mycobionts from the lichens *Cladonia macrophylla, Cladonia caespiticia,* and *Physcia stellaris* (Lecanorales, Ascomycetes). *Mycol. Res.*, submitted.

Jahns, H. M. 1988a. The lichen thallus. In: Galun, M., ed., *Handbook of Lichenology*. CRC Press.

Jahns, H. M. 1988b. The establishment, individuality and growth of lichen thalli. *Bot. J. Linn. Soc.* 96: 21–29.

James, P.W., and Henssen, A. 1976. The morphological and taxonomic significance of cephalodia. In: Brown, D. H., Hawksworth, D. L., and Bailey, R. H., eds., *Lichenology: Progress and Problems*. Academic Press.

Kidston, R., and Lang, W. H. 1921. On old red sandstone plants showing structure, from the Rhynie chert bed, Aberdeenshire, Part V. *Trans. Roy. Soc. Edinb.* 52: 855–902.

Larson, D. W. 1983. The pattern of production within individual Umbilicaria lichen thalli. *New Phytol.* 94: 409–419.

Law, R., and Lewis, D. H. 1983. Biotic environments and the maintenance of sex: Some evidence from mutualistic symbioses. *Biol. J. Linn. Soc.* 20: 249–276.

Lewis, D. H. 1987. Evolutionary aspects of mutualistic associations between fungi and photosynthetic organisms. In: Rayner, A. D. M., Brasier, C. M., and Moore, D., eds., *Evolutionary Biology of the Fungi*. Cambridge University Press.

Lines, C. E. M., Ratcliffe, R. G., Rees, T. A. V., and Southon, T. E. 1989. A ^{13}C NMR study of photosynthate transport and metabolism in the lichen *Xanthoria calcicola* Oxner. *New Phytol.* 111: 447–482.

Moreau, F., and Moreau, Mme. 1919. Recherches sur les lichens de la famille des Peltigeracées. *Ann. Sci. Nat.*, 10e serie Bot. 1: 29–137.

Ott, S. 1987a. The juvenile development of lichen thalli from vegetative diaspores. *Symbiosis* 3: 57–74.

Ott, S. 1987b. Sexual reproduction and developmental adaptations in *Xanthoria parietina*. *Nord. J. Bot.* 7: 219–228.

Ott, S. 1987c. Reproductive strategies in lichens. *Bibl. Lichenol.* 25: 81–93.

Pirozynski, K. A., and Malloch, D. W. 1975. The origin of land plants: A matter of mycotrophism. *BioSystems* 6: 153–164.

Plessl, A. 1963. Über die Beziehungen von Pilz und Alge im Flechtenthallus. *Oest. Bot. Z.* 110: 194–269.

Poelt, J. 1970. Das Konzept der Artenpaare bei den Flechten. *Ber. Deutsch. Bot. Ges., N.F.* 4: 77–81.

Poelt, J. 1986. Morphologie der Flechten— Fortschritte und Probleme. *Ber. Deutsch. Bot. Ges.* 99: 3–29.

Poelt, J., and Doppelbauer, H. 1956. Über parasitische Flechten. *Planta* 46: 467–480.

Poelt, J., and Mayrhofer, H. 1988. Über Cyanotrophie bei Flechten. *Plant Syst. Evol.* 158: 265–281.

Smith, D. C. and Douglas, A. 1987. *The Biology of Symbiosis*. Edward Arnold.

Tschermak, E. 1941. Untersuchungen über die Beziehungen von Pilz und Alge im Flechtenthallus. *Oest. Bot. Z.* 90: 233–307.

Tschermak-Woess, E. 1988. The algal partner. In: Galun, M., ed., *Handbook of Lichenology*. CRC Press.

23

Symbiosis, Interspecific Gene Transfer, and the Evolution of New Species: A Case Study in the Parasitic Red Algae

Lynda J. Goff

It is the customary fate of new truths to begin as heresies and to end as superstitions.

Thomas Henry Huxley

Horizontal (or lateral) gene transfer is a process by which genetic information from one species is transferred to another, where it may be integrated into the genome of the recipient cell. This phenomenon is well documented among prokaryotes and plays an important role in their evolution (Clark and Warren 1979; Trieur-Cuot et al. 1988). That horizontal gene transfer occurs between prokaryotes and eukaryotes was demonstrated by the ability of bacterial conjugative plasmids to mobilize DNA transfer between bacteria and yeast (Heinemann and Sprague 1989). The phytopathogenic bacterium *Agrobacterium* transmits bacterial genes to many plant hosts. The plasmid DNA enters plant cells, where it is integrated into the nuclear genome and codes for products required by the bacterium. An ancestral I_i-DNA transfer event from *Agrobacterium* may have played a role in the evolution of *Nicotiana* species, since many of them harbor a copy of the I_i-DNA plasmid that affects the plant's development (Furner et al. 1986).

With few exceptions (Busslinger et al. 1983; Spolsky and Uzzell 1984; Syvanen 1984, 1985, 1986; Gould 1986; Landsmann et al. 1986; Krieber and Rose 1986; Lavitrano, et al. 1989), horizontal gene transfer between eukaryotes is thought to occur rarely and to have had little significance in their evolution. Stachel and Zambryski (1989) stated: "As any student of mythology can attest, conjugal mating between disparate species will bring forth wonderful and novel creatures. . . but Nature has wisely set barriers to such horizontal gene transfer in eukaryotes. . . ." Incompatibility barriers that inhibit cells of one species from fusing with another ("trans-species sex") limit the gene transfer between species; trans-species gene transfer rarely occurs via this mechanism. However, genes may move from one species to another via viruses (Went 1971; Jeppsson 1986; McClure et al. 1987) or during the intimate associations that occur between organisms involved in symbiosis.

Although it has not been experimentally documented, the symbiosis literature is full of tantalizing evidence suggesting that trans-species gene transfer has played a major role in the evolution of the Protoctista. Numerous cases of symbioses are known among the protoctists in which one or more of the genomes of a symbiont resides in the cytoplasm of a host. For example, in *Peridinium balticum* (a dinomastigote) a eukaryotic nucleus, presumably originating from a chrysophyte alga, resides in cells along with the characteristic mesokaryotic, dinomastigote nucleus (Dodge 1971; Tomas and Cox 1973). The two nuclei divide synchronously before cell replication, and a synchronous sexual reproduction cycle has evolved for the endosymbiont and its host (Chesnick and Cox 1987, 1989). The nucleomorphs of the cryptophytes and the chlorarachnions provide additional examples of organisms that contain more than one genetic type of nucleus. The DNA-containing nucleomorph is postulated to be the vestigial remains of an ancestral cryptophyte cell whose nucleus entered along with its plastid (Ludwig and Gibbs 1987, 1989).

Plastids from one eukaryotic species may reside in cells of unrelated species (Gibbs 1981). Presumably, either these plastids must be genetically autonomous (i.e., not requiring gene products from the host nucleus) or the gene products required by the plastid must have been previously transferred into the host's nucleus from a resident plastid or from a previous plastid or nuclear endosymbiont. In either case, the sequential symbiont acquisition of organelles and their genomes by one organism from another has apparently given rise to a plethora of new protoctistan species, and in some cases to entirely new divisions of organisms (Whatley and Whatley 1981, 1984).

Once intracellular genomes become synchronous with the cell cycle of the host, the opportunity exists for genes to move from one organism's genome to the other. The mechanism remains obscure, but it may be analogous to the mechanism involved in the transfer of genes between organelles that occurs regularly in eukaryotic cells (Lonsdale et al. 1983; Stern and Palmer 1984; Timmis and Scott 1984; Lonsdale 1985; Kemble et al. 1983, 1985; Gellissen and Michaelis 1987).

The one symbiotic system discussed here is that of the parasitic red algae, which demonstrates horizontal gene transfer in symbiosis and evolution of species.

The Organisms

Nearly a quarter of all known genera of red algae exist as parasites of other red algae (Goff 1982; figures 1 and 2 here). Genera of parasitic red algae have evolved in all orders of red algae in which the ability to form secondary pit connections[1] is characteristic (Goff and Coleman 1985). Although most parasitic red algae have few or none of the pigments characteristic of the division Rhodophyta, historically they have been classified as red algae for the following reasons: they exhibit a triphasic life history that is characteristic of the florideophycean red algae; their developmental morphology (typically filamentous, pseudoparenchymatous) is similar to that of other red algae, particularly their hosts; pit plugs (pit connections) occur between their cells; the proplastids of the parasites, although highly reduced in size and structural complexity, may exhibit characteristic features unique to the red algae (i.e., parallel, single traversing thylakoids, a parietal encircling thylakoid, and infrequently a few phycobilisomes); parasites have typical saccate red algal mitochondria directly associated with the golgi complex; and the characteristic carbon storage compound, glycogen, occurs within their cytoplasm.

Until recently, the development of the parasitic red algae was thought to progress as described in the red algae *Harveyella* and *Choreocolax* (Goff and Coleman 1985). Spores (either haploid or diploid), produced by the region of the parasite thallus that protrudes from the tissue of the host, attach to and germinate on the surface of the host in these organisms (figure 3a). The cytoplasm of the spore migrates into the germ tube that emerges from the germinating spore, and this tube penetrates through the outer wall surface of the host. Within the intercellular wall region of the host, a cell is cut off from the penetrating germ tube and divides to form a branching filament of cells that grows throughout a localized region of host tissue. Within these regions, the parasite's filaments may coalesce into a pseudoparenchymatous mass, which will eventually rupture through the outer wall layer of the host to form the characteristic thallus of the

Figure 1
Plocamiocolax pulvinata (2N) and its red algal host *Plocamium cartilagineum*. The arrow indicates the mass of heterokaryotic tissue in which nuclear transfer, from parasite to host, has occurred during development.

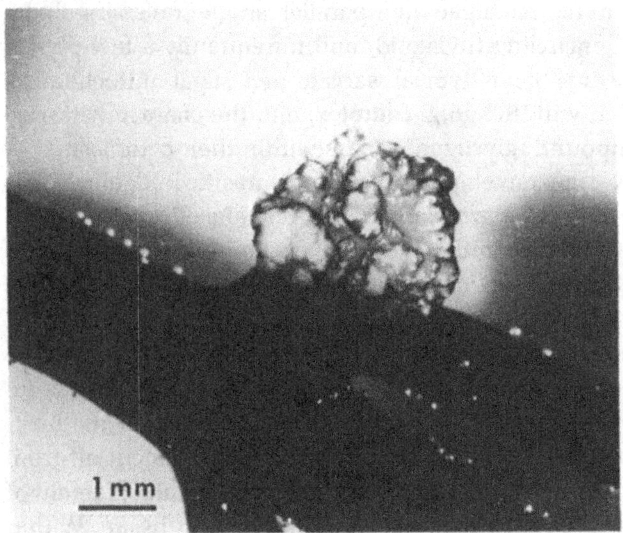

Figure 2
Gardneriella tuberifera and its red algal host *Sarcodiotheca gaudichaudii*. The pigmented tissue of the tumor mass is composed of host cells which have not yet received parasite nuclei and in which plastids are still functional photosynthetically.

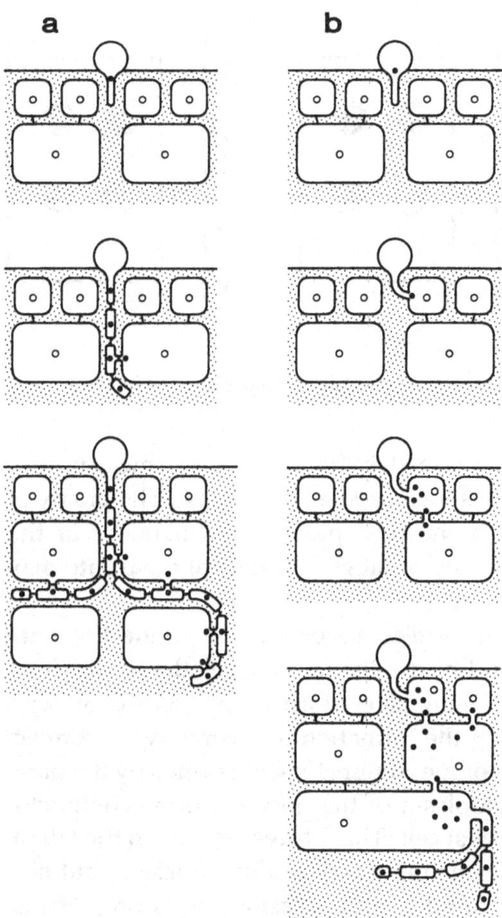

Figure 3
Initial infection of host plants by spores of parasitic red algae. In the first type (a), the parasite injects nuclei into host cells but these nuclei do not undergo DNA synthesis or nuclear division. The parasite forms invasive filaments of cells that grow intercellularly within tissues of its host. In the second type (b), the parasite injects nuclei into host cells and these undergo DNA synthesis and division. The parasite genome is spread intracellularly from host cell to host cell.

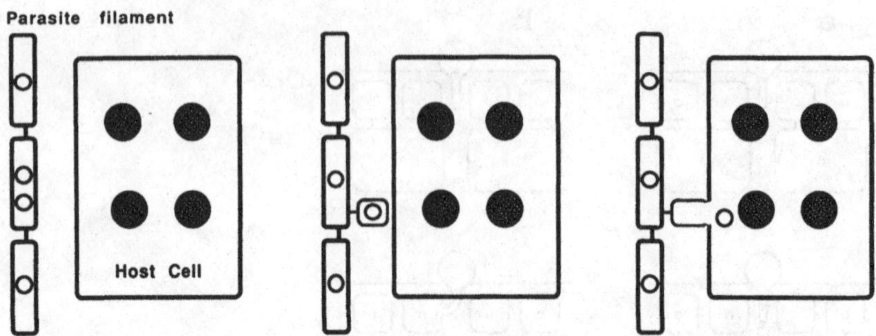

Figure 4
The formation of a secondary pit connection (plug) between parasite and host cells.

parasite that produces either gametes, haploid spores (meiospore = tetraspores), or diploid spores (carpospores = zygotes). The ploidy of the generation produced on the host plant is independent of the host's ploidy, and therefore any or all generations of a parasite may be found on a single host individual.

In both *Harveyella* and *Choreocolax*, secondary pit connections are formed between parasite and host cells; the first cell that forms from the initial penetration tube immediately forms a pit connection with an adjacent host cell. During the formation of secondary pit connections (figure 4), the nucleus of the conjunctor cell, formed by the parasite, is injected into the cytoplasm of the host when the conjunctor cell fuses with an adjacent host cell. This process results in the lateral transfer of not only the genetic information of the nucleus, but also that of the parasite's mitochondria and occasionally its proplastids, which are also injected into the host's cytoplasm. In *Choreocolax* and *Harveyella*, the parasite nuclei do not undergo DNA synthesis or karyokinesis within the host's cytoplasm, but they are likely instrumental in redirecting the physiology of the host cell for the non-photosynthetic parasite tissues.

The examination of several other red algal parasites, particularly in the adelphoparasites (those closely related systematically to their hosts), has revealed an entirely different pattern of parasite development, one previously unknown in any other eukaryotic organisms (Goff and Coleman 1987). In these taxa, little or no somatic tissue is produced directly by the germ tube of the parasite. Rather, the parasite spore, upon attaching to a host thallus, sends out a germ tube that will fuse directly with an underlying host epidermal or subepidermal cell (figure 3b). Thus, all the cytoplasmic contents of the parasite spore, including its nucleus, its proplastids and its mitochondria, enter into the cytoplasm of the host cell. Microspectrophotometric analysis of the DNA of the parasite nuclei reveals that these nuclei undergo DNA synthesis within the host's cytoplasm and rapidly divide, eventually repopulating the cytoplasm of the host cell with parasite nuclei. The host nuclei often appear to degenerate as the parasite nuclei synthesize DNA and divide.

From these "parasitized host cells" (figure 5), which are transitionally heterokaryotic, parasite nuclei may be dispersed across the host-host intercellular connection that may result upon the dissolution of the pit plug (primary pit connection), or they may be "packaged" into cells that form from the parasitized host cell and which in turn divide to give rise to filaments of cells derived from the heterokaryotic host cell but containing primarily or solely the nuclear genome of the parasite. These filaments grow intrusively throughout localized tissue regions within their hosts, where they may fuse directly with host cells or form interconnections via the formation of secondary pit connections (figures 6a,b and 7a,b). In each case, the process results in spreading the parasite's genome(s?) to previously uninfected host cells, from which new filaments containing parasite nuclei will arise.

The characteristic reproductive thallus of the parasite that protrudes from the host is formed by an interaction of the parasite and the host. In the case of the parasite *Gracilariophila oryzoides*, parasitism induces host-cell proliferation in a localized region immediately surrounding the infection region. This forms a tumor of host tissue in which the developing "parasite" is contained. Eventually, the parasite genome will be spread to all this surrounding tissue; from this mass, gametes, haploid spores, and diploid spores will be produced and disseminated to effect additional host infections.

Host tissue immediately surrounding the developing parasite, *Gardneriella tuberifera*, is stimulated to divide, resulting in a mass of callus tissue that protrudes from the host (figure 2). Parasitized filaments produced from the initial heterokaryotic cells proliferate rapidly and invade this tissue, transferring parasite nuclei to many host cells. Often regions of host tissue in the "tumor" do not receive parasite nuclei. Remaining pigmented and photosynthetically active, these cells transfer photosynthate to the nonpigmented cells that contain the parasite genome (Goff 1982).

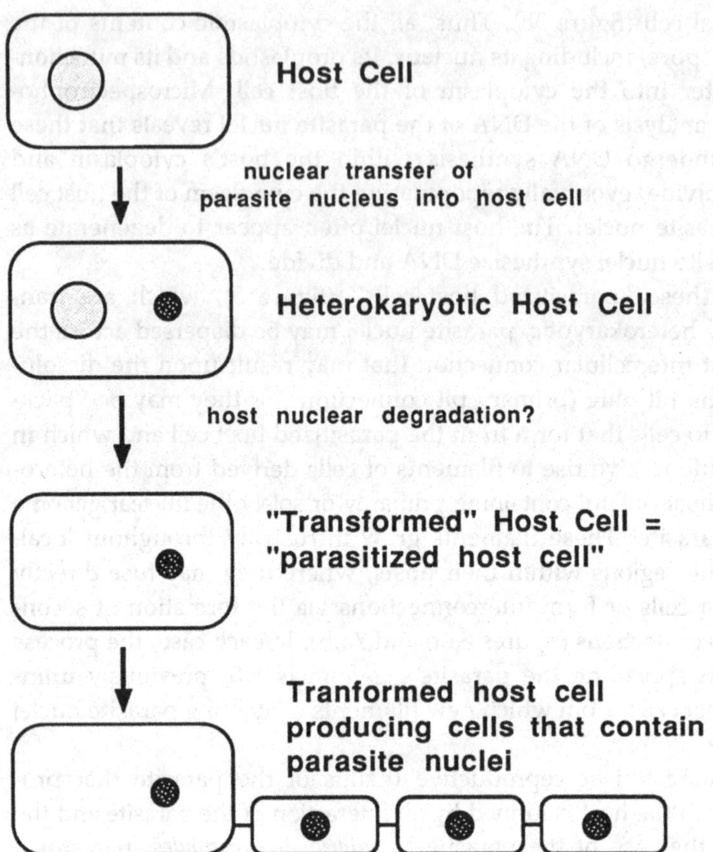

Figure 5
The process by which a red algal host cell is "transformed" by a parasite nucleus.

One of the most remarkable aspects of nearly all the parasitic red algae is how their developmental morphology "mimics" that of its host. In his study of a species of the parasite *Janczewskia* from Japan and one from California, Saito (1972) observed that there was much greater morphological similarity between each parasite and its respective host than there was between the two *Janczewskia* species. These similarities suggest that parasitic red algae may have evolved directly from their hosts.

It is also plausible, partioularly in view of the unusual way that these organisms parasitize the host cell, that the parasitic red algae are totally unrelated to their hosts, and that the expression of the host's growth and reproductive features in the parasite is really a manifestation of the host, via the presence and expression of long-lived host mRNA in parasitized host cells, or is due to intraspecific transfer of genes from the host to the parasite genome. In fact, if the cell-biological and morphological features we use to classify these organisms are really those determined by host genes, parasitic red algae may not even be red algae. Cytoplasm glycogen, red algal-like proplastids, and mitochondrial/golgi complexes in parasite cells may merely be those of their hosts.

Parasite Evolution: Comparisons of Host and Parasite Genomes

The genomes of parasites and hosts are being analyzed and compared to determine the possible origins of the parasitic red algae. One hypothesis is that these organisms evolved directly from the life histories (as "rogues") of their host via some mutation that resulted in a loss of the ability to live autonomously. A second hypothesis is that these organisms originated from a very different group of organisms but have secondarily acquired some of the developmental and morphological characteristics of their host through the transfer (and expression) of host genes into the genome of the parasite.

Nuclear Genome Comparisons

Microspectrofluorometric measurements of DNA levels in host and parasite nuclei reveal the two genomes clearly differ in DNA content. Although these analyses are complicated by the occurrence of developmental polyploidy in both the parasite and the host, generally

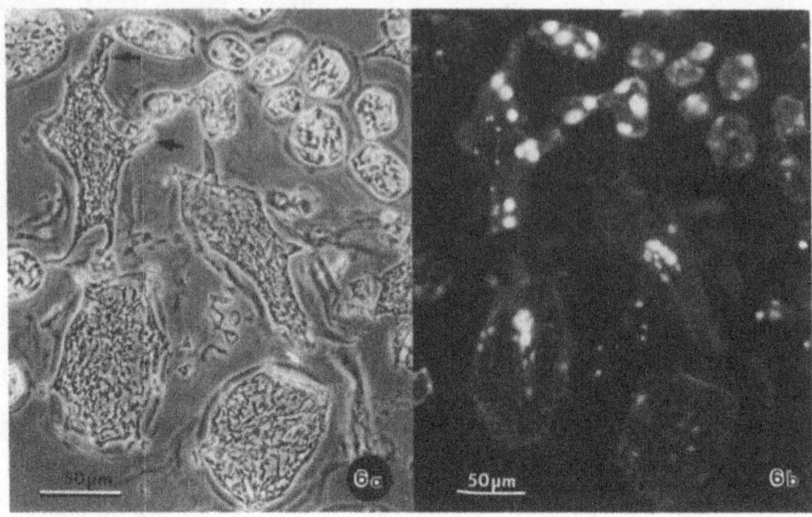

Figure 6
Heterokaryotic cells of *Laurencia spectabilis* that has received nuclei from its 1N parasite, *Janczewskia gardneri*. In figure 6b, the smaller host nuclei and the larger parasite nuclei are evident. The nuclei have been transferred into the host cell via secondary pit connection formation. The regions where conjunctor cells have fused with host cells, thereby delivering parasite nuclei, are indicated by arrows in figure 6a.

the overall genome size of the parasite is slightly smaller than that of its host.

Comparisons of restriction fragment length patterns of the nuclear genome of parasites and their hosts reveal differences in organization between the two (figures 8a,b and 9a,b). However, the rate of change of unconstrained DNA regions such as occur in nuclear genomes suggests that these differences may only be due to nuclear-genome divergence since the parasite's divergence from host.

A good way to compare these genomes is by using the sequence data of some evolutionarily conserved gene, such as the DNA coding for the 16S-like ribosomal DNA gene. Such studies of *Gracilaria lemaneiformis* (Bhattacharya et al. 1990) and its parasite *Gracilariophila* are underway; no conclusions are yet available.

Plastid Genome Comparison

The restriction fragment length patterns of the plastid DNAs of several hosts and their parasites have also been compared. As in flowering

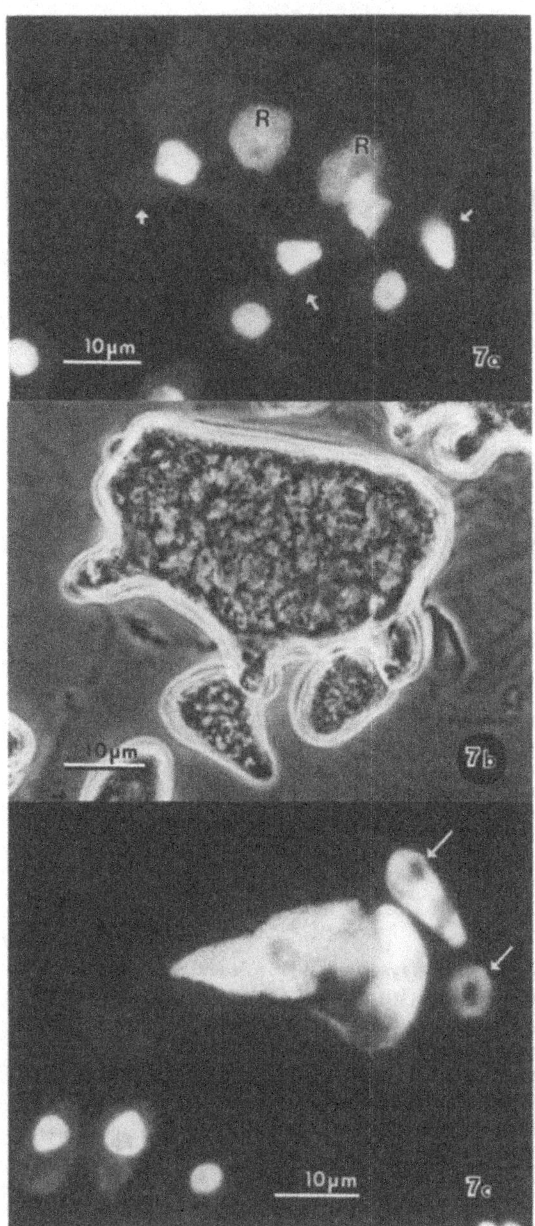

Figure 7
Nuclear transfer from the parasite *Dawsoniocolax bostrychiae* to its red algal host *Bostrychia radicans*. A binucleate pericentral cell of the host (7b) has received four nuclei from the parasite via fusion of several nucleated conjunctor cells with the host cell (7a, arrows). The two resident nuclei (R) of the host cell are indicated in figure 7a. In figure 7c, nucleoli are evident (holes in the DAPI-stained nuclei, indicated by arrows), indicating that RNA synthesis is occurring in these G2 nuclei.

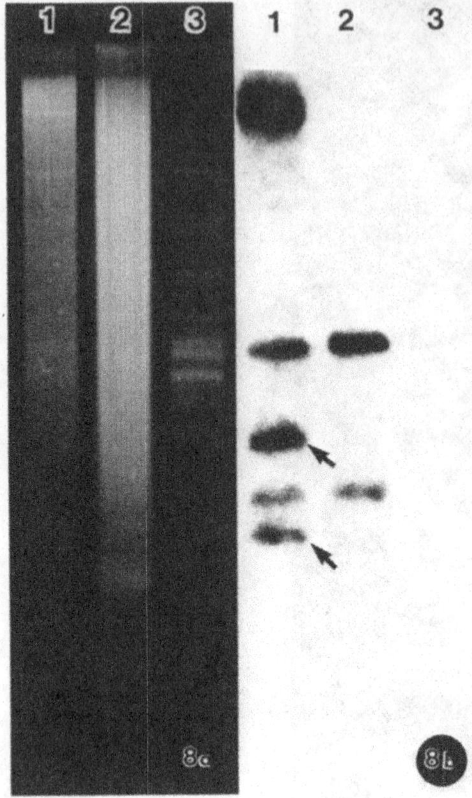

Figure 8
Nuclear DNA of the parasite *Gracilariophila oryzoides*(lane 1 in 8a and 8b) and its host *Gracilaria lemaneiformis* (land 2 in 8a and 8b). In figure 8a, nuclear DNA has been digested with Eco RI and stained with ethidium bromide. Lane 3 is DNA isolated from upper band of cesium chloride gradient; this fraction contains plastid and mitochondrial DNA and plasmid DNA. In figure 8b, the DNA from figure 8a has been transferred to nitrocellulose and probed with Pha-2 (nuclear ribosomal probe from pea). Note that in the parasite lane two unique bands occur (arrows). The upper band in lane 1 likely represents undigested DNA.

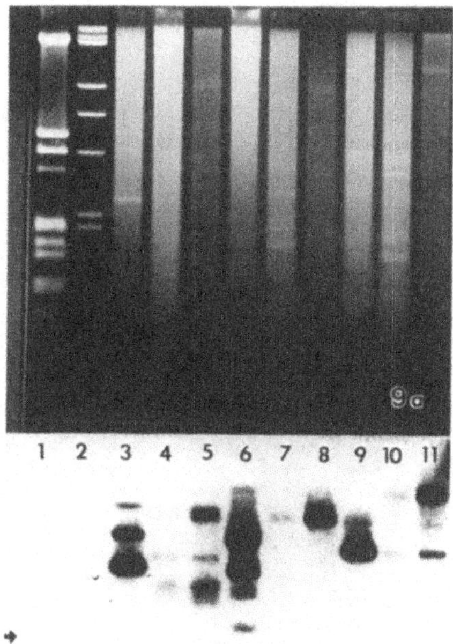

Figure 9
Nuclear DNA of the parasite *Plocamiocolax pulvinata* (lanes 3, 6, 9) and its host *Plocamium cartilagineum* (lanes 4, 5, 7, 8, 10, 11), digested with Bgl II (lanes 3, 4, 5), Eco RI (lanes 6, 7, 8) and Pst I (lanes 9, 10, 11). Lanes 4, 7, and 10 are from one fraction of nuclear DNA (lower fraction of Hoechst 33258 cesium chloride); lanes 5,8, and 11 are from a different fraction that is enriched in nuclear ribosomal DNA. Lanes 1 and 2 are DNA standards.

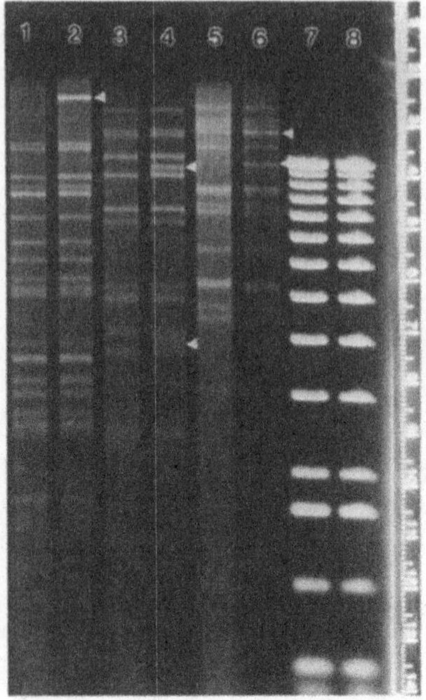

Figure 10
DNA from host *Plocamium cartilagineum* (lanes 1, 3, 5) and parasite *Plocamiocolax pulvinata* (lanes 2, 4, 6). The DNA fractions from each are those from the upper portion of the Hoechst cesium chloride gradients and contain both plastid and mitochondrial DNAs. Lanes 1 and 2 have been cut with Bgl II, lanes 3 and 4 with Eco RI, and lanes 5 and 6 with Pst I. Arrowheads indicate DNA bands unique to parasite DNA. Lanes 7 and 8 are Lambda 1Kb DNA standard.

plants, the plastid genome of red algae is very highly conserved at the species level (Goff and Coleman 1988). By comparing the plastid genome of the host with that of the proplastid of the parasite, we sought to determine either independent evolution of the parasite or derivation of the parasite from its host.

The plastid genomes of several parasites and their hosts show identical plastid DNA patterns (figure 10). These data are not decisive in determining the origins of the parasites; the parasites evolved relatively recently directly from their host. But it is also entirely possible that the proplastid DNA isolated from parasite cells is actually that of the host, since the cells making up the parasite thallus are derived initially from the host and may therefore possess host organelles.

Microscopic observations (fluorescence and electron) of heterokaryotic (parasite + host) cells during the early stages of parasite nuclear invasion and proliferation suggest that the proplastids in the parasite cells originate from host plastids. Fluorescence microscopy reveals that upon the injection and proliferation of parasite nuclei into host cells, the pigmentation of the host's plastids is degraded (the autofluoresence of the chlorophylls and phycobiliproteins disappear; see Goff and Coleman 1987). Electron-microscopic observations of these cells reveal the presence of both plastids and proplastids (figure 11), as well as plastids actively budding and/or dividing to form proplastids (a very unusual process in red algae) (figures 11, 12). The resulting proplastid is a small, structurally simple organelle that, although it contains plastid DNA and an occasional thylakoid, lacks phycobilisomes and appears to be devoid of pigmentation and photosynthetic capability. These proplastids are transmitted to the parasite cells that develop from the heterokaryon, are incorporated into spores, and are transferred back into the host cell upon subsequent infection.

The proplastids of the parasite cells do not regain the capabilities of photosynthesis, and the parasites remain colorless. However, in a few exceptional cases (e.g., the parasite *Janczewskia spectabilis*; see Court 1980) the mature parasite thallus is photosynthetically active and pigmented (though less so than the host). The proplastids of this parasite remain devoid of pigmentation throughout its development from the time of heterokaryon formation to the production of a small white tumor that protrudes from the host tissue. This tumor then begins to express the polysiphonious development characteristic of its host: while the tissue changes from white to pink, typical red-algal plastids differentiate from the proplastids.

The interconversion of photosynthetic plastids to nonphotosynthetic proplastids, and in some cases back to photosynthetic plastids, is interesting in the context of nuclear-plastid interactions. The loss of photosynthetic capabilities once the parasite nucleus invades that host cytoplasm results from the degeneration of the host's nuclear genome or from a cessation of host nuclear RNA transcription. Some nuclear-encoded proteins, such as phycobilisome-linker proteins required for light-harvesting and photosynthesis, might not be available to the plastid. All nonpigmented linker proteins involved in phycobilisome synthesis are nuclear encoded in red algae (Egelhoff and Grossman 1983). This is being tested by probing parasite and host nuclear DNA with heterologous probes to the linker genes from *Cyanophora*.

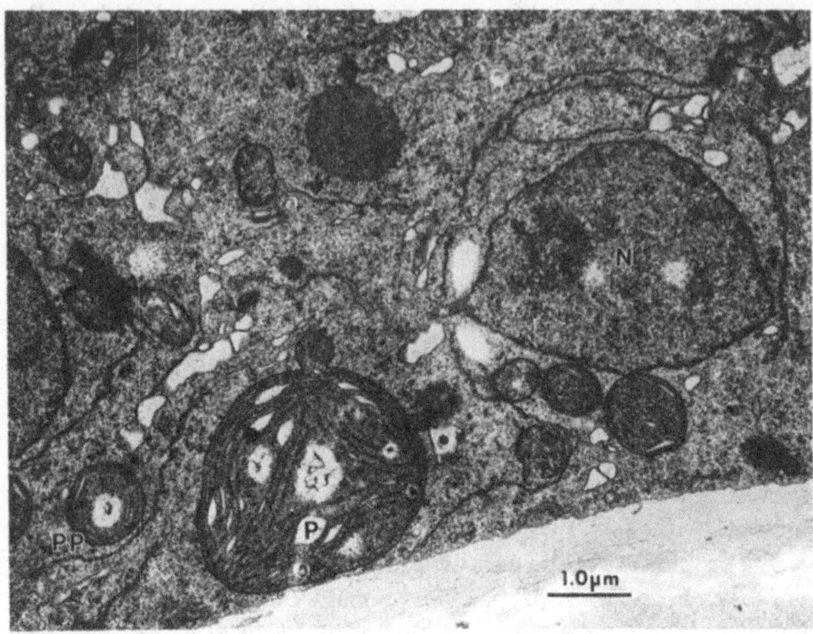

Figure 11
Host cell of *Sarcodiotheca gaudichaudii* that has recently received a parasite nucleus from *Gardneriella tuberifera*. Both plastids (P) and proplastids (PP) are evident. The plastid is budding in two regions.

Mitochondrial Genome Comparisons

Comparison of the restriction-fragment patterns of the parasitic red alga *Plocamiocolax* and its host *Plocamium cartilagineum* revealed fragments in the parasite that are absent from the host. The stoichiometry of DNA fragments stained with ethidium bromide indicates that the additional fragments are due to a genome present in the parasite but absent in the host (figure 10, arrowheads). The extraneous genome, ca. 25–30 kbp (figure 13a) as determined by reverse pulse electrophoresis, might be mitochondrial. Since no mitochondrial probes from other red algae are available yet to confirm the hypothesis, and no mitochondrial probes from other photosynthetic organisms cross-reacted with the putative mitochondrial DNA of red algae (Goff and Coleman 1988), this point is unresolved. An abundance of mitochondrial genomic DNA is also expected since parasite cells are generally highly enriched in mitochondria.

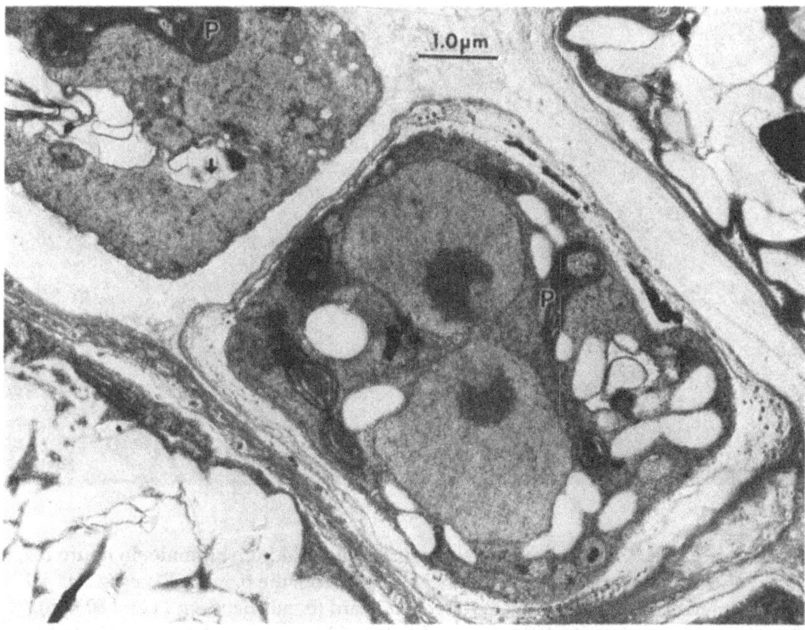

Figure 12
Host cell of *Gracilaria lemaneiformis* that has recently received a parasite nucleus from *Gracilariophila oryzoides*. The plastids (P) are dividing up into small proplastids that contain conspicuous DNA nucleoids.

If the DNA were indeed mitochondrial, we reasoned, lack of evidence (apparent absence) of this genome in the host tissue might only be due to problems of detection, since the number of mitochondria is several orders of magnitude lower in host cells than in parasites. Therefore, we isolated DNA from host gonimoblast, the 2N tissue that results from fertilization and produces the zygotes (= carpospores) and is known to have abundant mitochondria. Pulse-field electrophoresis and electron-microscopic comparisons of DNA isolated from the host gonimoblast tissue (figure 13b) and that from the parasite indicated that both contain a circular genome of similar size (ca. 25–30 kbp; figures 13c and 14a,b).

After cloning several regions of the parasite's putative mitochondrial DNA, we conclude that the restriction patterns of the host and parasite "mitochondrial" genomes differ. However, the similarity in overall size of these circular genomes of both parasite and host, and the ability of regions of this molecule cloned from the parasite to

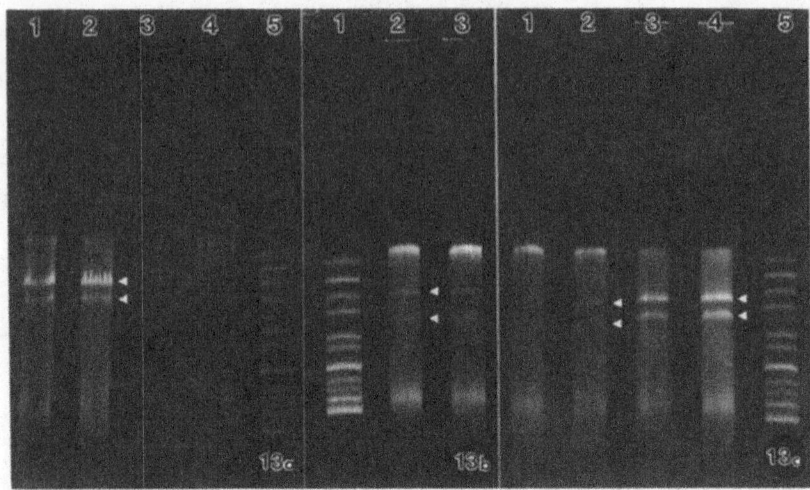

Figure 13
Reverse pulse-field electrophoresis gels, stained with ethidium bromide. In figure 13a, lanes 1 and 2 contain mitochondrial-enriched DNA from the parasite *Plocamiocolax pulvinata*. Lane 5 is the high-molecular-weight standard (bands between 11 and 50 kpb). The DNA is from the upper fraction of a Hoechst 33258 cesium chloride gradient of *Plocamium cartilagineum* cystocarpic tissue (mitochondrial enriched); the arrowheads indicate the putative mitochondrial DNA (circular = upper band vs. linearized = lower band?). In figure 13c, the putative mitochondrial DNAs from the host (lanes 1, 2) are compared against those of the parasite (lanes 3, 4). Lane 1 of 13b and lane 5 of 13c are high-molecular-weight standards.

hybridize with the genome of the host, indicate that host and parasite may be quite closely related.

Conclusion

A coherent story emerges from these investigations once we recognize that the plastids of hosts and their red algal parasites are genetically identical, and if we assume that plastid identity is due to the parasitic red algal plastids' actually originating from their host. Parasitic red algae may be able to survive as nutritional parasites of their hosts only because they have lost, or never acquired, their own functional plastids.

The observations that the mitochondrial genomes of host and parasites differ in their restriction-fragment pattern, but not in their overall size or general sequence homologies, indicate that parasitic red

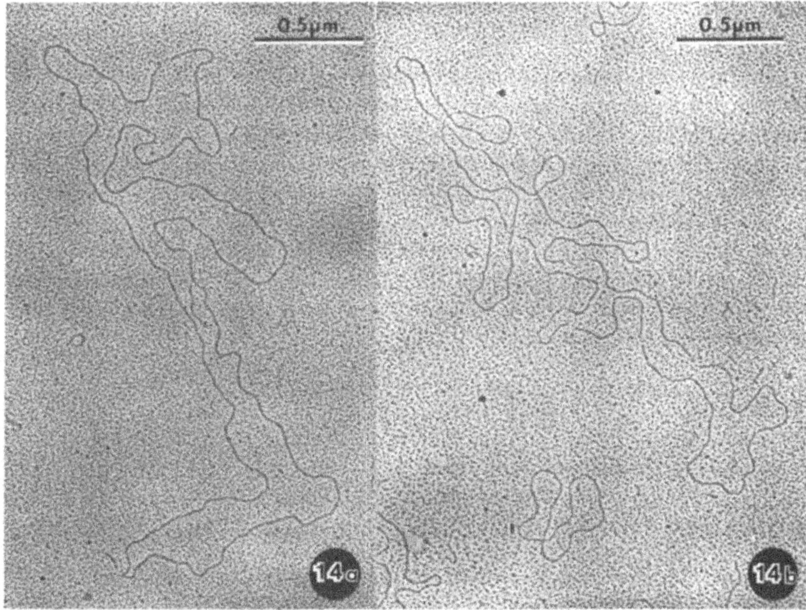

Figure 14
Cytochrome spreading of DNA from the uppermost portion of a Hoechst 33258 cesium chloride gradient from the parasite *Plocamiocolax pulvinata* (14a) and its host *Plocamium cartilagineium* (14b). The small plasmid in figure 14b (arrow) is a 4755-bp plasmid standard (PJMC 110).

algae may have evolved directly from their modern-day or previous red algal host. The mechanism by which a host gives rise to these most remarkable parasites remains obscure, but certainly this case is one in which symbiosis and the horizontal transfer of genomes has played a major role in the evolution of new species.

Acknowledgments

My sincere thanks are extended to the Bellagio Conference organizers, Lynn Margulis and Ken Nealson, for the opportunity of presenting this paper, and to the Rockefeller Foundation for supporting the conference. I thank K. Jeon, R. Fester, A. Coleman, and L. Liddle for their careful reviewing and editorial help. This research was supported by the National Science Foundation (BSR 8415760 and BSR B709239).

Note

1. Secondary pit connections form between contiguous photosynthetic cells of some red algae when one cell (the donor) divides unequally to produce a small, nucleated conjunctor cell that is interconnected to its parent cell by a pit plug. The pit plug is deposited in the septal pore which results from incomplete centripetal infurrowing of the plasmalemma during cytokinesis. During secondary pit connection formation, the conjunctor cell formed by the donor cell fuses with the adjacent recipient cell. By this mechanism, the parent cell of the conjunctor and the recipient cell become interconnected via the pit plug that initially had been deposited between the parent and the short-lived conjunctor cell.

References

Bhattacharya, D., Elwood, H. J., Goff, L. J., and Sogin, M. L. 1990. The phylogeny of *Gracilaria lemaneiformis* (Rhodophyta) based on the sequence analysis of its small subunit ribosomal RNA coding region. *J. Phycol.* 26: 181–186.

Busslinger, M., Rusconi, S., Rohrer, U., and Birnstiel, M. L. 1983. A horizontal gene transfer between sea urchin species? In: Robberson, D. L., and Saunders, G. F., eds., *Perspectives on Genes and the Molecular Biology of Cancer*. Raven.

Chesnick, J. M., and Cox, E. R. 1987. Synchronized sexuality of an algal symbiont and its dinoflagellate host, *Peridinium balticum* (Levander) Lemmermann. *BioSystems* 21: 69–78.

Chesnick, J. M., and Cox, E. R. 1989. Fertilization and zygote development in the binucleate dinoflagellate *Peridinium balticum* (Pyrrhophyta). *Amer. J. Bot.* 76: 1060–1072.

Clark, A. J., and Warren, G. J. 1979. Conjugal transmission of plasmids. *Ann. Rev. Gen.* 13: 99–123.

Court, G. J. 1980. Photosynthesis and translocation studies of *Laurencia spectabilis* and its symbiont *Janczewskia gardneri* (Rhodophyceae). *J. Phycol.* 16: 270–279.

Dodge, J. D. 1971. A dinoflagellate with both a mesocaryotic and a eucaryotic nucleus. I. Fine structure of the nuclei. *Protoplasma* 73: 145–157.

Egelhoff, T., and Grossman, A. 1983. Cytoplasmic and chloroplast synthesis of phycobilisome polypeptide. *Proc. Nat. Acad. Sci.* 80: 3339–3343.

Furner, I. J., Juffman, G. A., Amasino, R. M., Garfinkel, D. J., Gordon, M. P., and Nester, E. W. 1986. An *Agrobacterium* transformation in the evolution of the genus *Nicotiana*. *Nature* 319: 422–427.

Gellissen, G., and Michaelis, G. 1987. Gene transfer: Mitochondria to nucleus. In: Lee, J. J., and Fredrick, J. F., eds., *Endocytobiology III*. New York Academy of Sciences.

Gibbs, S. P. 1981. The chloroplasts of some algal groups may have evolved from endosymbiotic eukaryotic algae. In: Fredrick, J. F., ed., *Origins and Evolution of Eukaryotic Intracellular Organelles*. New York Academy of Sciences.

Goff, L. J. 1982. The biology of parasitic red algae. In: Round, F., and Chapman, D., eds., *Progress in Phycological Research*, volume 1. Elsevier.

Goff, L. J., and Coleman, A. W. 1985. The role of secondary pit connections in red algal parasitism. *J. Phycol.* 21: 483–508.

Goff, L. J., and Coleman, A. W. 1987. Nuclear transfer from parasite to host. A new regulatory mechanism of parasitism. In: Lee, J. J., and Fredrick, J. F., eds., *Endocytobiology III*. New York Academy of Sciences.

Goff, L. J., and Coleman, A. W. 1988. The use of plastid DNA restriction endonuclease patterns in delineating red algal species and populations. *J. Phycol.* 24:357–368.

Gould; S. J. 1986. Linnaean limits. *Natural History* 95: 16–19.

Heinemann, J. A., and Sprague, G. F. 1989. Bacterial conjugative plasmids mobilize DNA transfer between bacteria and yeast. *Nature* 340: 205–209.

Jeppsson, L. 1986. A possible mechanism in convergent evolution. *Paleobiology* 12: 80–88.

Kemble, R. J., Gabay-Laughnan, S., and Laughnan, J. R. 1985. Movement of genetic information between plant organelles: Mitochondria-nuclei. In: Hohn, B., and Dennis, E. S., eds., *Plant Gene Research. Genetic Flux in Plants*. Springer.

Kemble, R. J., Mans, R. J., Gambay-Laughnan, S., and Saughnan, J. R. 1983. Sequences homologous to episomal mitochondrial DNAs in the maize nuclear genome. *Nature* 304: 744–747.

Krieber, M., and Rose, M. R. 1986. Molecular aspects of the species barrier. *Ann. Rev. Ecol. Syst.* 17: 465–485.

Landsmann, J., Dennis, E. S., Higgins, T. J. V., Appleby, C. A., Kortt, A. A., and Peacock, W. J. 1986. Common evolutionary origins of legume and nonlegume plant haemoglobins. *Nature* 324: 166–169.

Lavritrano, M., Camaioni, A., Fazio, V. M., Dolci, S., Farace, M. G., and Spadafora, C. 1989. Sperm cells as vectors for introducing foreign DNA into eggs: Genetic transformation of mice. *Cell* 57: 717–723.

Lonsdale, D. M. 1985. Movement of genetic material between the chloroplast and mitochondrion in higher plants. In: Hohn, B., and Dennis, E. S., eds., *Plant Gene Research: Genetic Flux in Plants.* Springer.

Lonsdale, D. M., Hodge, T. P., Howe, C. J., and Stern, D. B. 1983. Maize mitochondrial DNA contains a sequence homologous to the ribulose-1,5-biphosphate carboxylase large subunit gene of chloroplast DNA. *Cell* 34: 1007–1014.

Ludwig, M., and Gibbs, S. P. 1987. Are the nucleomorphs of Cryptomonads and Chlorarachnion the vestigial nuclei of eukaryotic endosymbionts? In: Lee, J. J., and Fredrick, J. F., eds., *Endocytobiology III.* New York Academy of Sciences.

Ludwig, M., and Gibbs, S. P. 1989. Evidence that the nucleomorphs of *Chlorarachnion reptans* (Chlorarachniophyceae) are vestigial nuclei: Morphology, division and DNA-DAPI fluorescence. *J. Phycol.* 25: 385–394.

McClure, M. A., Johnson, M. S., and Doolittle, R. F. 1987. Relocation of a protease-like gene segment between two retroviruses. *Proc. Nat. Acad. Sci.* 84: 2693–2697.

Saito, Y. 1972. Two species of *Janczewskia* from Japan and their systematic relationships. In: Nisizawa, K., ed., *Proceedings of the Seventh International Seaweed Symposium.* University of Tokyo Press.

Spolski, C., and Uzzell, T. 1986. Natural interspecies transfer of mitochondrial DNA in Amphibians. *Proc. Nat. Acad. Sci.* 81: 5802–5805.

Stachel, S. E., and Zambryski, P. C. 1989. Generic trans-kingdom sex? *Nature* 340: 190–191.

Stern, D. B., and Palmer, J. D. 1984. Extensive and widespread homologies between mitochondrial DNA and chloroplast DNA in plants. *Proc. Nat. Acad. Sci.* 81: 1946–1950.

Syvanen, M. 1985. Cross-species gene transfer: Implication for a new theory of evolution. *J. Theor. Biol.* 112: 333–343.

Syvanen, M. 1986. Cross-species gene transfer: A major factor in evolution. *Trends in Genetics* 2: 63–66.

Timmis, J., and Scott, N. S. 1984. Promiscuous DNA: Sequence homologies between DNA of separate organelles. *Trends in Biochem. Sci.* 9: 271–273.

Tomas, R. W., and Cox, E. R. 1973. Observations on the symbiosis of *Peridinium balticum* and its intracellular alga. I. Ultrastructure. *J. Phycol.* 9: 304–323.

Trieur-Cuot, P., Carlier, C., and Courvalin, P. 1988. Conjugative transfer from *Enterococcus faecalis* to *Escherichia coli. J. Bacteriology* 170: 4388–4391.

Went, F. W. 1971. Parallel evolution. *Taxon* 20: 197–226.

Whatley, J. M., and Whatley, F. R. 1981. Chloroplast evolution. *New Phytologist* 87: 233–247.

Whatley, J. M., and Whatley, F. R. 1984. Evolutionary aspects of the eukaryotic cell and its organelles. In: Linskens, H. F., and Heslop-Harrison, J., eds., *Encyclopedia of Plant Physiology,* N.S., volume 17: *Cellular Interactions*. Springer-Verlag.

24 Galls, Flowers, Fruits, and Fungi

A. Pirozynski

This chapter speculatively explores some implications of the idea (Pirozynski 1988) that angiospermous flowers and fleshy fruits are derived from arthropod-induced galls via the incorporation of microbial or fungal DNA into the plant genome.

In the conventional view, flowers are derived from an archetype through progressive rearrangement and transformation of ancestral features, with pollinating animals accorded a major selective role. Likewise, fleshy fruits in angiosperm families are viewed as parallel or convergent end products of adaptive evolution, with seed-dispersing vertebrates contributing to the selection pressures.

However, recent paleobotanical evidence underscores rather than bridges the discontinuous innovation seen in Cretaceous flowers and, in particular, the 40-million-year gap that separates the initial radiation of flower types in the mid-Cretaceous from that of fleshy fruits in the late Cretaceous (Friis et al. 1987). Concomitantly, flowers and fruits resemble galls in their basic structure; "faithful" pollinating animals are derived from mycophagous and carnivorous rather than pollinating phytophagous ancestors; the origin and early radiation of fleshy fruits coincides with that of gall-making holometabolan insects, rather than that of frugivorous birds; and frugivorous birds are of carnivorous rather than herbivorous ancestry.

Ninety-eight percent of all known galls are on angiosperms (Mani 1964). The earliest fossils interpreted to be insect-induced galls resemble modern oak-leaf spangle galls found on angiospermous leaves of uncertain identity from early Cretaceous sediment (Larew 1986). Fossils comparable to modern galls date from the Eocene (Gagné 1984). What happened in the intervening 60 or 70 million years must be inferred from the reconstructed histories of gall-makers.

Organoid Galls

Organoid galls are condensed and modified shoots, e.g., psyllid-induced rosette galls or mite-induced witches' brooms. Channabasavanna and Nangia (1984) described the latter as "somewhat branched and compacted [shoot apices] with shortened internodes . . . bunches of discoloured and distorted leaves." The "discolouration" may be yellow, white, red, or violet (Mani 1964; Ananthakrishnan 1984). Rudimentary or young leaves may be multifariously modified, i.e., folded, fused, or transposed, to resemble structures found in flowers and fruits. Some fold and roll galls are swollen, spongy, fleshy, woody, or fibrous. A psyllid-induced hypertrophy of carpels of *Rumex* makes the normally small ovary into a "pod-shaped structure up to 2 cm long" (Hodkinson 1984). Some Homoptera that induce fleshy open-pit galls at the base of young leaves void honey-dew through the gall exit aperture (Beardsley 1984).

Organoid galls are often the outcome of the combined action of populations rather than individuals, and are considered primitive (Mani 1964). They are induced, in all classes of plants, by the more ancient groups of arthropods (mites, thrips, homopterans), whose association with plants probably predated the rise of angiosperms. Teratogenic prokaryotes and fungi (e.g., *Taphrina*) would have not only preceded but perhaps also fostered gall-making in those arthropods.

A fortuitous concatenation of homopteran and fungal galls of sporophylls could have provided morphogenetic novelty as well as an extraordinary opportunity for initiating coevolution with pollinators if it involved some combination of the following:

- a roll gall folding an ovulate sporophyll into a carpel-like enclosure
- secretion of honey-dew or scent, possibly attracting a free-ranging mate (Beardsley 1984), or opportunistic honey-dew feeders (Downes 1973) and parasitoids contaminated with pollen
- pigmentation attracting vectors (contaminated with pollen) of microbial symbionts deployed in gall formation (Nealson, this volume)
- sexual transmutation with accompanying hypertrophy, as envisaged in the evolution of the female ear of cultivated maize (Iltis 1983; Lamboy 1984), and even of the angiospermous flower (Meyen 1988).

Pollination

Apparently insect pollination long predates the origin of angiospermous flowers. The first pollinators were probably mycophagous or predatory beetles that adapted to protein-rich pollen, ovules, and seeds of Paleozoic and early Mesozoic plants: pollination was incidental (Crepet 1985) and remains largely so. A large gap separates pollinating beetles and "faithful" pollinators, a category that includes Hymenoptera (chiefly bees and wasps), some Diptera (flies), Lepidoptera (butterflies and moths), and vertebrates (birds and bats).

Like beetles, some Hymenoptera began as mycophages before adapting to phytophagy. Unlike beetles, they developed more intimate relationships with plants. In the late Cretaceous, the ancestors of bees probably fed on pollen and nectar but fed insect larvae to their own nest-bound progeny. The prey would have included gall-makers and their symbionts on flowers in which the predators foraged. Feeding transitions in Hymenoptera have often been preceded by cohabitation with future prey (Rasnitsyn 1980). Furthermore, examples of transition from feeding on gall insects to feeding on gall tissue or products are common throughout the order. Some pollinating Diptera feed on nectar only, others on nectar and pollen. Their association with flowers dates from the Tertiary and probably evolved from opportunistic nectar-feeding (Willemstein 1987). Downes (1973) suggested that the utilization of honey-dew by early Diptera would have preadapted their pollinating descendants for a diet of nectar. Lepidoptera first appear in the early Cretaceous as carnivores and saprophages. What is now a highly evolved or coevolved pollination relationship probably originated in the Tertiary from nectar thieving by adults (Willemstein 1987). Pollination by birds and bats is also of recent, opportunistic origin (Crepet 1985).

Therefore, exploitation of gall symbionts by Hymenoptera and, initially, of gall-makers' products (honey-dew) by Lepidoptera and Diptera provides a plausible starting point for the coevolution of flowers with "faithful" pollinators.

Histioid Galls

The histologically complex histioid galls are formed primarily on angiosperms by the Diptera (cecidomyiid or gall midges) and the Hymenoptera (cynipid and chalcid wasps). Jurassic ancestors of gall midges were mycophagous (Mamaev 1975); the gall-forming habit has evolved independently at least three times since the late Cretaceous. The subfamily Asphondyliini are chiefly deformers of flowers and fruits. Flowers, the preferred habitat for fly larvae (Mamaev 1975), are the presumed original site of cecidomyiid gall formation (Mani 1964). Feeding on honey-dew, floral nectar, flower-inhabiting fungi, or liquid products of the fungal degradation of flowers may have prepared gall midges for gall-making, as well as for the most recently derived "free" phytophagy (Mamaev 1975; Bissett and Borkent 1988).

At the time of the radiation of flowers in the early Cretaceous, the Apocritan Hymenoptera were chiefly carnivorous Ichneumonomorpha and Vespomorpha. By the end of the Cretaceous, both taxa diversified markedly and diverged from each other and from a common Jurassic ancestor (Rasnitsyn 1980). One lineage of the late-Cretaceous Vespomorpha were pollinating wasps and bees. Some Ichneumonomorpha changed from endoparasites of insects into endoparasites of seeds and gall-makers. The evolution in the second half of the Cretaceous of gall-forming chalcid and cynipid wasps, presumably from insect predators, would have involved a switch from laying eggs in parasitized insects to laying them in the reproductive parts of plants (Willemstein 1987). The switch culminated in the "reformation [of plant meristems] into a structure similar to a fruit and seed" (Cornell 1983). The Diptera and Hymenoptera that today induce histologically complex histioid galls, which in some cases involve hypertrophy and hyperplasia of the ovary wall and floral receptacles, did not radiate along with early flowers in the first half of the Cretaceous. Instead, their origin and diversification appear to more closely parallel the initial radiation of fleshy fruits.

Many soft galls resemble colorful, succulent fruits to a remarkable degree; others, sclerified, look like stony or woody fruits. The resemblance goes beyond the superficial features of shape, size, color, pubescence, and glands; it includes food reserves, secondary metabolites, tissue differentiation (including mechanical and conducting), dehiscence, and phenology (Slepyan 1961; Mani 1964). The dehiscence mechanisms deployed in histioid galls for the release of insects, including abscission, lacerated rupture, and dehiscence along predetermined lines (as in lids and plugs), are like those deployed in fruits for the release of mature seeds. A nut-like gall of *Pruthidiplosis* on *Mimusops* opens by irregular cracking of the hard outer rind to expose pupae "sticking out embedded in the spongy core" (Mani 1964). Hypertrophy and hyperplasia of ovaries normally suppress formation of ovules. However, even in grossly modified flowers—such as those of *Crataea,* which the gall-midge Aschistonyx reduces to a fleshy, yellowish gall—normal ovules sometimes form (Mani 1964). Furthermore, modifications to fertilized ovaries, even as radical as the fleshing out of an incipient dry fruit, often spare the developing seeds. An incorporation of indigestible or unpalatable seed in succulent galls attracting vertebrate consumers could have presented an extraordinary ecological advantage to the host plant.

Parasitoids, Inquilines, and Successori

The biology of histioid galls may be molded by complexly interrelated successions of symbionts (Mani 1964; Raman 1984), including the following:

- parasitoids of gall makers
- inquilines, i.e., insects and mites that cohabit or take over living galls and sometimes induce (perhaps synergistically with primary gall-makers) extensive gall modifications
- cecidophages, i.e., arthropods and vertebrates that feed on gall occupants or tissues
- successori that occupy vacated galls, mainly ants and their symbionts.

Together these communities elicit morphogenetic responses in plants that sometimes amount to more than the sum of the components.

Escape by gall-formers from parasitoids "proves the major selective force molding complex and diverse gall morphology" (Cornell 1983). Secretions, both gummy and sugary, protect galls: the former may deter arthropods physically; the latter attract ants. Gall-attending ants attack potential competitors of gall occupants, and the association has led to the evolution of mutualisms between gall-makers and ants (Cornell 1983; Abe 1988). Do the prevalence and diversity of symbioses between plants and ants (Beattie 1985) and the overwhelming presence of ants in the canopy of Amazonian rain forests (May 1989) betray their heritage as gall successori?

Some strategies for escaping parasitoids accentuate the "fruitness" of a gall. It has been argued (Weis et al. 1988, 1989; but see Waring and Price 1989) that larger galls place the gall-maker beyond the reach of parasitoids' ovipositors, but are selectively eaten by vertebrates. Color, odor, and succulence may either prevent detection by or deter competitors (Rothschild 1975; Pellmyr and Thien 1986). Although such traits may attract gall-eating vertebrates, they could retain their selective advantage by protecting a greater number of galls from parasitic arthropods. What once protected gall-makers from parasitoid arthropods may now attract seed-dispersing vertebrates to fruits. Fig synconia harboring chalcid wasp inquilines become larger, softer, juicier, and almost black, and remain on trees longer (Askew 1984). I interpret the fig synconium as a controlled gall that depends on agaonid wasp inquilines for pollination.

Seed Dispersal

Contrary to ornithologists' conventional opinion (Snow 1981), succulent fruits of the kind eaten by birds did not evolve early in the history of angiosperms. Until the late Cretaceous, fruits were small, more or less clustered dry follicles, capsules, and nutlets (Friis and Crepet 1987). The appearance of fleshy fruits in the late Cretaceous and the progressive increase in fruit size thereafter may reflect the increased importance of specialized seed dispersal by small frugivorous vertebrates (Wing and Tiffney 1987), initially birds (Regal 1977). However, there is no evidence for the major radiation of frugivorous birds in the Cretaceous postulated by Regal (Martin 1983). Modern groups of frugivores evolved in the Tertiary from unknown, presumably insectivorous ancestors. Many extant frugivores supplement a diet of invertebrates with fruit (Snow 1981), or eat fruit but feed invertebrates to their nest-bound young. Southwood (1985) views the adaptation to a mixed diet as transitional along the insectivore's route to phytophagy via plant parts (fruits and seeds) that are rich in proteins. The critical step, however, would have been the initial switch to plant proteins.

The elaboration of histioid gall-making in the second half of the Cretaceous would have established communities of gall-forming Diptera and Hymenoptera (with their extensive coteries of parasitoids, inquilines, and successori) in the canopies of trees, and would have amplified selective pressure on insectivorous vertebrates for the evolution of arboreal adaptations. The ability to fly would have offered an initial advantage. Concomitantly, the "packaging" of insects inside the increasingly fruit-like histioid galls could have initiated a transition to frugivory. Some modern birds and small rodents, as well as insects, eat histioid galls. Insects and woodpeckers often cut through galls to feed on gall-makers and inquilines, but there are also insects and birds that feed exclusively on gall tissues. The two extremes are bridged by a full range of intermediate trophic adaptations (Mani 1964).

Mechanisms of Gall Formation

Heterochronic, heterotopic, and homeotic shifts during embryonic development (Guerrant 1988; Sattler 1988), amplified into abnormalities or terata (Van Steenis 1969), are held responsible for crucial evolutionary innovations in the history of plants (Takhtajan 1976; Doyle 1978; Asama 1982; Graham 1985; Rothwell 1987; Meyen 1988; DiMichelle et al. 1989). Like other terata, plant galls have long been considered to be phenotypic expressions of silent genes from the plants' repertoire of plasticity, triggered by an environmental shock (which need not be biological) through the relaxation of developmental constraints (Braun 1978). Yet an insect gall is not merely a more or less profound externally triggered departure from the morphogenetically constrained expression of innate variability. As has been noted repeatedly (Cornell 1983; Shorthouse and Lalond 1986), some galls differ fundamentally in anatomy and histology from the plant organs in which they arise. Some gall-makers introduce morphologies foreign to the host plant, its ancestors, or its close relatives, or they "reproduce" on one plant morphologies of a distantly related species. For example, the gall midge *Kiefferia pimpinellae* replaces the small, dry, spiny fruit of a carrot with a colorful, enormously inflated, utricular gall with thick fleshy walls. Fusions, dissections or transpositions of floral parts, and hairs or glands that arise *de novo* on galls are frequently normal traits of other plants (Mani 1964 and references therein). In short, gall-makers can abruptly modify plant organs into complex structures resembling flowers or fleshy fruits.

Slepyan (1961) demonstrated that, as plant organs, galls and fruits form a parallel series, but he could not invoke homology. Instead, he interpreted the similarities as due to the analogous patterns of tissue growth and differentiation produced by plant growth regulators. Slepyan's model pivots on the existence of a transferable package of growth regulators sequestered from plant organs on which gall-makers fed and then redeployed elsewhere to replicate the morphogenesis of the donor organ. In my model the package differs in both contents (DNA) and destination (from gall-maker to plant). The former prediction is prompted by the current opinion (Guern 1987) that plant growth regulators alone do not account for the full spectrum of controlled developmental phenomena: changes result not only from levels of growth regulators but also from sensitivity of cells to these substances. Implied in the latter prediction is the transfer of gall genetic determinants to plants.

Horizontal Gene Transfer?

The molecular genetics of hormone-mediated morphogenesis of insect galls is not yet known. The development of the gall phenotype is currently viewed as a product of two genotypes: the insect's coding for stimulus and the plant's coding for response (Weis and Abrahamson 1986). Genetic transformation of plants by plasmid-borne bacterial DNA (Zambryski et al. 1989) prompted speculation on the enigmatic gall-inducing factor (Cornell 1983; Pirozynski 1988). However, unlike the agrobacterial neoplastic tumors that result from a one-time transfer of the Ti plasmid, insect galls require the presence of living larvae to continue development (Cornell 1983). Consequently, with reference to cynipid wasp galls, Cornell rejected the Ti plasmid-like interaction in favor of a virus "forging a permanent ecological link."

From plant pathology comes evidence of endogenous symbionts in what were previously considered spontaneous ontogenetic aberrations resembling organoid galls. For example, mycoplasma-like organisms inhabit some hypertrophied and hyperplastic inflorescences (Westphal 1980), witches' brooms, and other developmental abnormalities of flowers. The symbiont "could be either a source of genes that affect the expression of host plant hormone metabolism or a source of prokaryotic DNA coding for genes involved in the synthesis of gibberellins" (Golino et al. 1988).

An agrobacterial gall is a tumorous response to parasitic T-DNA (Tempé et al. 1984) which, integrated into host chromosomes, codes for the production of plant growth regulators and novel metabolites (opines). The reputedly primitive *Pontania* saw-fly galls on willow, which are neoplasms induced by adult females at oviposition (Mani 1964; Mamaev 1975), may be similarly induced. The precisely controlled, species- or organ-specific complex histioid galls are conceivably more like the "refined parasitism" of legumes by *Rhizobium* (which leads to the formation of a novel organ, the nitrogen-fixing nodule); the functioning of the nodule requires the presence of the bacteria, but some genes controlling the interaction are incorporated in the plant chromosomes (Palacios and Verma 1988). Indeed, Weis et al. (1988) rationalized that if exogenous DNA were involved in gall development, its role would likely be regulation of the expression of duplicate silent genes in the host plant.

Gall-making is transient; new associations evolve and old ones dissolve. The mélange of modern galls provides some evidence for a historical trend identified by Slepyan (1961): gall-makers increasingly leave control of gall morphogenesis to plants. Some plants control the gall-makers' emergence by inducing physiological senescence of gall tissues, and may even provide elaborate means of escape. The fate of vacated fleshy galls is not always abscission and decay. Some are reabsorbed; others remain to serve the insect as a pupation site after larvae have stopped feeding, or to accommodate predatory successori, such as ants (Mani 1964). A cynipid wasp gall on oak continues to secrete honey-dew, which attracts defending ants even after the gall-maker has been killed by parasitoids (Washburn 1984).

The production of food bodies by *Piper cenocladum* depends on the presence of *Pheidole* ants; on other ant plants, food bodies form whether ants are present or not. Genetic determinants of the swollen stipular thorns of ant acacias (claimed to be of gall ancestry; see Pirozynski 1988 and references therein) can be transferred to, and expressed in, non-ant acacias. The series of more conventional plant organs begins with organs of storage, some of which not only resemble galls superficially but also appear to be relatively recent intrusions on the phylogeny of the bearer (e.g., tubers in *Solanum*).

Whence the Genes?

Many plant-animal interactions are mediated by microbial symbionts (Southwood 1985) (e.g., autonomous communities of microorganisms in guts of herbivores) permitting exploitation of unbalanced or inaccessible resources (Schwemmler, this volume).

The ability to induce histioid galls is closely linked in invertebrate gall-formers with extraintestinal digestion, i.e., degradation of food outside the animal's body. In gall midges, whose larva is the sole feeding phase in the life cycle, digestive juices ejected during feeding originate in the highly developed salivary glands and the mid-gut. The rest of the digestive tract is simplified by, among other changes, the elimination of outgrowths which in ancestral forms house gut symbionts. There appears to be a concomitant progressive subordination of gut microorganisms as intracellular symbionts (Mamaev 1975). Ultimately, the functional symbiont could be a less readily detectable mobile package of genetic gall determinants.

Although there is nothing to suggest that fungi are uniquely adapted (or even equal to viruses and prokaryotes) as potential mediators of insect-plant gall interaction, there are several reasons why they should be considered as the morphogenetic agents employed by gall-forming insects:

• Mycophagy appears to be the primitive mode of feeding in many groups of insects, including gall midges (Bissett and Borkent 1988), and to have paved the way to phytophagy (Wheeler and Blackwell 1984; Willemstein 1987).

• Fungi mediate exploitation of otherwise unusable substrates, e.g., wood (Anderson et al. 1984: Wilding et al. 1989), especially by Hymenoptera (Rasnitsyn 1980).

• Fungi synthesize plant growth regulators (Sokolovskaya and Kuznetsov 1984; Gay et al. 1989) and induce neoplasia (Wenzler and Meins 1987), hypertrophy, hyperplasia, and morphological and functional transposition and transformation (Mani 1964). In short, they duplicate the basic morphogenetic arsenal of gall-forming insects.

• Fungi and hyphae-forming protoctists were ancient tumorigens of plants and are, next to aphids and gall wasps, the most important gall-makers. The oldest recognizable gall, on a root of Carboniferous lepidodendron approximately 315 million years old, is attributed to an *Urophlyctis*-like chytrid (Larew 1986).

Furthermore, in their often intimate and prolonged contact with plants, and with all potentially reproductive cells, fungi are well suited to accept and transmit alien genes directly or via plasmids, viruses, or mycoplasmas (Tzean et al. 1983; Modjo and Hendrix 1986; Scannerini and Bonfante-Fasolo, this volume).

Implications of Gall Symbioses

"Flowers" and fleshy "fruits" exploited by animals evolved repeatedly in seed plants (Dilcher 1979; Herrera 1989). Therefore, the appearance of true flowers and fruits, merely as new combinations of old traits, seems insufficient to precipitate the "angiosperm revolution."

As plant growth, galls are saltatory innovations. However, unlike a "monster" conceived from an internal ontogenetic accident, a gall is a product of the ecosystem. Gall-exploiting parasitoids, inquilines, cecidophages, and successori may have utilized plant modifications in a way that improved the function of the affected organ (e.g., feeding on honey-dew and cross-pollination by insects, or feeding on fleshy tissue and seed dispersal by birds), or conferred new advantages (e.g., structures catering to defending arthropods; see O'Dowd and Willson 1989). The transition of a gall into a plant organ would require the acquisition of the genetic determinants of galls from gall-makers by the plant, their integration into the plants' heritable genome, and their diffusion by sex and selection. The indeterminate growth habit of plants and their propensity for vegetative propagation would have facilitated the heritability of exogenous DNA, and incorporation in pollen could have ensured its rapid diffusion (Ellstrand 1988). The modification of plant reproductive features by gall symbionts could have permitted or contributed to the following:

- Advanced (constant or "faithful") pollination.
- concomitant dispersal of seeds with spores of VA mycorrhizal fungi. There could be a functional distinction between the fleshy fruits of Mesozoic seed plants and those of angiosperms. Fleshiness in non-angiospermous fruits (originally microbial tumors?) would have been selected for by herbivorous terrestrial vertebrates, e.g., seed-dispersing frugivorous reptiles and seed-eating multituberculates (Wing and Tiffney 1987; Del Tredici 1989). Brightly colored, shoot-borne, fleshy angiospermous fruits (whose late appearance might have awaited the acquisition, by gall-making insects, of the genetic control of tumor-inducing microbial symbionts) appear to have been originally adapted for birds, which were not initially herbivorous. Exclusive frugivory is rare even among extant birds: a diet of fruit is usually supplemented with invertebrates. According to Izhaki and Safiel (1989), bird-adapted fruits do not satisfy the nutritional needs of the animal for which they are intended. This they see as the plant's "dispersal tactics" to force birds to seek other kinds of food and so

disperse seeds more effectively. Indeed, such "tactics" employed by a mycorrhizal plant (as all early woody angiosperms probably were) would permit the harvesting of fruit-borne seed and soil-invertebrate-borne fungal spores with concomitant dispersal and subsequent rapid geographical radiation (Pirozynski and Malloch 1988).

• The appearance of novel phytochemicals (pigments, scents, toxic secondary metabolites)—in particular, complex molecules shared by plants with animals or microorganisms (e.g., terpenoids in floral scents that act as insect sex pheromones).

• A "horizontal" spread of novel traits by symbiosis (i.e., parallel development of modifications cross-induced by related gall-makers; see also Went 1971 and Jeppsson 1986). There is no unequivocal evidence that fleshy, colorful, shoot-borne seeds in Mesozoic gymnosperms evolved before fleshy fruits in angiosperms (Thomas and Spicer 1987; Del Tredici 1989). The brightly colored seeds of modern *Juniperus, Podocarpus*, and even *Taxus* that appear to be adapted for dispersal by frugivorous birds and the analogous fruits of angiosperms could be contemporaneous structures induced by related gall-makers. Indeed, Mamaev (1975) has argued persuasively the derived status of midges forming histioid galls on conifers.

The implications of gall symbiosis as an independent factor in plant morphogenesis reach beyond the historical question of angiosperm radiation and impinge on the conventional tenets of reproductive morphology and their phylogenetic and taxonomic interpretation.

Acknowledgments

I gratefully acknowledge the advice, constructive criticism, and encouragement offered by David L. Dilcher, Anthony J. Downes, Lynda Goff, David H. Lewis, Peter W. Price, Ernest Small, and the editors of this volume. The shortcomings of the contents and the presentation reflect on me alone. The paper is a heuristic experiment intended to generate questions rather than provide answers. It makes no other pretenses.

References

Abe, Y. 1988. Trophobiosis between the gall wasp, *Andricus symbioticus*, and the gall-attending ant, *Lasius niger*. *Applied Entomology and Zoology* 23: 41–44.

Ananthakrishnan, T. N. 1984. Adaptive strategies in cecidogenous insects. In: Ananthakrishan, T. N., ed., *Biology of Gall Insects*. Edward Arnold.

Anderson, J. M., Rayner, A. D. M., and Walton, D. W. H., eds. 1984. *Invertebrate-Microbial Interactions*. Cambridge University Press.

Asama, K. 1982. Evolution and phylogeny of vascular plants based on the principles of growth retardation. Part 6. Triphyletic evolution of vascular plants. *Bulletin of the National Science Museum, Tokyo, Series* C 8: 94–115.

Askew, R. R. 1984. The biology of gall wasps. In: Ananthakrishnan, T. N., ed., *Biology of Gall Insects*. Edward Arnold.

Beardsley, J. W., Jr. 1984. Gall-forming coccoidea. In: Ananthakrishnan, T. N., ed., *Biology of Gall Insects*. Edward Arnold.

Beattie, A. J. 1985. *The Evolutionary Ecology of Ant-Plant Mutualisms*. Cambridge University Press.

Bissett, J., and Borkent, A. 1988. Ambrosia galls: The significance of fungal nutrition in the evolution of the Cedicomyiidae (Diptera). In: Pirozynski, K. A., and Hawksworth, D. L., eds., *Coevolution of Fungi with Plants and Animals*. Academic Press.

Braun, A. C. 1978. Plant tumors. *Biochimica et Biophysica Acta* 516: 167–191.

Channabasavanna, G. P., and Nangia, N. 1984. The biology of gall mites. In: Ananthakrishnan, T. N., ed., *Biology of Gall Insects*. Edward Arnold.

Cornell, H. V. 1983. The secondary chemistry and complex morphology of galls formed by the Cynipinae (Hymenoptera): Why and how? *American Midland Naturalist* 110: 225–234.

Crepet, W. L. 1985. Advanced (constant) insect pollination mechanisms: Pattern of evolution and implications vis-à-vis angiosperm diversity. *Annals of the Missouri Botanical Garden* 71: 607–630.

Del Tredici, P. 1989. Ginkgos and multituberculates: Evolutionary interactions in the Tertiary. *BioSystems* 22: 327–339.

Dilcher, D. L. 1979. Early angiosperm reproduction: An introductory report. *Review of Palaeobotany and Palynology* 27: 291–328.

DiMichelle, W. A., Davis, J. I., and Olmstead, R. G. 1989. Origins of heterospory and the seed habit: The role of heterochrony. *Taxon* 38: 1–11.

Downes, J. A. 1973. Endopterygote insects and the origin of the angiosperm flower. Abstract, First International Congress of Systematic and Evolutionary Biology, Boulder, Colorado.

Doyle, J. A. 1978. Origin of angiosperms. *Annual Review of Ecology and Systematics* 9: 365–392.

Ellstrand, N. C. I988. Pollen as a vehicle for the escape of engineered genes? *Trends in Ecology and Evolution* 3: 30–32.

Friis, E. M., and Crepet, W. L. 1987. Time of appearance of floral fetures. In: Friis, E. M., Chaloner, W. G., and Crane, P. R., eds., *The Origins of Angiosperms and Their Biological Consequences*. Cambridge University Press.

Friis, E. M., Chaloner, W. G., and Crane, P. R., eds. 1987. *The Origins of Angiosperms and Their Biological Consequences*. Cambridge University Press.

Gagné, R. J. 1984. The geography of gall insects. In: Ananthakrishnan, T. N., ed., *Biology of Gall Insects*. Edward Arnold.

Gay, G., Rouillon, R., Bernillon, J., and Favre-Bonvin, J. 1989. IAA biosynthesis by the ectomycorrhizal fungus *Habeloma hiemale* as affected by different precursors. *Canadian Journal of Botany* 67: 2235–2239.

Golino, D. A., Oldfield, G. N., and Gumpf, D. J. 1988. Induction of flowering through infection by beet leafhopper transmitted virescence agent. *Phytopathology* 78: 285–288.

Graham, L. E. 1985. The origin of the life cycle of land plants. *American Scientist* 73: 178–186.

Guern, J. 1987. Regulation from within: The hormone dilemma. *Annals of Botany* 60 (Supplement 4): 75–102.

Guerrant, E. O., Jr. 1988. Heterochrony in plants. The intersection of evolution, ecology and ontogeny. In: McKinney, M. L., ed., *Heterochrony in Evolution*. Plenum.

Herrera, C. M. 1989. Seed dispersal by animals: A role in angiosperm diversification? *American Naturalist* 133: 309–322.

Hodkinson, I. D. 1984. The biology of ecology of the gall-forming Psylloidea (Homoptera). In: Ananthakrishnan, T. N., ed., *Biology of Gall Insects*. Edward Arnold.

Iltis, H. H. 1983. From teosinte to maize: The catastrophic sexual transmutation. *Science* 222: 886–894.

Izhaki, I., and Safiel, U. N. 199. Why are there so few exclusive frugivorous birds? Experiments on fruit digestibility. *Oikos* 54: 23–32.

Jeppsson, K. 1986. A possible mechanism in convergent evolution. *Paleobiology* 12: 80–88.

Lamboy, W. T. 1984. Evolution of flowering plants by fungus-to-host horizontal gene transfer. *Evolutionary Theory* 7: 45–51.

Larew, H. G. 1986. The fossill gall record: A brief summary. *Proceedings of the Entomological Society of Washington* 88: 385–388.

Mamaev, B. M. 1975. *Evolution of Gall-forming Insects: Gall Midges*. British Library Board, Boston Spa.

Mani, M. S. 1964. *Ecology of Plant Galls*. W. Junk, The Hague.

Martin, L. 1983. The origin and early radiation of birds. In: Bush, A. H., and Clark, G. A., Jr., eds., *Perspectives in Ornithology*. Cambridge Univeristy Press.

May, R. M. 1989. An inordinate fondness for ants. *Nature* 341: 386–387.

Meyen, S. V. 1988. Origin of the angiosperm gynoecium by gamoheterotopy. *Botanical Journal of the Linnean Society* 97: 171–178.

Modjo, H. S., and Hendrix, J. W. 1986. The mycorrhizal fungus *Glomus macrocarpum* as a cause of Tobacco Stunt disease. *Phytopathology* 76: 688–691.

O'Dowd, D. J., and Willson, M. F. 1989. Leaf domatia and mites on Australasian plants: Ecology and evolutionary implications. *Biological Journal of the Linnean Society* 37: 191–236.

Palacios, R., and Verma, D. P. S., eds. 1988. *Molecular Genetics of Plant-Microbe Interactions*. American Phytopathological Society Press.

Pellmyr, O., and Thien, L. B. 1986. Insect reproduction and floral fragrances: Keys to the evolution of the angiosperms? *Taxon* 35: 76–85.

Pirozynski, K. A. 1988. Coevolution by horizontal gene transfer: A speculation on the role of fungi. In: Pirozynski, K. A., and Hawksworth, D. L., eds., *Coevolution of Fungi with Plants and Animals*. Academic Press.

Pirozynski, K. A., and Malloch, D. W. 1988. Seeds, spores and stomachs: Coevolution in seed dispersal mutualisms. In: Pirozynski, K. A., and Hawksworth, D. L., eds., *Coevolution of Fungi with Plants and Animals*. Academic Press.

Raman, A. 1984. Gall insect-host relationships: An ecological perspective. *Proceedings of the Indian Academy of Sciences (Plant Sciences)* 93: 293–300.

Rasnitsyn, A. P. 1980. *The Origin and Evolution of Hymenoptera*. Trudy Paleotologicheskogo Instituta, "Nauka," USSR Academy of Sciences, Moscow.

Regal, P. J. 1977. Ecology and evolution of flowering plant dominance. *Science* 196: 622–629.

Rothschild, M. 1975. Remarks on carotenoids in the evolution of signals. In: Gilbert, L. E., and Raven, P. H., eds., *Coevolution of Animals and Plants*. University of Texas Press.

Rothwell, G. W. 1987. The role of development in plant phylogeny: A paleobotanical perspective. *Review of Palaeobotany and Palynology* 50: 97–114.

Sattler, R. 1988. Homeosis in plants. *American Journal of Botany* 75: 1606–1617.

Shorthouse, J. D., and Lalonde, R. G. 1986. Formation of flowerhead galls by the Canada thistle gall-fly, *Urophora cardui* (Diptera: Tephritidae), under cage conditions. *Canadian Entomologist* 118: 1199–1203.

Slepyan, E. I. 1961. Comparison of galls and terates caused by insects with fruits and seeds. *Botanickiy Zhurnal* 46: 1702–1717 (in Russian).

Snow, D. W. 1981. Coevolution of birds and plants. In: Forey, P. L., ed., *The Evolving Biosphere*. Cambridge University Press.

Sokolovskaya, I. V., and Kuznetsov, L. V. 1984. Gibberellin-like substances in the mycelium of haploid and diploid strains of the smut-fungus *Ustilago zeae* (DC.) Ung. *Prikladnaya Biokhimiya i Mikrobiologiya* 20: 484–489.

Southwood, T. R. E. 1985. Interactions of plants and animals: Patterns and processes. *Oikos* 44: 5–11.

Takhtajan, A. L. 1976. Neoteny and the origin of flowering plants. In: Beck, C. B., ed., *The Origin and Early Evolution of Angiosperms*. Columbia University Press.

Tempé, J., Petit, A., and Ferrand, S. K. 1984. Induction of cell proliferation by *Agrobacterium tumefaciens* and *A. rhizogenes*: A parasite's point of view. In: Verma, D. P. S., and Hohn, T., eds., *Genes Involved in Microbe-Plant Interactions*. Springer.

Thomas, B. A., and Spicer, R. A. 1987. *The Evolution and Palaeobiology of Land Plants* (Ecology, Phytogeography and Physiology Series, volume 2). Dioscorides Press.

Tzean, S. S., Chu, C. L., and Su, H. J. 1983. Spiroplasmalike organisms in vesicular-arbuscular mycorrhizal fungus and its mycoparasite. *Phytopathology* 73: 989–991.

Van Steenis, C. G. G. J. 1969. Plant speciation in Malesia, with special reference to the theory of non-adaptive saltatory evolution. *Biological Journal of the Linnean Society* 1: 97–133.

Waring, G. L., and Price, P. W. 1989. Parasitoid pressure and radiation of a gall-forming group (Cecidomyiidae: *Asphondylia* spp.) on creosote bush (*Larrea tridentata*). *Oecologia* 79: 293–299.

Washburn, J. O. 1984. Mutualism between a cynipid gall wasp and ants. *Ecology* 65: 654–656 .

Weis, A. E., and Abrahamson, W. G. 1986. Evolution of host-plant manipulation by gall makers: Ecological and genetic factors in the *Solidago-Eurosta* system. *American Naturalist* 127: 681–695.

Weis, A. E., Walton, R., and Crego, C. L. 1988. Reactive plant tissue sites and the population biology of gall makers. *Annual Review of Entomology* 33: 467–486.

Weis, A. E., Wolfe, C. L., and Gorman, W. L. 1989. Genotypic variation and integration in histological features of the goldenrod ball gall. *American Journal of Botany* 76: 1541–1550.

Went, F. W. 1971. Parallel evolution. *Taxon* 20: 197–226.

Wenzler, H., and Meins, F., Jr. 1987. Persistant changes in the proliferative capacity of maize leaf tissues induced by *Ustilago* infection. *Physiological and Molecular Plant Pathology* 30: 309–319.

Westphal, E. 1980. On the anatomy and cytology of the virescence of *Beta vulgaris* L. (Chenopodiaceae). *Cecidologia Internationale* 1: 9–15.

Wheeler, Q., and Blackwell, M., eds. 1984. *Fungus-Insect Relationships*. Columbia University Press.

Wilding, N., Collins, N. M., Hammond, P. M., and Webber, J. F., eds. 1989. *Insect-Fungus Interactions*. Academic Press.

Willemstein, S. C. 1987. An evolutionary basis for pollination ecology. *Leiden Botanical Series* 10: 1–425

Wing, S. L., and Tiffney, B. H. 1987. The reciprocal interaction of angiosperm evolution and tetrapod herbivory. *Review of Palaeobotany and Palynology* 50: 179–210.

Zambryski, P., Tempé, J., and Schell, J. 1989. Transfer and function of T-DNA genes from *Agrobacterium* Ti and Ri plasmids in plants. *Cell* 56: 193–201.

25

Luminous Bacterial Symbiosis in Fish Evolution: Adaptive Radiation among the Leiognathid Fishes

Margaret Jean McFall-Ngai

The bony fishes (Osteichthyes) are the most diverse and widely distributed group of vertebrates (Nelson 1984). Thus, they are an important group in which to analyze the extent to which symbiotic associations have been involved in wide-scale radiations among the vertebrates. Although our knowledge of symbiotic phenomena among the fishes is relatively poor, researchers have amassed a considerable amount of data on symbioses between fishes and marine luminous bacteria (Herring and Morin 1978; Hastings et al. 1987). Three types of marine luminous bacteria, *Photobacterium leiognathi*, *Photobacterium phosphoreum*, and *Vibrio fischeri*, occur in specific, extracellular light-organ associations with fishes. These bacteria also occur in other ecological niches, as enteric and parasitic symbionts of marine animals and as part of the free-living bacterial community (Hastings and Nealson 1981). A number of associations exist—ceratioids (deep-sea anglerfishes) and anomalopids (flashlight fishes)—from which the bacterial symbionts have not yet been cultured and for which identifications have therefore not been made (Hastings et al. 1987). As part of an examination of the role of luminous symbioses in evolutionary events in fishes, this paper includes an examination of the phylogenetic occurrence of these associations, with an analysis of the extent to which such associations appear to be taxon-specific; an analysis of the radiation of the most evolutionarily successful group of shallow-water fishes having such a symbiosis, the Leiognathidae, emphasizing their contribution to our understanding of the evolution of symbiotic associations; and a consideration of the importance of the morphological characteristics of these relationships as they reflect the evolution of light-organ symbioses in this family of fishes.

Occurrence of Light-Organ Symbioses within the Vertebrates

Specific light-organ associations are phylogenetically restricted among the vertebrates (Herring and Morin 1978). (The taxonomic scheme used throughout this paper follows Lauder and Liem 1983.) Light-organ symbioses occur only in the teleosts, the most recently evolved of the extant ray-finned fishes (actinopterygiians); they are not present in the elasmobranchs, or in the agnathans (figure 1).

The other groups of ray-finned fishes, including the sturgeons and paddlefishes (Chondrostei) and the gars (Ginglymodi), occur primarily in fresh water (Nelson 1984). Most species of luminous bacteria require 50 mM sodium for growth (approximately 1/10 the amount found in full-strength sea water; Reichelt and Baumann 1974). The absence of bacterial light organs in freshwater fishes may be as much a function of the availability of symbionts as a function of the phylogenetic position of the animals. It should be mentioned that, although luminous bacteria can now be isolated from freshwater habitat only very rarely, it is not possible to speculate on the abundance of such bacteria in ancient lakes and rivers.

Among teleost fishes, light-organ symbioses again are restricted to the more recently evolved taxa, probably occurring only in the Euteleostei, although there are reports of light organs, hypothesized to contain luminous bacteria, in certain eel species (Teleostei:Elopomorpha) (Nielsen and Bertelsen 1985; Castle and Paxton 1984). Many other extant groups of teleosts, such as the Clupeomorpha (anchovies, herring, and their relatives), have significant radiations within marine habitats and occasionally have nonbacterial luminescence, but no representatives with bacterial light organs are known.

Within the eight major groups of the Euteleostei, light-organ associations are spread across four taxa: the Protacanthopterygii, the Aulopiformes, the Paracanthopterygii, and the Percomorpha. No stomiiforms or myctophiforms, which are primarily mesopelagic marine fishes, have luminous bacterial symbioses, although they usually have autogenic (nonbacterial) photophores. Thus, an overview of the phylogenetic occurrence of light-organ associations reveals that they are confined to the more recently evolved taxa of the fishes.

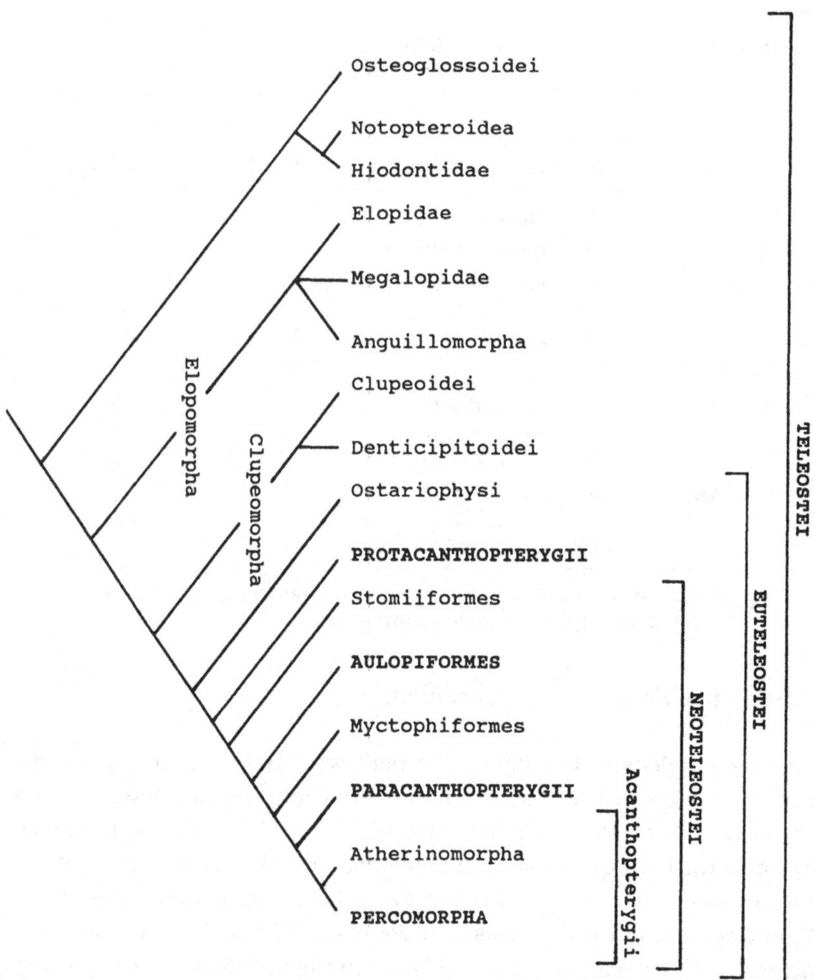

Figure 1
Phylogenetic occurrence of bacterial light organs in fishes. Those groups with bacterial light organs are listed in boldface capitals. (Modified from Lauder and Liem 1983.)

Table 1
The bacterial light organ as a taxon-defining character.

Host order	Host family (total genera)	Bacterial light organ present	
		No. genera	No. species[1]
Salmoniformes	Opisthoproctidae (5)	3	4
Aulopiformes	Chlorophthalmidae (7)	1	7
Gadiformes	Steindachneriidae (1)	1	1
	Moridae (17)	4	40
	Macrouridae (30)	10	146
Lophiiformes	Ceratioidei (34)	32	83
Beryciformes	Trachichthyidae (5)	1	5
	Anomalopidae (4)	4	5
	Monocentridae (2)	2	3
Perciformes	Leiognathidae (3)	3	21
	Apogonidae (26)	1	6
	Percichthyidae (20)	1	3

[1]The presence of a bacterial light organ is a characteristic of all examined species within every light-organ-containing genus.

Taxon Specificity of the Associations

Among euteleosts that have members with light-organ symbioses, the occurrence of these associations is not random but instead often characterizes all the members of a given family or genus (table 1). Because the light organs are often internal, the fishes in many of these groups have been assigned to the given taxon not because they have light organs, but on the basis of independent (usually external) characters. Such correlations give credence to the idea that within a group the evolution of a symbiosis may have been an important, and perhaps "key," innovation (Liem 1974), or may at least have been one of a suite of characters that were crucial in its radiation. The following analysis of the phylogenetic occurrence of all known light-organ symbioses serves to illustrate this point.

1. The protacanthopterygiian euteleosts include five orders of fishes, only one of which, the Salmoniformes, has representatives with bacterial light organs. Within this order, only the family Ophisthoproctidae has members with such organs; they occur in all four species of three of the five opisthoproctid genera (Ahlstrom et al. 1984; Bertelsen and Munk 1964; Bertelsen et al. 1965).

2. The aulopiform euteleosts are represented by eleven families of fishes, only one of which, the family Chlorophthalmidae, has members with bacterial light organs (Somiya 1977). This family has seven genera (Sulak 1977), with only the genus *Chlorophthalmus* having luminous bacterial symbioses. Although only two of the seven species within this genus are reported in the literature as having light organs, there are unpublished observations of light organs in other species, and it is suspected that all members have such associations (K. J. Sulak, personal commnication). If a close examination of this group bears out this suspicion, luminous bacterial symbioses will characterize the genus *Chlorophthalmus*.

3. The paracanthopterygiian euteleosts contain four orders, with a total of over 200 genera. Within this group, only two orders, the Gadiformes and the Lophiiformes, have members with luminous bacterial associations.

a. The gadiforms. The taxonomy of this order is in flux (Cohen 1989), but it contains roughly eight families, only three of which have bacterially luminous representatives: the Steindachneriidae, the Moridae (morid cods), and the Macrouridae (the rattails) (Herring and Morin 1978). The family Steindachneriidae is monotypic, containing only the bacterially luminescent species *Steindachneria argentea* (Cohen 1964a; Haneda 1968). The morid cods include seventeen genera, of which all species of four genera (Paulin 1989) have bacterial light organs (Haneda 1951; Marshall and Cohen 1973). Similarly, the rattail family has thirty genera, with all species in ten of these genera characterized by light-organ symbioses (Marshall 1965; Marshall and Iwamoto 1973; Okamura 1970a, b). Thus, there exists a taxon-specific set of associations in the gadiforms. Although the light organ is used as a taxonomic character in most of these associations, these taxonomic designations would stand independent of the use of the light organ in taxonomic assignments.

b. The lophiiforms. This order, the marine anglerfishes, contains sixteen families, eleven of which are in the suborder Ceratioidei, the bathypelagic anglerfishes. The females of all but two of the 34 genera in this suborder are characterized by a bacterial light-organ association (Bertelsen 1984).

4. The percomorph euteleosts. This group, which contains twelve orders with 10,000 species, represents nearly half of all fish species. They are the most recently evolved of all of the fish groups. Two of the orders, the Beryciformes and the Perciformes, have bacterially luminescent representatives (Herring and Morin 1978).

a. The beryciforms. This order of fishes has fourteen families, three of which—the Trachichthyidae, the Anomalopidae, and the Monocentridae—have light organs. These three families probably form a monophyletic assemblage (Zehren 1979). The family Trachichthyidae contains five genera, among which all species of one (*Paratrachichthys*) are bacterially luminous. According to the recent revision of Johnson and Rosenblatt 1988, the family Anomalopidae comprises five known species in four genera, all bearing a large subocular light organ. Although the details of the light-organ system have provided an important character in the classification of the species within this family, other taxonomic characteristics, particularly of the skeleton, define the family (Zehren 1979; Johnson and Rosenblatt 1988). Similarly, numerous characters independent of the presence of a light organ in the lower jaw, a feature common to all three monocentrid species (in two genera), distinguish the monocentrids from other beryciforms (Zehren 1979). Thus, among the Beryciformes, there are two families in which all the members have bacterial light organs.

b. The perciforms. Of the 150 families in this order, only three have representatives with luminous bacterial light organs: the Percichthyidae, the Apogonidae, and the Leiognathidae. Within the family Percichthyidae, which contains twenty genera, the three species of the genus *Acropoma* all have bacterial light organs (Nelson, 1984). Similarly, the family Apogonidae has 26 genera, only one of which (*Siphamia*) has bacterial light organs. All six species of this genus have these symbioses (Iwai 1958, 1959, 1971; Haneda 1966). The family Leiognathidae is the most speciose of all the perciform families with luminous bacterial associations. The approximately 21 species of this family all have circumesophageal light organs (Haneda and Tsuji 1976; McFall-Ngai and Dunlap 1984). The internal light organ of the leiognathids has been largely overlooked as a taxonomic character (Kühlmorgen-Hille, 1974; Dunlap and McFall-Ngai 1984). Thus, all the members of the group have been assigned on the basis of characters entirely independent of the light organ.

To summarize: The occurrence of bacterial light organs is limited to the more recently evolved teleost species. In all cases the presence of the symbiosis defines a taxon, characterizing all the individuals of a genus or family. In teleosts in which the taxonomic relationships are well established, the members of a taxon that are characterized by a luminous bacterial symbiosis have been assigned to that taxon on the basis of numerous other independent traits.

Age of the Associations

The phylogenetic patterns presented above provide evidence that luminous bacterial light-organ associations represent a derived, rather than a primitive, character state in fishes. In addition, information obtained from the fossil record indicates that most of these associations are relatively recent. Although the teleosts emerged around 200 million years ago (Carroll 1988) in the late Triassic or the early Jurassic, many forms do not appear until much later (Lauder and Liem 1983). The taxa in which most bacterially luminous fishes occur arose after the beginning of the Tertiary; for example, the gadiformes (Nolf and Steurbaut 1989), ceratioids (Carroll 1988), and perciformes (Carroll 1988) all arose during the Cenozoic, beginning approximately 65 Ma. The beryciforms are probably among the oldest, with a fossil record that dates from the Cretaceous (Patterson 1968; Zehren 1979; Woods and Sonoda 1973).

Specificity of the Associations

In addition to the presence of light-organ associations as a defining character of certain taxa, the symbiotic associations themselves show species specificity on the part of the host fish. That is, although three species of luminous bacteria are known to occur in light organs, a particular group of fishes will always have the same symbiont. For example, *Photobacterium phosphoreum* is cultured in all gadiform (Haneda and Yoshiba 1970; Singleton and Skerman 1973; Ruby and Morin 1978), chlorophthalmid aulopiform (Hastings et al. 1987), and trachichthyid beryciform light organs (P. V. Dunlap, personal communication); *Vibrio fischeri* occurs in the light organs of all monocentrids (Fitzgerald 1977; Ruby and Nealson 1976); and *P. leiognathi* occurs in the perciform light organs (Dunlap 1984; Fukasawa et al. 1988). The as-yet-unculturable light-organ symbionts of the ceratioids and the anomalopids have not been identified.

The forces that define and maintain this specificity are not yet known. It has been postulated that this specificity is a reflection of the host habitat; that is, deep-living fish species have the psychrophilic (cold-loving) luminous bacteria *P. phosphoreum*, temperate fish species have *Vibrio fischeri*, and tropical species have *P. leiognathi* (Hastings et al. 1987). However, neither biogeographic data on the bacteria nor details of the life history of the fish host support this hypothesis (table 2). Although the deep-water gadiforms have *P. phosphoreum*, some

Table 2
Characteristics of light-organ symbioses in fishes with culturable bacterial symbionts.

Host family	Symbiont[1]	Depth range (m)	Habitat[2]	Geographical distribution[3]	Anatomical location
Opisthoproctidae	Pp	100–890[a]	Meso/bathy	Trop & temp	Hindgut
Moridae	Pp	40–1500[b]	Benthypel	Worldwide	Hindgut
Steindachneriidae	Pp	180–370[c]	Benthypel	Trop	Hindgut
Macrouridae	Pp	200–2000[d]	Benthypel	Worldwide	Hindgut
Trachichthyidae	Pp	90–500[e]	Benthypel	Trop & temp	Hindgut
Monocentridae	Vf	5–185[f]	Reef	Trop & temp	Superficial (lower jaw)
Leiognathidae	Pl	0–300[g]	Demersal	Trop & temp	Foregut
Apogonidae	Pl	0–30[h]	Reef	Trop	Midgut
Percichthyidae	Pl	40–450[i]	Benthypel	Trop & temp	Hindgut

[1] Pp = *Photobacterium phosphoreum*; Vf = *Vibrio fischeri*; Pl = *Photobacterium leiognathi*.
[2] meso = mesopelagic; bathy = bathypelagic; benthypel = benthypelagic.
[3] trop = tropical; temp = temperate.
[a] Cohen 1964b; [b] Paulin 1989; [c] Cohen 1964a; [d] Nelson 1984; [e] Woods and Sonoda 1973; [f] Haneda 1966; [g] Pauly, pers. comm.; [h] Fraser 1972; [i] Haneda 1950.

species live in shallower, warmer waters of the temperate zones and tropics as larvae, during which stage the developing light organ probably becomes inoculated with bacteria (Cohen 1984; Fahay and Markle 1984). Furthermore, both leiognathids and monocentrids are found in the shallow waters of both temperate and tropical areas of the Indo- West Pacific (Nelson 1984). In fact, most of the luminous fishes have representatives in waters where all three symbiotic luminous bacteria probably occur, at least during some periods of the year (Ruby et al. 1980).

Another correlation invoked to explain the patterns of occurrence of light-organ associations has been the anatomical location of the light organ. Drawing upon the bacterial identifications available to them at the time, Hastings et al. (1987) noted that the light organs of fishes with hindgut associations contained *Photobacterium phosphoreum*, whereas those with either foregut or midgut associations contained *P. leiognathi*. However, percichthyids of the genus *Acropoma*, which occur on the continental shelves of the temperate zones at depths of 150–400 m (Haneda 1950), have a hindgut association that contains *P. leiognathi* (Fukasawa et al. 1988). Thus, the anatomical location of the light organs does not correlate perfectly with the species of bacterium and, in itself, provides little insight into the nature of the symbioses.

Thus, the present evidence best supports the conclusion that light-organ associations between fishes and luminous bacteria are taxon-specific rather than related to the habitat or to the anatomical location within the animal. Further support for this idea comes from an analysis of the occurrence of luminous bacteria in the only other group of animals with culturable bacterial symbionts, the cephalopod mollusks. The loliginid cephalopods have light organs with *P. leiognathi* (Fukasawa and Dunlap 1986), and the sepiolid cephalopods have light organs with *Vibrio fischeri* (Herring et al. 1981; Boettcher and Ruby 1989). This pattern of occurrence holds even though associations of luminous bacteria with species of these two cephalopod groups are found in both temperate and tropical waters.

The Symbiosis between Perciform Fishes of the Family Leiognathidae and the Luminous Photobacterium Lieognathi

All leiognathid fishes have a circumesophageal light organ containing a dense culture of *Photobacterium leiognathi* (figure 2). The leiognathids

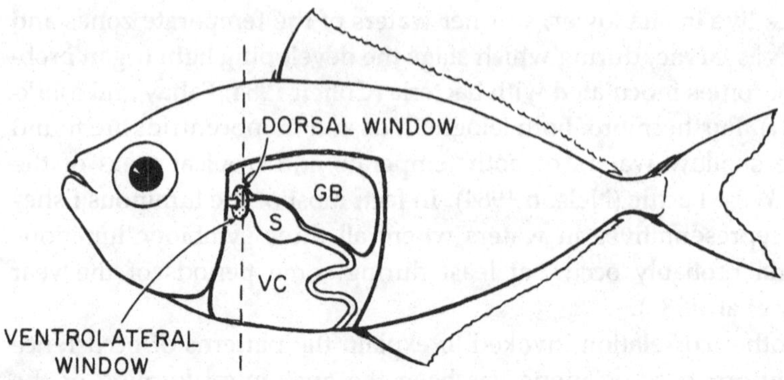

Figure 2
The general appearance of the leiognathid light-organ system, as seen in female fishes and in fishes such as *Leiognathus equulus* that bear nondimorphic light organs. Note the location of the dorsal and ventrolateral shutters. S = stomach, GB = gas bladder, VC = visceral cavity. (Reprinted from McFall-Ngai and Dunlap 1984.)

comprise approximately 21 species in three genera (Nelson 1984). The evolutionary radiation of the leiognathid fishes offers a rare opportunity for the study of inter- and intraspecific differences in the expression of symbiosis within a family. In a broader sense, the leiognathids are an ideal group to study in regard to the evolution and biogeography of symbiosis, because the family is speciose, is biogeographically widespread and ecologically diverse, is easily collected in large numbers, has easily identifiable, planktonic larvae, is characterized by a complex symbiosis that not only affects the structure and function of proximal tissues but also incorporates tissues distant from the association, and has intraspecific variations in the symbiotic light organ that correlate with the geographical distribution of the population and with the sex of the individual (McFall-Ngai 1983a).

Geographical Distribution and Habitat Variation in the Occurrence of Leiognathids

Members of this family occur throughout the Indo-West Pacific. The center of their distribution (Kühlmorgen-Hille, 1974), as is the case in many other fish families and among corals and mollusks, is in the "fertile triangle" area bounded by the Philippines, Malaysia, and Indonesia (Briggs 1974). The periphery of their distribution includes three species found in Japan's temperate waters, the few species reported from the Solomon Islands of the central South Pacific, and several species found along the east coast of Africa. A single species, *Leiognathus klunzingeri*, has migrated through the Suez Canal into the Mediterranean Sea (Ben-Tuvia 1966). This species is of particular interest because it has been gradually moving west through the Mediterranean since its introduction in the late 19th century. Thus, it offers the opportunity to study the influence of the introduction of a host on the dynamics of free-living symbiont populations.

Leiognathids have been reported from a variety of habitats, but are almost always associated with areas having a sandy or muddy substrate. Most reports of their distribution confine them to shallow, muddy coastal bays and estuaries. However, at least three species have been caught in the surf zone along high-energy, ocean-facing coastline (personal observation). Further, although most reports cite their depth range as 0–30m (Kühlmorgen-Hille 1974), ichthyologists at the University of the Philippines have caught them along the continental shelf at depths to 300 m (D. Pauly, personal communication). With such a wide geographical distribution and diversity of habitats, the leiognathid family as a whole can be expected to be exposed to all the species of luminous bacteria known to form symbiotic associations (Ruby et al. 1980).

Where they occur, leiognathids are usually found in large, dense schools that often contain several species of the family (Kühlmorgen-Hille 1974). The density of leiognathids in the shallow waters of the Orient has resulted in their heavy exploitation by the fishing industry (Pauly 1976; Rau and Rau 1980); freshly caught specimens can be obtained in almost any fish market in the Orient. Thus, for studies of luminous bacterial symbioses, they are the most readily accessible and available of the fish hosts.

The Leiognathid Light-Organ System

The tubules of the light organ of leiognathids arise from a deepening and a gradual pinching off of the folds of the esophagus (M. J. McFall-Ngai, unpublished data). It has been assumed that the infection begins anew with each generation because *Photobacterium leiognathi* can be cultured from the seawater where leiognathids are found (Reichelt et al. 1977), because the eggs have not been shown to carry the bacteria, and because the light organ forms from the gut and is therefore exposed to the symbiont from the first feeding. In 5-mm larvae, bacteria can be seen within these folds (personal observation); however,

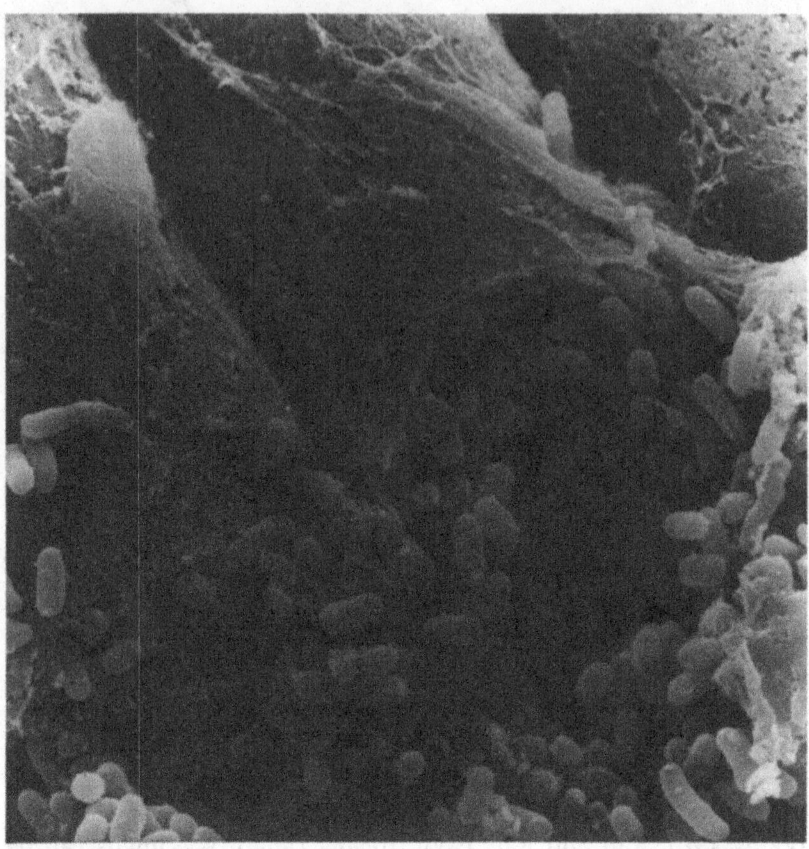

Figure 3
The luminous bacterial symbionts of leiognathids occur in high densities in the numerous parallel tubules of the light organ. b = bacterium; bar = 5 μm. (Reprinted from Dunlap and McFall-Ngai 1987.)

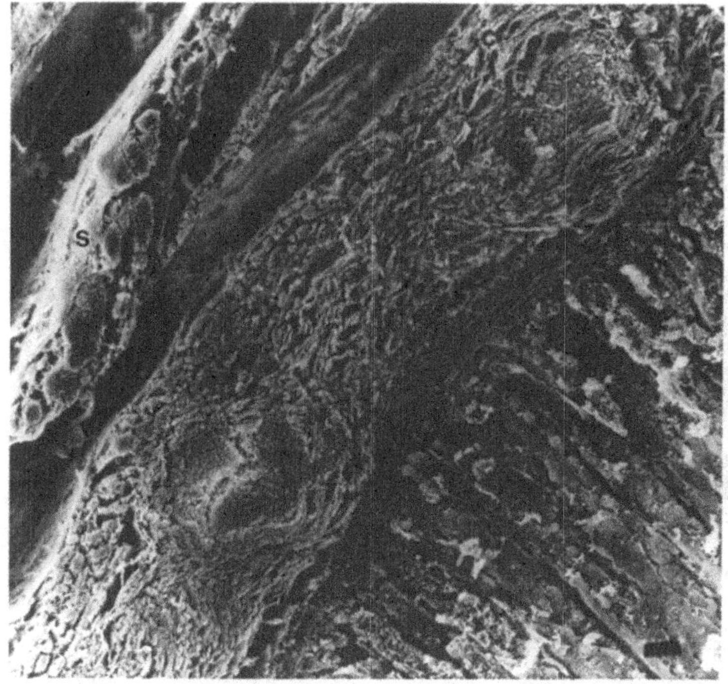

Figure 4
Tissues in many portions of the fish's body serve to impose some control over bacterial light output. One set of such tissues, including a chromatophore-bearing layer and a light-impervious muscular shutter, is found directly over the light-organ tubules. t = bacterial tubules, m = condensed melanin, c = chromatophore layer; s = muscular shutter; bar = 10 μm.

it has not yet been determined whether the bacteria in these early stages occur as a pure culture of *Photobacterium leiognathi*. Further, no information is available on the morphogenic events that give rise to the adult light organ.

In the adults, the light organ is composed of bacteria-containing tubules that surround the esophagus (figure 3; see McFall-Ngai 1983a). The surface of the light organ is covered by muscular shutters and chromatophores that partially control light emission (figure 4). In addition, almost the entire body of the fish is recruited into the light-organ system, the components of which are involved to various degrees in the wide variety of luminescent behaviors observed in leiognathids (McFall-Ngai and Dunlap 1983; Dunlap and McFall-Ngai 1987). The anteroventral surface of the gas bladder abuts the dorsum

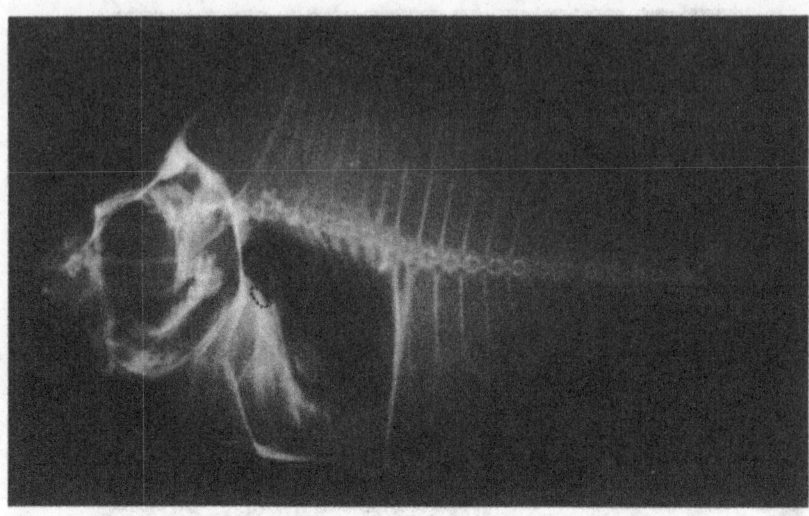

Figure 5
X-ray photograph of *Leiognathus equulus*. The central dark region is the gas-bladder space. The approximate location of the light organ is stippled in black. (Reprinted from McFall-Ngai 1983b.)

of the light organ and acts as the primary reflector (McFall-Ngai 1983b). The translucent ventral muscles act as diffusers; guanophores in the skin give the light direction (McFall-Ngai 1983a); chromatophores in all components of the system serve to modulate intensity or prohibit light emission in a particular direction. All these levels of control provide a luminescence system in leiognathids that is extensive, complex, and dynamic.

The Host's Gas Bladder as a Central Morphological Feature of the Symbiotic System

The gas bladder of leiognathids serves two critical functions as a component of the light-organ system (McFall-Ngai 1983b; Dunlap and McFall-Ngai 1987): (1) as the primary reflector of bacterially produced light in counterillumination, which is antipredatory behavior (figures 5–7); (2) as a source of oxygen for the luminescent reaction of the bacterial symbionts (figure 8). In the former function, the gas bladder aids in the emission of light, which is produced by the bacteria of the internal light organ, into the environment. Purines, the compounds responsible for reflective tissues in fishes, are deposited in the leiog-

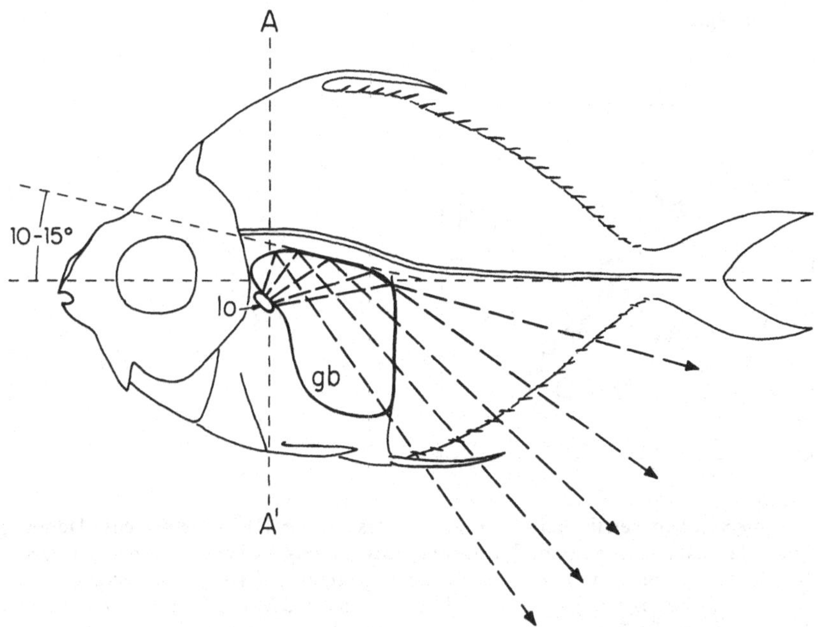

Figure 6
Tracing of figure 5, showing the relationship of the light organ to the gas bladder and the path of bioluminescent light through the gas bladder (indicated by dashed lines and arrows.)

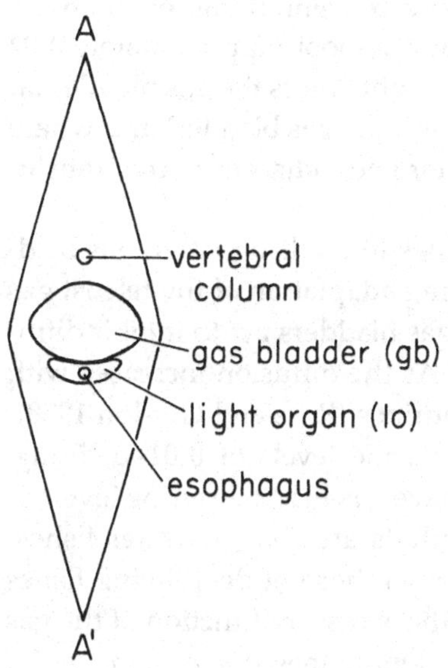

Figure 7
Cross-section of A–A′ in figure 6, illustrating the cupping of the light organ around the anteroventral bladder surface. (Reprinted from McFall-Ngai 1983b.)

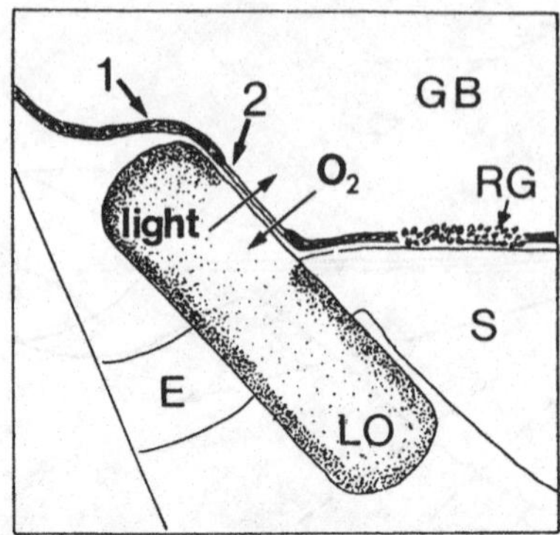

Figure 8
The interface between the light organ and the gas bladder. (1) Purine-rich gas-bladder lining. (2) Thin, oxygen-permeable membrane separating the light organ from the gas bladder and the red gland, a counter-current exchange system for the secretion of oxygen into the gas bladder. (Reprinted from Dunlap and McFall-Ngai 1987.)
LO = light organ, GB = gas bladder, RG = red gland, E = esophagus.

nathid gas bladder in high concentrations to form an efficient reflector. Further, the pattern of purine deposition in the gas bladder is consistent with the light path of the bioluminescence. The dorsum of the gas bladder has particularly heavy concentrations of guanine, averaging 2.83 mg/cm^2. In contrast, purine-poor (approximately 0.02 mg/cm^2) translucent areas occur where light enters the gas bladder (at the interface between the light organ and the gas bladder) and where light leaves the bladder (at the posterior end, adjacent to the translucent ventral musculature) (table 3).

The presence of high levels of purines in the leiognathid gas bladder represents selection on a preexisting adaptation of the teleost gas bladder. Purines in the walls of fish gas bladders act to inhibit diffusion of gas into surrounding tissues. As the diffusion increases with depth, so does the concentration of purines (Ross and Gordon 1978). Shallow-water fishes typically have purine levels of 0.01–0.05 mg/cm^2, whereas fishes living at 2000 m have gas-bladder purine levels of 2–3 mg/cm^2. Thus, although leiognathids are shallow-water fishes, deposition of levels of guanine similar to those of deep-living fishes appears to be a result of selection for the accessory function of the gas bladder as the primary reflector of the bioluminescent system.

The low levels of guanine at the interface between the gas bladder and the light organ permit the diffusion of oxygen into the light

Table 3
Purine content of areas of the *Leiognathus equulus* gas bladder ($n = 8$). Reprinted from McFall-Ngai 1983b.

Area	Purine content ($mg/cm^2 \pm SE$)	% guanine	% hypoxanthine	G + H (% of wet mass tissue ± SE)	G + H (% of dry mass tissue ± SE)
Dorsal	2.80 ± 0.16	95.2 ± 0.6	4.8	16.0 ± 1.6	50.6 ± 4.9
Lateral	1.81 ± 0.16	93.5 ± 0.6	6.5	14.0 ± 1.5	44.4 ± 3.8
Ventral	1.22 ± 0.10	91.0 ± 0.8	9.0	—[1]	—
Posterior	0.19 ± 0.01	59.3 ± 2.8	40.7	—	—
"Window"	0.09 ± 0.01	75.0 ± 2.7	25.0	—	—

[1] Not measured.

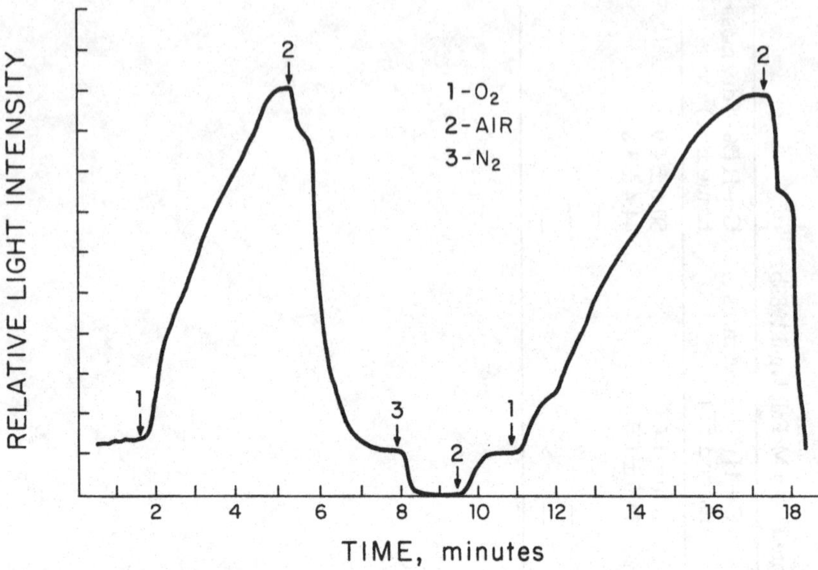

Figure 9
The light organ's luminescence levels in response to changes in the gas bladder's gas composition. The leiognathid light organ is separated from the gas-bladder space by a thin, transparent membrane. Experiments with changing the gas composition of the bladder showed that oxygen in the bladder is available through this membrane for use by the bacteria in their luminescence. Gas was exchanged through fine-gauge needles, and luminescence was monitored with a light guide (1 mm in diameter) inserted into the gas-bladder space. Pure oxygen increased luminescence output between 5 and 10 times over that of air, and pure nitrogen yielded no detectable luminesecence from the light organ. Controls were done to ensure that the light organ's shutters were completely open throughtout the experiment.

organ. In the anesthetized fish, the light organ's shutters remain open, and the bacterial tubules of the dorsum of the light organ interface directly with the purine-poor membrane of the gas bladder. In experiments in which the gas quality of the bladder was changed by exchanging the gas in the space for nitrogen, oxygen, or combinations of these two gases, the light emission of the bacterial culture could be manipulated (figure 9). Thus, the gas bladder has the potential to supply to the bacteria a significant portion of the oxygen required for the luminescent reaction. The fish is able to control bacterial light emission by pulling up the purine- and melanophore-rich shutters that are directly on the light organ. This has the effect of decreasing the light emission of the bacteria by two mechanisms: absorption of any light before it leaves the light organ and restriction of gas-bladder oxygen, quenching the luminescence.

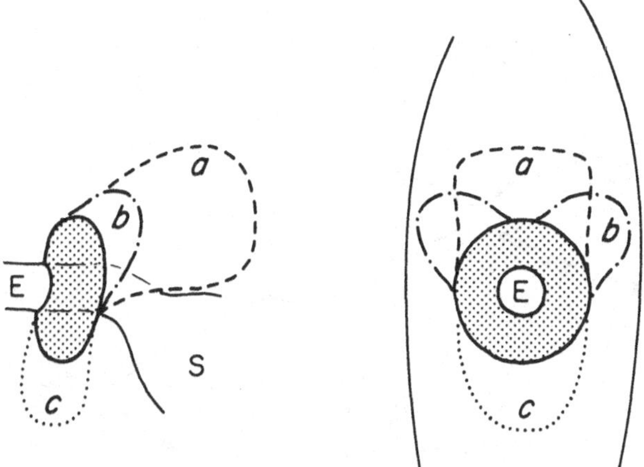

Figure 10
Lateral (left) and cross-sectional views of the general light organ (stippled) and the sexually dimorphic hypertrophy of dorsolateral and ventrolateral lobes seen in males of (a) *Leiognathus elongatus* and *L. leuciscus,* (b) *L. bindus* and *L. aureus,* (c) *Gazza* and *Secutor* species. E = esophagus, S = stomach. (Reprinted from McFall-Ngai and Dunlap 1984.)

Selection on the gas bladder as a central component of the light-organ system must have been strong (McFall-Ngai 1983b; Dunlap and McFall-Ngai 1987), because both tissues of the bladder spatially associated with the symbionts and bladder tissue some distance from the bacteria have been modified. But light can be considered analogous to any translocated metabolite from a symbiont, in that it is the symbiont product that apparently drives selection of tissue traits.

Variation in the Morphology of the Light Organ

Two aspects of the morphology of leiognathid light organs are of particular interest in relation to the symbiotic association: sexual dimorphism (McFall-Ngai and Dunlap 1984) and the geographical variation in the light organs within a species. Most of the leiognathids that have been examined show distinct sexual I dimorphism. The male's light organ usually is significantly larger than that of the female (figure 10), which, in all species examined, forms a relatively small, symmetrical, donut-shaped light organ around the esophagus. In contrast, specific areas of the male's light organ often will be hypertrophied, thus bringing the face of the light organ to just beneath the surface of the fish on both sides. These enlarged, lateral lobes usually

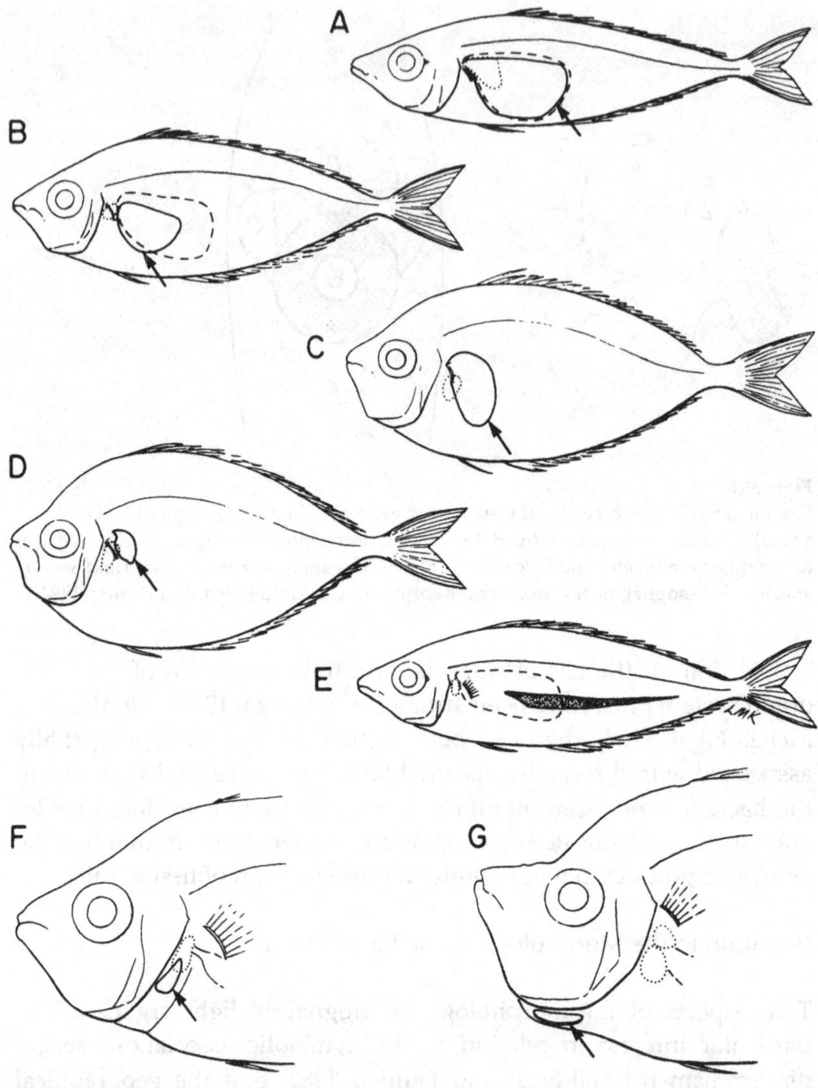

Figure 11
Relationship between external and internal sexual dimorphism in leiognathid fishes. Arrow at solid line indicates extent of clear skin patch on flank (A–D) or at opercular margin (F,G). Dotted line indicates internally located light organ, and dashed line indicates internally located gas bladder. (1) *Leiognathus elongatus* (similar to *L. leuciscus*), (B) *L. rivulatus*, (C) *L. aureus* (gas bladder not shown), (D) *L. bindus* (gas bladder not shown), (E) *L. stercorarius* (males bear flank stripe rather than clear patch), (F) *Gazza* sp., (G) *Secutor* sp.

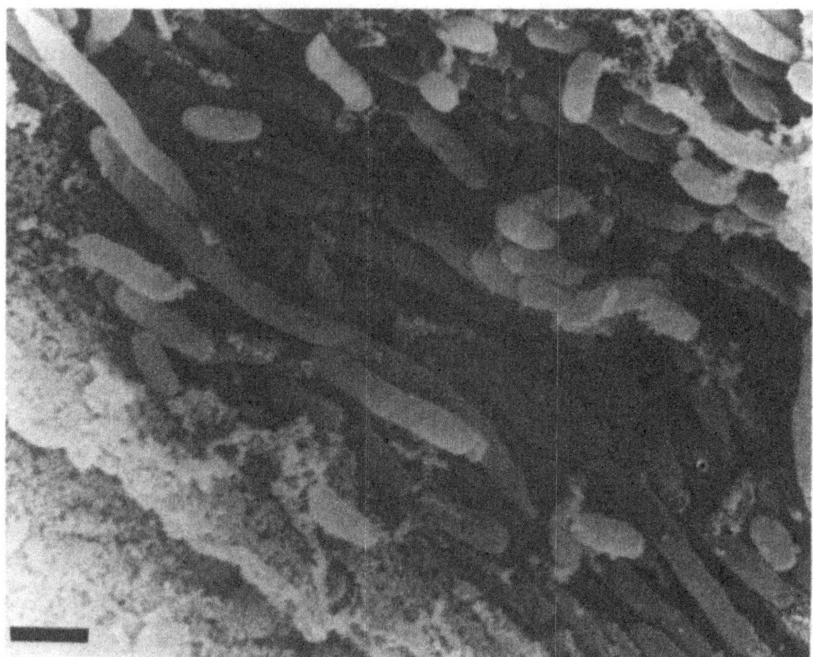

Figure 12
The light-organ bacteria of male *Leiognathus elongatus* appear as long thin cells, unlike the short rods that are typically found in the light organs of leiognathids. (bar = 2 μm.)

correspond with transparent patches on the body surface of the male (figure 11), which probably indicates sex-specific light emission through the body walls of the males.

The influence of the morphology of the male light organ on the physiology and morphology of the bacterial symbionts is of interest in the context of light-organ symbiosis (figure 12 here; Dunlap and McFall-Ngai 1984; Dunlap 1984). The luminous bacteria of light organs of females are coccobacilloid in shape and easily cultured. However, in species in which the male has a distinctly large light organ, the bacteria are markedly elongated and have required special culturing conditions (Dunlap and McFall Ngai, 1984). Because these observations have been made in only a few species, gradations in bacterial morphology and ease of culturing correlating with degree of sexual dimorphism have not yet been documented. Further, it is not known whether the male and the female of each fish species culture different strains of *Photobacterium leiognathi* from early in their life history, or whether the bacteria are instead the same strain in both sexes and the milieu of the male light organ influences the form and function of the symbionts.

In addition to variation between the sexes within a particular species, differences within a species occur over its geographical distribution. That is, for a given body size (for either males or females) the volume of the light organ of a given species differs among populations. Certain species of leiognathids occur over a very wide range (e.g., from the Phillippines to India). It is not known whether differences in bacterial strain or differences in habitat are responsible for this interesting example of intraspecific variation.

Conclusion

A central issue of this symposium concerned whether or not symbiosis has been important in determining the patterns of adaptive radiation we see among organisms. Although it is difficult to find direct evidence for the involvement of symbiosis, some supportive data may be available in patterns of phylogenetic occurrence. The complete coincidence of the presence of a symbiosis with the taxonomic alignment of a group of organisms, particularly if such an alignment has been made without regard to the symbiosis, is strong presumptive evidence that such a symbiotic association was important in the radiation of that group.

Bacterial light-organ symbioses in fishes occur as a derived character in a number of families. As presented in this paper, where they do occur they are characteristic of all the individuals of a particular species, a particular genus, and sometimes a particular family. This pattern of occurrence is important because the classification of fishes into these taxa has been based almost entirely on characters that are independent of the presence of a symbiotic association. Because the light organ of many fishes is internal, and ichthyologists traditionally use skeletal features as the most significant taxonomic characters, the luminous fishes provide a particularly compelling case for symbiosis as an important innovation. However, future analysis of other groups of symbiotic organisms, such as corals, leguminous plants, lichens, and hydrothermal-vent animals, may show similar correlations of the presence of a symbiosis and taxonomic affinities. A careful review of symbiotic groups is required to assess the extent to which these factors are coincident.

Within the context of the present contribution, it is relevant to ask how selection may have occurred to produce bacterially luminous glands, the light of which is used by the host in its behavior, and to ask how this innovation promoted radiation within a group of fishes. In other symbioses, such as those mentioned above, the microbial symbiont is a source of scarce nutrients. In contrast, present-day light-organ associations provide strong evidence that light is the primary bacterial product used by the host. For the luminous organ to be functional (i.e., visible), not only must the bacteria be concentrated and in high numbers; in addition, other portions of the fish must be adapted to permit and control light emission. Devising a scenario by which selection would occur to generate these adaptations is reminiscent of Darwin's problem with evolution of the vertebrate eye. However, many of the light-organ symbioses are gut-associated, and luminous bacteria (members of the Enterobacteriaceae) form a significant component of the gut microbiota of many fishes. It has been reported that up to 100 percent of the culturable aerobic gut microbionts are luminous bacteria (Nealson and Hastings 1979; Ruby and Morin 1979). Thus, it is possible to envision selection for the elaboration of a gut diverticulum that contained enterobacteria that happen to be luminous. For bacterial symbioses that are not gut-associated, such as those in the anomalopids and the monocentrids, no readily supportable hypothesis is yet available. Superficial pockets of bacteria seem to be rare in vertebrates.

The morids, the macrourids, the ceratoidei, and the leiognathids are the four groups in which members with luminous bacteria symbioses have undergone extensive radiation. In order to study the impact that symbiosis may have had on the evolution of these groups, the ecology and behavior of these fishes must be critically analyzed. The leiognathids are the only group among these four through which the question can be readily approached. Often they occur sympatrically in mixed-species schools in bays and estuaries. Their evolution might be studied in a manner similar to that applied to the sympatric cichlid fishes in the lakes of Africa, among which the key innovation driving selection and adaptive radiation appears to have been a change in the feeding mechanisms of the fish (Liem 1974). Extensive knowledge of the ecology and behavior of leiognathids, and an understanding of the role of the symbiosis in their biology, are areas that must be explored if we are to obtain a coherent picture of the importance of these associations in the evolution of this group.

Acknowledgments

I thank the organizers of the Bellagio symposium, L. Margulis, K. Nealson, and R. Fester, for an exceptional intellectual experience. I thank D. Cohen, R. Rosenblatt, E. Ruby, and J. Webb for critical comments on the manuscript, and W. Park for technical assistance.

References

Ahlstrom, E. H., Moser, H. G., and Cohen, D. M. 1984. Argentinoidei development and relationships. In: Moser, H. G., Richards, W. J., Cohen, D. M., Fahay, M. R., Dendall, A. W., Jr., and Richardson, S. L., eds., *Ontogeny and Systematics of Fishes*. Alan Press.

Ben-Tuvia, A. 1966. Red Sea fishes recently found in the Mediterranean. *Copeia* 1966: 254–275.

Bertelsen, E. 1984. Ceratioidei: Development and relationships. In: Moser, H. G., Richards, W. J., Cohen, D. M., Fahay, M. R., Dendall, A. W., Jr., and Richardson, S. L., eds., *Ontogeny and Systematics of Fishes*. Alan Press.

Bertelsen, E., and Munk, O. 1964. Rectal light organs in the argentinoid fishes, *Opisthoproctus* and *Winteria*. *Dana Rep*. 62: 1–189.

Bertelsen, E., Theisen, B., and Munk, O. 1965. On a post-larval specimen, anal light organ and tubular eyes of the argentinoid fish *Rhyncholhyalus natalensis* (Gilchrist and von Bonde). *Vidensk. Meddr. Dansk. Naturh. Foren.* 128: 357–371.

Boettcher, K. J., and Ruby, E. G. 1989. A physiological basis for depressed light emission by a symbiotic luminous bacterium. *Abstracts of the Annual Meeting of the American Society for Microbiology* 87: 257.

Briggs, J. C. 1974. *Marine Biogeography*. McGraw-Hill.

Carroll, R. 1988. *Vertebrate Paleontology and Evolution*. Freeman.

Castle, P. H. J., and Paxton, J. R. 1984. A new genus and species of luminescent eel (Pisces:Congridae) from the Arafura Sea, Northern Australia. *Copeia* 1984: 172–81.

Cohen, D. M. 1964a. Bioluminescence in the Gulf of Mexico anacanthine fish *Steindachneria argentea*. *Copeia* 1964: 406–409.

Cohen, D. M. 1964b. Suborder Argentinoidea. In: Bigelow, H. B., Cohen, D. M., Dick, M. M., Gibbs, R. H., Jr., Grey, M., Morrow, J. E., Jr., Schultz, C. P., and Walters, V. M., eds., *Fishes of the Western North Atlantic*, part 4: *Isospondyli*. Sears Foundation for Marine Research.

Cohen, D. M. 1984. Gadiformes: Overview. In: Moser, H. G., Richards, W. J., Cohen, D. M., Fahay, M. R., Dendall, A. W., Jr., and Richardson, S. L., eds., *Ontogeny and Systematics of Fishes*. Alan Press.

Cohen, D. M. 1989. *Papers on the Systematics of Gadiform Fishes*. Natural History Museum, Los Angeles.

Dunlap, P. V. 1984. The Ecology and Physiology of the Light Organ Symbiosis between *Photobacterium leiognathi* and Ponyfishes. Ph.D. thesis, University of California, Los Angeles.

Dunlap, P. V., and McFall-Ngai, M. J. 1987. Initiation and control of the bioluminescent symbiosis between *Photobacterium leiognathi* and leiognathid fish. In: Lee, J. J., and Fredrick, J. F., eds., *Endocytobiology III*. New York Academy of Sciences.

Dunlap, P. V., and McFall-Ngai, M. J. 1984. *Leiognathus elongatus* (Perciformes:Leiognathidae): Two distinct species based on morphological and light organ characters. *Copeia* 1984: 884–892.

Fahay, M. P., and Markle, D. F. 1984. Gadiformes: Development and relationships. In: Moser, H. G., Richards, W. J., Cohen, D. M., Fahay, M. R., Dendall, A. W., Jr., and Richardson, S. L., eds., *Ontogeny and Systematics of Fishes*. Alan Press.

Fitzgerald, J. M. 1977. Classification of luminous bacteria from the light organ of the Australian pinecone fish, *Cleiopus gloriamaris*. *Arch. Microbiol.* 112: 153–156.

Fraser, T. H. 1972. Comparative osteology of the shallow water cardinal fishes (Perciformes:Apogonidae) with reference to the systematics of the family. *Ichthyological Bull. Rhodes University* 34: 1–105.

Fukasawa, S., Suda, T., and Kubota, S. 1988. Identification of luminous bacteria isolated from the light organ of the fish, *Acropoma japonicum*. *Agric. Biol. Chem.* 52: 285–286.

Fukasawa, S., and Dunlap, P. V. 1986. Identification of luminous bacteria isolated from the light organ of the squid, *Doryteuthis kensaki*. *Agric. Biol. Chem.* 50: 1645–1646.

Haneda, Y. 1950. Luminous organs of fish which emit light indirectly. *Pacific Science* 4: 214–277.

Haneda, Y. 1951. The luminescence of some deep-sea fishes of the family Gadidae and Macrouridae. *Pacific Science* 5: 372–378.

Haneda, Y. 1966. Luminous apogonid fish from the Moreton Bay, Brisbane. *Scientific Reports of the Yokosuka City Museum* 12: 1–3.

Haneda, Y. 1968. On the luminous organ of the anacanthine fish, *Steindachneria argentea*, from the Gulf of Mexico. *Scientific Reports of the Yokosuka City Museum* 14: 7–11.

Haneda, Y., and Tsuji, F. 1976. The luminescent system of ponyfishes. *J. Morphology* 150: 539–552.

Haneda, Y., and Yoshiba, S. 1970. On a luminous substance of the anacanthine fish *Steindachneria argentea*, from the Gulf of Mexico. *Scientific Reports of the Yokosuka City Museum* 16: 1–4.

Hastings, J. W., and Nealson, K. H. 1981. The symbiotic luminous bacteria. In: Starr, M. P., Stolp, H., Truper, H. G., Balows, A., and Schlegel, H. G., eds., *The Prokaryotes: A Handbook of Habitats, Isolation and Identification of Bacteria*. Springer-Verlag.

Hastings, J. W., Makemson, J., and Dunlap, P. V. 1987. How are growth and luminescence regulated independently in light organ symbionts? *Symbiosis* 4: 3–24.

Herring, P., and Morin, J. G. 1978. Bioluminescence in fishes. In: Herring, P., ed. *Bioluminescence in Action*. Academic Press.

Herring, P. J., Clarke, M. R., von Boletzky, S., and Ryan, K. P. 1981. The light organs of *Sepiola atlantica* and *Spirula spirula* (Mollusca:Cephalopoda): Bacterial and intrinsic systems in the order Sepioidea. *J. Marine Biological Association of the United Kingdom* 61: 901–906.

Iwai, T. 1958. A study of the luminous organ of the apogonid fish, *Siphamia versicolor* (Smith and Radcliffe). *J. Washington Academy of Science* 48: 267–270.

Iwai, T. 1959. Notes on the luminous organ of the apogonid fish, *Siphamia majimai*. *Annals Mag. Nat. Hist.* 13: 545–550.

Iwai, T. 1971. Structure of luminescent organ of apogonid fish, *Siphamia versicolor*. *Japanese J. Ichthyology* 18: 125–127.

Johnson, G. D., and Rosenblatt, R. H. 1988. Mechanisms of light organ occlusion in flashlight fishes, family Anomalopidae (Teleostei: Beryciformes), and the evolution of the group. *Zoological J. of the Linnean Society* 94: 65–96.

Kühlmorgen-Hille, G. 1974. Leiognathidae: FAO Species Identification Sheets for Fisheries Purposes, Eastern Indian Ocean (Fishing Area 57) and Western Central Pacific (Fishing Area 71). FAO for the United Nations, Rome.

Lauder, G. V., and Liem, K. F. 1983. The evolution and interrelationships of the actinopterygian fishes. *Bull. Museum of Comparative Zoology* 150: 95–197.

Liem, K. 1974. Evolutionary strategies and morphological innovations: Cichlid pharyngeal jaws. *Systematic Zoology* 20: 425–441.

Marshall, N. B. 1965. Systematic and biological studies of the macrourid fishes (Anacanthini—Teleostei). *Deep-Sea Research* 12: 299–322.

Marshall, N. B., and Cohen, D. M. 1973. Order Anacanthini (Gadiformes): Characters and synopsis of families. In: Cohen, D. M., Marshall, N. B., Ebeling, A. W., Roser, D. E., Iwamoto, T., Sunoder, P., McDowell, S. B., Reed, W. H., III, and Woods, L. P., eds., *Fishes of the Western North Atlantic*. Sears Foundation for Marine Research.

Marshall, N. B., and Iwamoto, T. 1973. Family Macrouridae. In: Cohen, D. M., Marshall, N. B., Ebeling, A. W., Roser, D. E., Iwamoto, T., Sunoder, P., McDowell, S. B., Reed, W. H., III, and Woods, L. P., eds., *Fishes of the Western North Atlantic*. Sears Foundation for Marine Research.

McFall-Ngai, M. J. 1983a. Patterns, Mechanisms and Control of Luminescence in Leiognathid Fishes. Ph.D. thesis, University of California, Los Angeles.

McFall-Ngai, M. J. 1983b. Adaptations for reflection of bioluminescent light in the gas bladder of *Leiognathus equulus* (Perciformes:Leiognathidae). *J. Experimental Zoology* 227: 23–33.

McFall-Ngai, M. J., and Dunlap, P. V. 1983. Three new modes of luminescence in the leiognathid fish *Gazza minuta*: Discrete projected luminescence, ventral body flash, and buccal luminescence. *Marine Biology* 73: 227–237.

McFall-Ngai, M. J., and Dunlap, P. V. 1984. External and internal sexual dimorphism in leiognathid fishes: Morphological evidence for sex-specific bioluminescent signaling. *J. Morphology* 182: 71–83.

Nealson, K. H., and Hastings, J. W. 1979. Bacterial bioluminescence: Its control and ecological significance. *Microbiological Reviews* 43: 496–518.

Nelson, J. S. 1984. *Fishes of the World*. Wiley-Interscience.

Nielsen, J. G., and Bertelsen, E. 1985. The Gulper-Eel Family Saccopharyngidae (Pisces:Anguilliformes). Steenstrupia Zoological Museum, University of Copenhagen.

Nolf, D., and Steurbaut, E. 1989. Importance and restrictions of the otolith-based fossil record of gadiform and ophidiiform fishes. In: *Papers on the Systematics of Gadiform Fishes*. Natural History Museum, Los Angeles.

Okamura, O. 1970a. *Fauna Japonica: Nacrourina (Pisces)*. Academic Press of Japan.

Okamura, O. 1970b. Studies on the macrourid fishes of Japan: Morphology, ecology and phylogeny. *Reports of the Usa Marine Biology Station of Kochi University* 17: 1–179

Patterson, C. 1968. The caudal skeleton in mesozoic acanthopterygian fishes. *Bulletin of the British Museum of Natural History* 17: 47–102.

Paulin, C. D. 1989. Review of the morid genera *Gadella, Physiculus,* and *Salilota* (Teleostei:Gadiformes) with descriptions of seven new species. *New Zealand J. Zoology* 16: 93–133.

Pauly, D. 1976. The Leiognathidae (teleosts): Their stock and fishery in Indonesia, with notes on the biology of *Leiognathus splendens*. In: *Ecology and Management of Some Tropical Shallow Water Communities*. Indonesian Institue of Sciences.

Rau, N., and Rau, A. 1980. *Commercial Fisheries of the Central Phillipines*. Deutsche Gesellschaft für Technische Zusammenarbeit, Eschborn.

Reichelt, J. L., and Baumann, P. 1974. Effect of sodium chloride on growth of heterotrophic marine bacteria. *Archives of Microbiology* 97: 329–345.

Reichelt, J. L., Nealson, K. H., and Hastings, J. W. 1977. The specificity of symbiosis: Pony fish and luminous bacteria. *Archives for Microbiology* 112: 157–161.

Ross, L. G., and Gordon, J. D. M. 1978. Guanine and permeability of swimbladders of slope-dwelling fish. In: *Proceedings of the 12th European Symposium on Marine Biology, Stirling*. Pergamon.

Ruby, E. G., and Morin, J. G. 1978. Specificity of symbiosis between deep-sea fishes and psychrotrophic luminous bacteria. *Deep-Sea Research* 25: 161–167.

Ruby, E. G., and Morin, J. G. 1979. Luminous enteric bacteria of marine fishes: A study of their distribution, densities, and dispersion. *Applied and Environmental Microbiology* 38: 406–411.

Ruby, E. G., and Nealson, K. H. 1976. Symbiotic association of *Photobacterium fischeri* with the marine luminous fish *Monocentris japonica*: A model of symbiosis based on bacterial studies. *Biological Bulletin* 151: 574–586.

Ruby, E. G., Greenberg, E. P., and Hastings, J. W. 1980. Planktonic marine luminous bacteria: Species distribution in the water column. *Applied and Environmental Microbiology* 39: 302–306.

Singleton, R. J., and Skerman, T. M. 1973. A taxonomic study by computer analysis of marine bacteria from New Zealand waters. *J. Royal Society of New Zealand* 3: 129–140.

Somiya, H. 1977. Bacterial bioluminescence in chlorophthalmid deep-sea fish: A possible relationship between the light organ and the eyes. *Experimentia* 33: 906–909.

Sulak, K. J. 1977. The systematics and biology of *Bathypterois* (Pisces, Chlorophthalimidae) with a revised classification of benthic myctophiform fishes. *Galathea Report* 14: 109–122.

Woods, L. P., and Sonoda, P. M. 1973. Order Berycomorphi (Beryciformes). In: Cohen, D. M., Marshall, N. B., Ebeling, A. W., Roser, D. E., Iwamoto, T., Sunoder, P., McDowell, S. B., Reed, W. H., III, and Woods, L. P., eds., *Fishes of the Western North Atlantic*. Sears Foundation for Marine Research.

Zehren, S. J. 1979. The comparative osteology and phylogeny of the Beryciformes (Pisces: Teleostei). *Evolutionary Monographs* 1: 1–389.

26 Symbiogenesis and the Evolution of Mutualism: Lessons from the *Nephromyces*-Bacterial Endosymbiosis in Molgulid Tunicates

Mary Beth Saffo

Mutualism has been thought to death; what we need are solid descriptions of how organisms actually interact, experiments with what happens when a potential mutualist is removed.

—*D. Janzen, 1985*

The evolution of mutualism is a topic which has traditionally been rich in dialectic and fascinating anecdotes but impoverished by critical experimental data. Mutualistic symbiosis has a simple theoretical definition: an intimate inter-species interaction in which both (or all, if there are more than two) species partners benefit. Mutualistic symbionts, by definition, are better off together than they are apart. Though a simple definition in theory, mutualism is difficult to demonstrate experimentally; it is also difficult to construct biologically realistic theories of the evolution of mutualist interactions.

One limitation to the development of theories of the evolution of mutualism has been the historic emphasis among theoretical evolutionary biologists on examples of mutualism drawn from external species interactions, such as pollination ecology (Janzen 1985; Templeton and Gilbert 1985), ant-plant relationships (Heithaus et al. 1980), damselfish-sea anemone symbioses (Roughgarden 1975), and phoretic interactions (Wilson 1983). Until recently (Vandermeer and Boucher 1978; Law and Lewis 1983; Law 1985, 1988), endosymbiotic mutualisms have been virtually ignored by theoretical ecologists, except for occasional, and sometimes astute, secondary mention (Addicott 1981; Keeler 1985). Conversely, endosymbioses have been studied most intensively by cell biologists and physiologists focusing on the mechanics of cellular and metabolic interactions, rather than the evolutionary consequences, of such interactions. This specialization of scholarly activity has led to caricatures of the ecology and evolution of mutualism drawn largely from considerations of animals and plants from tropical, terrestrial communities:

- Mathematical models (May 1973, 1976) and other analyses (Williams 1966) have predicted that mutualism is an unstable or implausible interaction, and consequently rare (May 1973) or ecologically and evolutionarily unimportant (Williamson 1972).
- Mutualistic interactions have been considered more prevalent in stable, nonseasonal, nonfluctuating environments, and thus (for instance) more common in the tropics (Vermeij 1983; May 1976, 1982, 1984) than in temperate habitats.
- For any given pair of species, mutualism has been treated as a physiologically permanent condition, a discrete evolutionary step, which is considered capable of evolving to a parasitic interaction (or vice-versa) over several generations, but not of changing its nature within a single generation.
- Voicing a commonly held view, May (1982) has asserted that in endosymbiotic mutualisms the distinction between one organism and two mutualistic partners is so blurred that the symbiotic partners "no longer have a life of their own." Because of such blurred identities, mutualistic endosymbioses have been considered analytically intractable and also fundamentally different from other mutualisms.

The generality of these perspectives is tested provocatively by experimental consideration of a number of "chronic" endosymbioses (that is, of symbioses in which all individuals of a given host species, rather than merely a small percentage of a host population, are inhabited by a particular endosymbiotic taxon or group of taxa for a substantial portion of the host life cycle). I examine one of these—a tripartite symbiosis among molgulid tunicates, heterotrophic protists, and Gram-negative bacteria—and suggest some ways in which chronic endosymbioses imply a significant impact of symbiotic interactions on species radiation and evolutionary change, some ways in which chronic but cyclically reestablished endosymbioses can provide experimentally tractable systems for addressing evolutionary and ecological questions, and some ways in which the study of endosymbioses can inspire fresh, more complex perspectives on the evolution and the ecology of mutualism.

The Impact of Symbiosis on Molgulid Ascidians

Molgulid ascidian tunicates are a diverse, widely distributed family of filter-feeding, marine invertebrate chordates. Globally distributed in the marine benthos, their species diversity is markedly greater in higher latitudes than in the tropics (van Name 1945; Kott 1969; Monniot 1969). All molgulid species thus far examined—all adult individuals of at least seven species of two molgulid genera (*Molgula* and *Bostrichobranchus*; Saffo 1978, 1982, 1988) from a diverse array of habitats along the Atlantic, Pacific, and Gulf Coasts of the United States, and numerous molgulid species from the coasts of Europe and elsewhere (de Lacaze-Duthiers 1874; Giard 1888; Harant 1931; Buchner 1965)—are infected by a heterotrophic protist, "*Nephromyces*" (Giard 1888).[1]

Variously classified as a gregarine (de Lacaze-Duthiers 1874), a chytridiomycete (Giard 1888; Harant 1931), and an idiotypic "lower fungus" (Buchner 1965), *Nephromyces* remains a taxonomically recalcitrant protist of uncertain affinities. Though it resembles a fungus both in its hyphal-like trophic stages (figures 1–5) and in its chitinous walls (Saffo and Fultz 1986), many of its features are not typical of the kingdom Fungi or of other fungus-like protistan taxa. Its tubular mitochondrial cristae (figure 5), for instance, are more typical of protozoan cells than they are of chytridiomycetes or fungi, and its morphologically eclectic life cycle, especially the morphology of its reproductive stages (figure 6), does not resemble that of any other protoctistan group (Saffo 1981; Saffo and Fultz 1986).

Thus far, *Nephromyces* has been found only among molgulid tunicates. All developmental stages except the host transfer stages of *Nephromyces* are limited to the lumen of a molgulid organ called the renal sac (Saffo and Nelson 1983). Though the host transfer stage, an encysted spore, can be maintained in axenic culture for at least a month (Saffo and Davis 1982), the other life-history stages of *Nephromyces* have thus far not been cultured outside the animal host. All of these observations suggest that *Nephromyces* is an obligate symbiont of molgulid tunicates.

The renal sac (figures 1a, 2), *Nephromyces*' microhabitat, has been traditionally considered an excretory organ, largely because of its urate-containing concretions. However, the absence of ducts from the renal sac, the relatively unusual presence in a marine animal of uric acid as a major "excretory" product, the presence of other com-

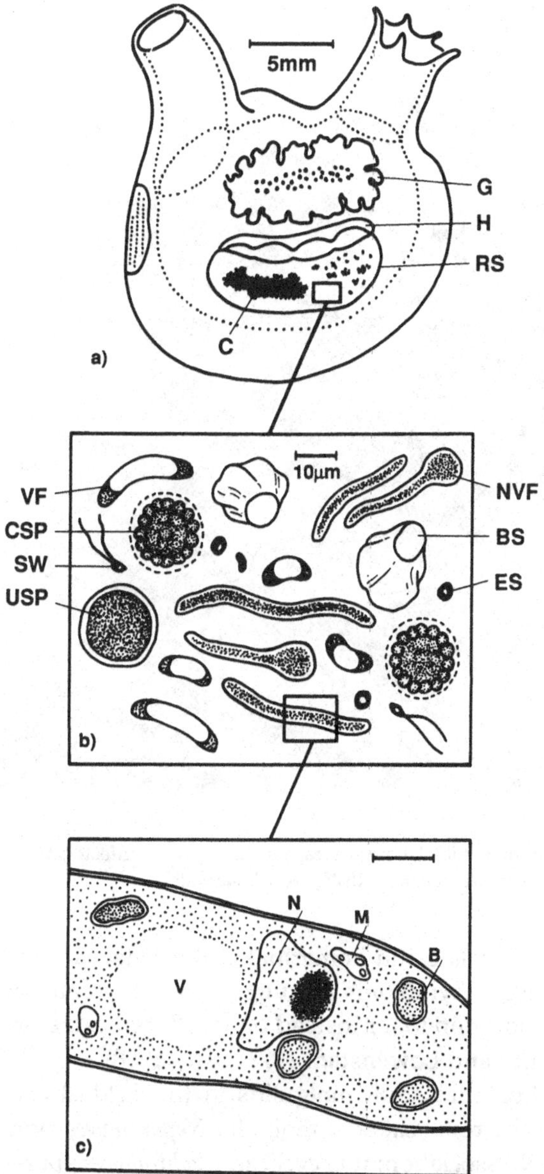

Figure 1
(a) Antomy of adult *Molgula manhattensis*, viewed from the right side. C = concretion; G = gonad; H = heart; RS = renal sac. (b) Cell types of *Nephromyces* isolated from renal-sac lumen of *M. manhattensis*. BS = "basket" (wall of discharged sporangium); CSP = cleaved sporangium; ES = encysted spore (host transfer stage); NVF = nonvacuolate filamentous cell; SW = bimastigote swarmer; USP = uncleaved sporangium; VF = vacuolate filamentous cell. (c) Filamentous cell of *Nephromyces*, drawn from electron micrograph. B = bacterium; M = mitochondrion; N = nucleus; V = vacuole.

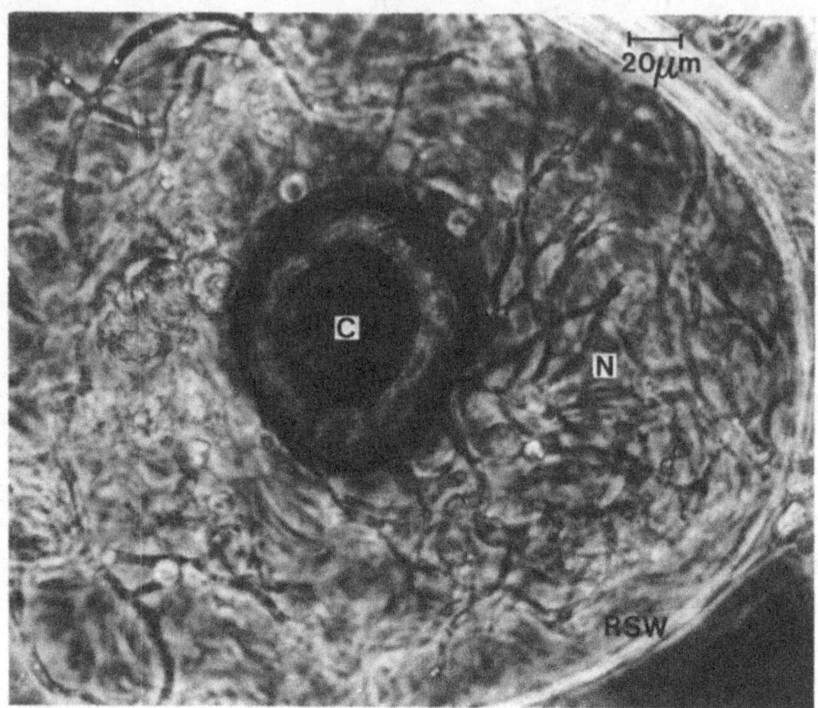

Figure 2
Renal sac of juvenile (postmetamorphic), laboratory-raised, *Nephromyces*-infected *M. manhattensis*. C = concretion; N = *Nephromyces*; RSW = renal-sac wall.

pounds (notably calcium oxalate and homarine) in the renal sac, and the conspicuous presence of symbionts all suggest that the renal sac may play other roles in addition to, or instead of, its alleged excretory function (Saffo 1978; Saffo and Lowenstam 1978; Saffo 1988).

Despite the observation that adult molgulids in the field always contain *Nephromyces* in the renal sac, the molgulid-*Nephromyces* symbiosis is not a hereditary association; it is cyclical. *Nephromyces* spores infect post-metamorphic zooids of *Molgula manhattensis* anew each generation (Saffo and Davis 1982). Thus, in nature and in the laboratory, larvae of *M. manhattensis* are *Nephromyces*-free, and post-metamorphic field-grown juveniles can vary seasonally in their rate of infection (Saffo and Davis 1982). However, 100 percent of the adult population of *M. Manhattensis* in the field contains *Nephromyces*.

Recent studies (Saffo 1987, 1990) have revealed a third partner in this symbiosis. In both molgulid species investigated (*M. manhattensis*

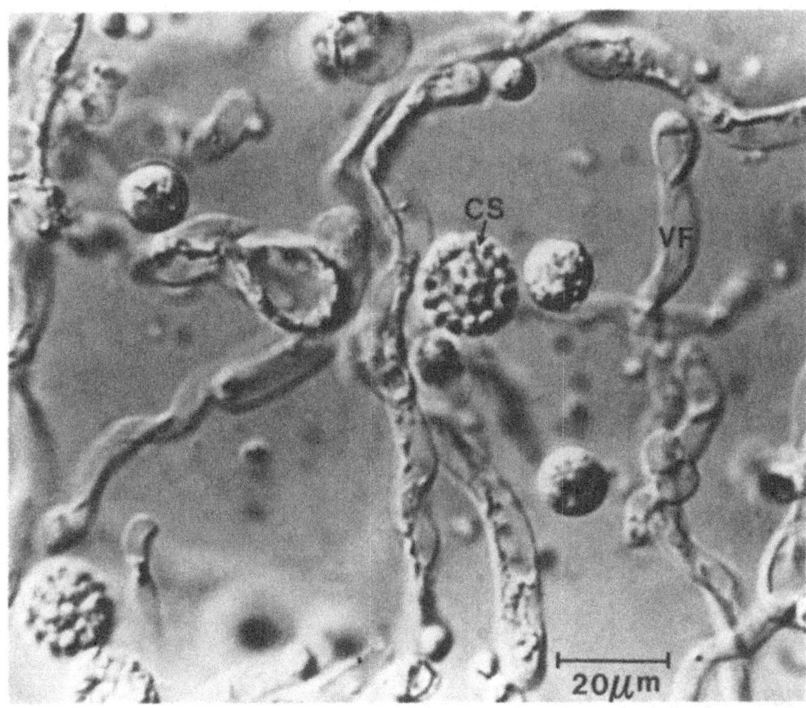

Figure 3
Nephromyces-infected renal-sac fluid from *M. manhattensis*. CS = cleaved sporangium; VF = vacuolate filamentous cell.

and *M. occidentalis*), *Nephromyces* cells harbor Gram-negative, intracellular bacteria (figures 1c, 5). Bacteria are present in both trophic and reproductive stages of *Nephromyces*. Ultrastructural data suggest that, at least in the trophic stages, the intracellular bacteria may lie directly in the *Nephromyces* cytoplasm, without an intervening host membrane. In the renal sac, bacteria are present only in *Nephromyces*. They are never present in renal-sac cells of *Nephromyces*-infected *Molgula* spp., and they are completely absent from both tissues and extracellular renal-sac fluid of experimental populations of *Nephromyces*-free *M. manhattensis*.

As a persistent, intimate, and hereditarily transmitted component of *Nephromyces*, these intracellular bacteria add new dimensions to the molgulid-*Nephromyces* symbiosis. The nonhereditary, cyclically reestablished symbiosis between *Molgula* spp. and *Nephromyces* is underlain, for instance, by a hereditary symbiosis between *Nephromyces*

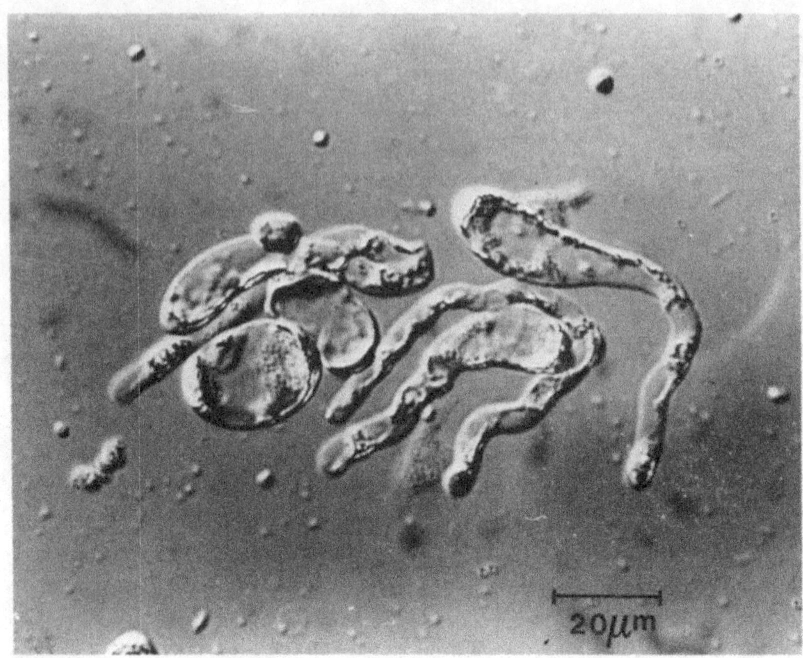

Figure 4
Vacuolate filamentous cell of *Nephromyces* from *M. manhattensis*.

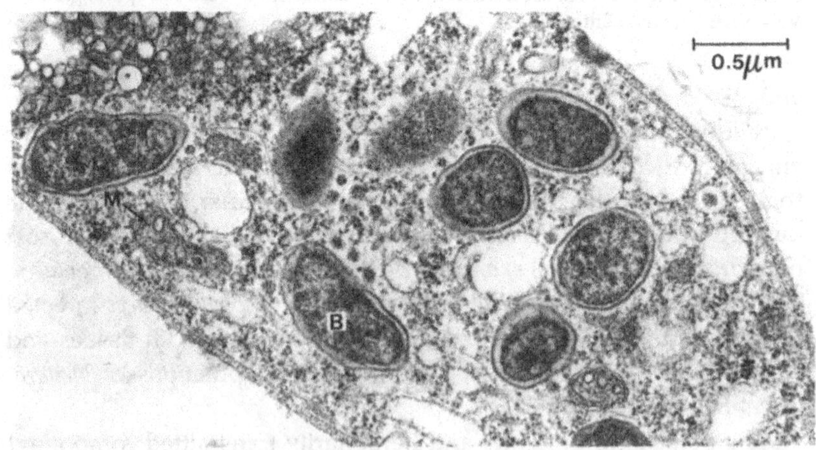

Figure 5
Transmission electron micrograph of filamentous cell of *Nephromyces*, isolated from *M. occidentalis*. B = bacterium; M = mitochondrion.

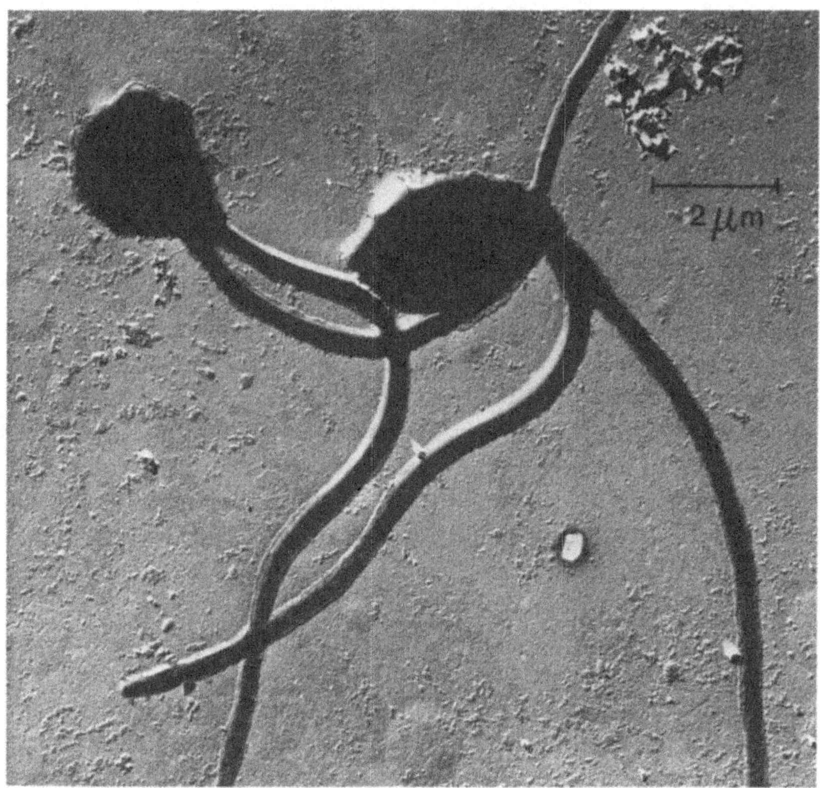

Figure 6
Swarmer cells of *Nephromyces*, from the renal sac of *M. manhattensis*, with whiplash undulipodia.

and its bacteria. The metabolic interactions between molgulids and *Nephromyces* now incorporate an unexpected prokaryotic component. Finally, to the ecological impact of *Nephromyces* on molgulids is added the complex dynamic of a third-partner interaction.

From a number of perspectives, it can be argued that the evolution of molgulid ascidians is bound up with that of *Nephromyces* (and its bacteria). First, *Nephromyces* is probably present in all molgulid species; conversely, *Nephromyces* has been found only in molgulids. Second, evidence suggests it is a coevolved association. For instance, the morphology of *Nephromyces* cells varies between host species, even when individuals of different host species, such as *M. manhattensis* and *M. citrina*, occasionally co-occur (Giard 1888; Saffo 1982). Also, cross-inoculations of laboratory-raised, symbiont-free *M. man-*

hattensis with *Nephromyces* from another host species, *M. occidentalis*, do not result in successful infection (M. B. Saffo, unpublished data), whereas parallel inoculation of *M. manhattensis* with symbionts from *M. manhattensis* results in 100-percent infection. Both observations are suggestive of host-species-correlated differentiation in *Nephromyces*.

Third, molgulids are distinct in a number of striking ways from other ascidians—including their closest relatives, the Pyuridae, from which the Molgulidae probably evolved. The Molgulidae have been regarded as the "most specialized" (van Name 1945) and "without doubt the most highly developed" (Berrill 1950) family of ascidians. Molgulids exploit an unusually wide range of marine habitats. They are found in all latitudes, but they exhibit greatest species diversity in boreal and polar latitudes (Kott 1969; van Name 1945). There are molgulids in both shallow and abyssal waters, and in both rocky and soft-bottom habitats (Monniot 1969). Though molgulids are by no means the only ascidians to inhabit soft-bottom substrates (van Name 1945; Kott 1972), they are unusual in their family-wide tendency to do so. At least two molgulid genera, *Eugyra* and *Bostrichobranchus*, are found exclusively in soft-bottom habitats; a large percentage of species of the largest genus, *Molgula*, also live "unattached" in sand and mud (van Name 1945; Berrill 1950). *M. manhattensis*, one of the few tunicates to tolerate low salinities, is abundant in many harbors and estuaries, and has been increasing its range within historic times (Tokioka and Kado 1972).

In view of their ecological and species diversity, molgulids are surprisingly homogeneous in some aspects of their morphology (Monniot 1969). They are usually rounded or elliptical in general body shape. The inner surface of the branchial sac (the filtering organ in ascidians) is always complex among molgulids (Monniot 1969); it is characterized by internal longitudinal vessels (usually borne on internal branchial folds) and spiral, ciliated slits (stigmata), which in some species are carried on conical infundibula penetrating into the branchial cavity.

In living animals, the most striking and diagnostic feature of molgulids is the presence of the renal sac on the right side of the molgulid body. There is no renal sac in the closely related pyurids. In the absence of ducts, and in the presence of urate-containing deposits, the renal sac bears some resemblance to other so-called renal tissues of other ascidian families, such as the renal vesicles of the ascidian families Ascidiidae and Corellidae. However, it also differs from such tis-

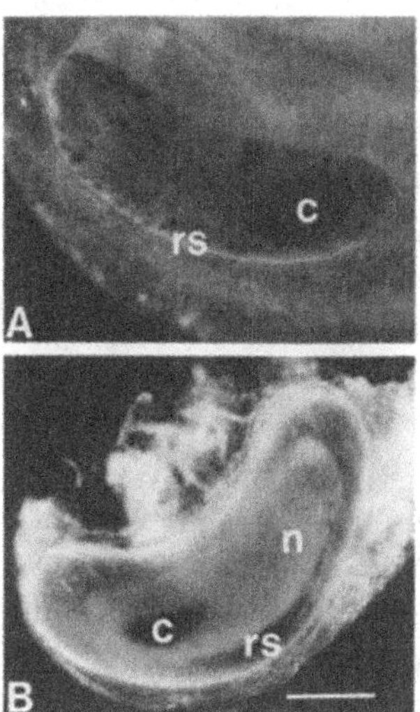

Figure 7
Renal-sac morphology in laboratory-raised *Molgula manhattensis*, comparing symbiont-free and *Nephromyces*-infected *M. manhattensis* individuals. In the renal sac of *Nephromyces*-infected *M. manhattensis* (B), concretions can be smaller, and the lumen fluid cloudier, than in that of symbiont-free counter parts (A). (A) Symbiont-free renal sac in *M. manhattensis* (entire animal, minus its tunic). (B) *Nephromyces*-infected renal sac, dissected from *M. manhattensis*. rs = renal-sac wall; n = *Nephromyces*; c = concretions. Scale bar = 0.5 mm.

sues in a number of ways. Easily one of the most conspicuous organs of molgulids, it is much larger than other ascidian urate-containing tissues. Also unlike these tissues, it is a single organ, and it closely abuts the heart. Some of its chemical contents, such as calcium oxalate and homarine (Saffo and Lowenstam 1978), have not been described from the "renal" tissues of other ascidians. Finally, the molgulid renal sac is apparently also unique among ascidian renal tissues (M. B. Saffo, unpublished observations on the Corellidae and Ascidiidae) in the presence of endosymbionts. Could the presence of *Nephromyces* be correlated with the evolution of the renal sac, and with the unusual ecological features of molgulids?

Fourth, the activities of *Nephromyces* and its bacteria markedly affect the activities of the renal sac. This impact is sometimes morphologically obvious (figure 7), and it is certainly demonstrable biochemically. While *M. manhattensis* produces and accumulates uric acid in the renal sac, even in the absence of *Nephromyces* (Saffo 1977, 1988), the metabolic fate of that uric acid differs profoundly in *Nephromyces*-

infected and symbiont-free *Molgula* spp. Whether or not they are infected with *Nephromyces*, molgulids themselves show no significant enzymatic ability to utilize the renal-sac urate. However, *Nephromyces* (with its bacteria) possesses striking urate oxidase activity (Saffo 1988), with specific activities among the highest published values for any organism (table 1). Relative quantities of homarine, calcareous deposits, and other compounds (Saffo, 1987 and unpublished) also differ in symbiont-free and *Nephromyces*-infected *M. manhattensis*. These symbiont-induced differences in the physiology of the renal sac must have an impact on the general biology of *Molgula* spp.

Similarly, the origin and the functional significance of the numerous distinctive features of *Nephromyces* are best considered in the context of the dual symbiotic associations of this protist. The striking urate oxidase activity of *Nephromyces* might well be the result of its own bacterial endosymbionts, which perhaps serve generally as peroxisomal analogues (Saffo 1990). Some of the taxonomically paradoxical morphological features of *Nephromyces*, such as the unusual posteriorly directed pair of whiplash undulipodia of its swarmer cell (figure 6), might be fruitfully interpreted in part as adaptations to endosymbiotic life in the viscous, chemically unusual lumen fluid of the renal sac.

Experimental Usefulness of Cyclically Reestablished Endosymbioses

The cyclical reestablishment of the molgulid-*Nephromyces* symbiosis (Saffo and Davis 1982) makes it possible to separate the animal and microbial symbionts without the use of artifact-inducing antibiotic or chemical treatments (Douglas and Smith 1983), allowing parallel laboratory culturing of *Nephromyces*-infected and symbiont-free *M. manhattensis*. Parallel-culture systems such as this can serve as a powerful tool for direct measurement of the impact of microbial symbionts on an animal host.

So far, this system has provided critical evidence that *Nephromyces* is an organism foreign to its ascidian host (Saffo and Davis 1982), made possible delineation of the protist's life cycle and the general infective process (Saffo and Nelson 1983), and contributed evidence suggesting that the intracellular bacteria of *Nephromyces* are transmitted hereditarily with their protistan host. It has also been used to localize urate oxidase activity in the symbiosis, and it has suggested

Table 1
Urate oxidase activity in molgula and its symbionts compared with other organisms.

Organism/tissue	Specific activity (μm/mg prot./min × 10^5)	Reference
Homo sapiens liver	0	Friedman et al. 1985
uninfected *Molgula manhattenensis* renal sac wall	0–1.1	Saffo 1988
chicken liver homogenates	3	Scott et al. 1969
chicken liver peroxisomes	16.5	Scott et al. 1969
frog kidney homogenate	120	Scott et al. 1969
frog kidney peroxisomes	600	Scott et al. 1969
old world monkeys: liver homogenate	241–536	Friedman et al. 1985
spinach leaf perioxisomes	1000	Huang et al. 1983
mouse liver	1130	Friedman et al. 1985
Nephromyces: field-collected *Molgula manhattensis*	713–1567	Saffo 1988
rabbit liver	1980	Friedman et al. 1985
purified urate oxidase (porcine liver)	4788	Saffo 1988
Nephromyces: laboratory-raised *Molgula manhattensis*	4800–6245	Saffo 1988
soybean nodule perixisomes	8900	Huang et al. 1983
fish peroxisomes	10,000	Hanks and Tolbert 1982
rat liver peroxisomes	13,000	Huang, et al. 1983
Nephromyces: Molgula occidentalis	4103–23,696	Saffo 1988

other chemical activities of the microbial symbionts. It is now being used to determine the partitioning of enzymes of the urate degradative pathway among the three symbionts, to measure the effects of endosymbionts on rates of urate synthesis in molgulids, and to assess the general metabolic effects on molgulids of colonization by *Nephromyces*.

Finally, parallel culture of symbiont-free and *Nephromyces*-infected *Molgula manhattensis* has been used to assess the effect of *Nephromyces* and its bacteria on the growth, survivorship, and reproductive success of molgulids. Experiments have yielded unexpectedly complex data. The consistent presence of *Nephromyces* among adult molgulids, coupled with the restriction of *Nephromyces* to molgulids, suggests strongly that the *Nephromyces*-molgulid symbiosis is a mutualistic association. However, laboratory data comparing the growth, survivorship, and reproductive output of *Nephromyces*-infected and symbiont-free *M. manhattensis* suggest a complex dynamic to the association. Although *Nephromyces* is an obligate symbiont of molgulids, the presence of *Nephromyces* is not physiologically obligate for *Molgula* spp., as symbiont-free *M. manhattensis* can reach sexual maturity both in laboratory conditions and in semi-natural field conditions (in ocean waters outside the natural distribution of *M. manhattensis*; Saffo, 1983, 1986, and unpublished). In some laboratory conditions (ascidians raised in seawater originating from their natural habitat; Saffo 1983), *Nephromyces*-infected *M. manhattensis* produce more offspring than their symbiont-free counterparts; in other conditions (animals raised in seawater from areas outside their natural distribution), there is no consistent difference between symbiont-infected and symbiont-free hosts in survivorship, growth, or reproduction. Early experiments with symbiont-infected and symbiont-free *M. manhattensis* grown in San Francisco Bay suggest that infected *M. manhattensis* survive markedly better than uninfected animals in natural field conditions. But in field conditions outside their natural range of distribution (in colder, more saline waters), juvenile-infected *M. manhattensis* have shown greater survivorship but slower growth rates than their uninfected counterparts.

The laboratory "habitat" of *M. manhattensis* differs from field conditions, most notably in the maintenance of laboratory animals in isolated aquaria rather than in constantly flowing sea water, in the use of filtered (0.45-μm pores) seawater to prevent infection of symbiont-free cultures by *Nephromyces* spores in incoming water, and in the reliance on six species of laboratory-grown algae as the chief food sources. These differences clearly affect the growth of *M. manhattensis* as well as the apparent relative growth rate of *Nephromyces*. In the laboratory, *Nephromyces*-infected *M. manhattensis* grows more slowly and is more heavily infected with *Nephromyces* than its counterparts grown in field conditions. Such differences could affect other aspects of symbiont-host interactions.

In other symbioses, differences between laboratory and field conditions do indeed influence interactions of symbiont and host. Hoegh-Guldberg et al. (1987) have noted that laboratory conditions markedly affect symbiotic interactions between the tropical sea anemone *Aiptasia* and its zooxanthellae. In the laboratory, the expulsion rate of zooxanthellae from *Aiptasia* is five times that of anemones in the field—representing a loss of carbon that is insignificant in the field but significant in the laboratory. Outcomes of experiments comparing the effect of mycorrhizae on plant growth can also differ between field and greenhouse conditions (Law 1988).

Complexity of Symbiotic Dynamics

Especially when combined with the compelling implications of the apparently universal distribution of *Nephromyces* among molgulids, the complex outcomes of differential survivorship, growth, and reproduction between symbiont-infected and symbiont-free *M. manhattensis* in different environmental conditions pose a paradox which can best be resolved by a more complex view of mutualism. From this perspective, "mutualism" is not necessarily a physiologically permanent condition, but a dynamic balance among symbiotic partners that can be shifted according to environmental conditions. Examples from other chronic endosymbioses bolster such a view. For instance, experiments on symbiotic and aposymbiotic green hydra (Douglas and Smith 1983) have documented a growth cost to hydra for maintenance of its algal symbionts. Symbiotic green hydra grown in the light can survive for long periods in the absence of organic food, while aposymbiotic green hydra do not survive. In the presence of an organic supplement as well as light, there is no difference in growth between symbiotic and aposymbiotic hydra. But in the absence of light and the presence of organic food, the growth rates of aposymbiotic hydra are the same as they are in the light, while those of symbiotic green hydra are depressed.

Experiments on other symbioses, notably vesicular-arbuscular mycorrhizae (VAM), suggest even more pointedly that chronic endosymbioses are not necessarily mutualistic in all circumstances. Bethlenfalvay, Brown, and Pacovsky (1982) have noted, along with other investigators, that the interaction between mycorrhizae and host plants varies with the level of phosphorus in the soil and with the developmental stage of the symbiotic interaction. At low levels of phosphorus, plant growth is enhanced by VAM. In the first 9 weeks after VAM inoculation of soybeans (with increasing fungal biomass), intermediate phosphorus levels result initially in growth reduction, then (as the concentration of available phosphorus is reduced and the fungal biomass peaks) in enhanced growth of the host plant (Bethlenfalvay, Brown, and Pacovsky 1982; Bethlenfalvay, Pacovsky, and Brown 1982). High levels of soil phosphorus are correlated with reduced plant growth as compared against uninfected controls; very high phosphorus levels inhibit colonization of the plant by VAM.

In still other endosymbioses, dissolution or inhibition of symbiotic interactions occurs in predictable environmental conditions; for instance, high nitrate levels inhibit rhizobial colonization of legumes, and lichen partners dissociate in high-nutrient, humid conditions. Various factors, including darkness and low temperatures, increase expulsion rates of zooxanthellae from cnidarian hosts (Smith and Douglas 1987).

Chronic endosymbioses serve to broaden and revise standard views of mutualism in a number of ways:

- Mutualistic endosymbioses (at least, chronic endosymbioses that are plausibly or demonstrably mutualistic) are ubiquitous. Many such endosymbioses—reef-building corals, mycorrhizae, lichens, nitrogen-fixing symbioses, chemosynthetic symbioses in deep-sea hydrothermal-vent animals, molgulid ascidians—have significant ecological impact, whether measured by abundance, diversity, and distribution of the hosts or by trophic impact on ecosystem cycling (Lewis 1973, 1982).
- Chronic endosymbioses are by no means limited to the tropics, but instead are a significant presence in nontropical, temporally variable habitats as well. Symbioses such as lichens and mycorrhizae are conspicuous members of boreal and alpine communities and of other pioneer or disturbed communities (Lewis 1982; Law 1988). Several such endosymbioses are clearly more prevalent in higher latitudes than in the tropics. Molgulid diversity, for instance, is lowest in tropic

habitats and greatest in boreal and polar latitudes. The distribution of dicyemid mesozoans in the excretory organs of benthic cephalopods, variously interpreted as a parasitic, commensal, or mutualistic symbiosis (Lapan and Morowitz 1972), is also instructive. Although dicyemids are found in 100 percent of benthic cephalopods in temperate and polar habitats, the rate of infection in subtropical waters is less than 100 percent. No diceymids have been found among tropical cephalopods (Hochberg 1983).

• Mutualism is a delicate, dynamic balance which can be shifted by environmental conditions. Just as environmental conditions can affect the outcomes of external interactions between microbes (McGee et al. 1972; Dean 1985) or animals (Paine 1979; Wilson 1983), so can they affect endosymbiotic interactions. In many symbioses, such as VAM, the alternative outcomes to mutualism include not only dissolution of the association but also parasitism. For endosymbioses, these outcomes can change not just in evolutionary time (Janzen 1985) but also in physiological time. Such dynamic complexity has at least two implications for theoretical implications for the evolution of mutualism. First, a physiologically and ecologically variable interaction means that the transition from parasitism to mutualism (and vice-versa) is not necessarily a profound evolutionary step. Second, the factors which affect that balance can be tested, at least in some symbioses, within experimentally realistic periods of time.

• Chronic endosymbioses are not experimentally intractable. In many cases—nitrogen-fixing symbioses, mycorrhizae, lichens, algal-invertebrate symbioses, molgulid ascidians—hosts and symbionts are separable (Law and Lewis 1983; Law et al. 1984). Cyclically reestablished, nonhereditary symbioses are especially amenable to comparative assessments of symbiotic with aposymbiotic host biology. Particularly in concert with the possible or demonstrated variability in interaction outcomes in some of these symbioses, these endosymbioses therefore provide a potentially powerful tool with which to test a number of mutualism models. Cost-benefit models seem particularly amenable to such treatment.

• Research has increased appreciation for the intricacy of endosymbioses, both in the intersymbiont dynamics and in the number and significance of the taxonomic, metabolic, ecological, and evolutionary dimensions. While becoming less prone to caricatures as charming fairy tales, they have taken on a more scientifically useful role through the instructiveness of their complexity.

Acknowledgments

I am grateful to Jenny Wardrip for assistance with illustrations, and to the American Philosophical Society, the National Science Foundation, the Research Corporation, the Whitehall Foundation, and the Institute of Marine Sciences and the Graduate Division of the University of California, Santa Cruz, for support of the research described here. Debra Felix and Jonathan Margolis stimulated development of the perspective of an environmentally alterable dynamic in mutualistic interactions.

Note

1. Giard named three species of *Nephromyces*: *N. molgularum* isolated form *M. socialis* (= *M. manhattensis*); *N. sorokini*, isolated from the moluglid *Lithonephyra eugyranda* (= *M. complanata*), and *N. roscovitanus*, isolated from the molgulid *Anurella roscotivana* (= *M. occulta*). Since these species descriptions were based solely on very sketchy verbal descriptions of *Nephromyces* morphology, I have avoided the use of species names for morphological variants of *Nephromyces* Giard until species identification can be supported by more complete taxonomic data. *Nephromyces piscinus* Plehn (1916), isolated from kidneys of freshwater carp, has conidia and a septate mycelium; it is clearly unrelated to *Nephromyces* Giard.

References

Addicott, J. F. 1981. Stability properties of two-species models of mutualism: Simulation studies. *Oecologia* 49: 42–49.

Berrill, N. J. 1950. *The Tunicata: With an Account of the British Species*. Ray Society.

Bethlenfalvay, G. F., Brown, M. S., and Pacovsky, R. S. 1982. Parasitic and mutualistic associations between a mycorrhizal fungus and soybean: Development of the host plant. *Phytopath.* 72: 889–893.

Bethlenfalvay, G. F., Pacovsky, R. S., and Brown, M. S. 1982. Parasitic and mutualistic associations between a mycorrhizal fungus and soybean: Development of the endophyte. *Phytopath.* 72: 894–897.

Buchner, P. 1965. *Endosymbiosis of Animals with Plant Microorganisms*. Interscience.

Dean, A. M. 1985. The dynamics of microbial commensalisms and mutualisms. In: Boucher, D. H., ed., *The Biology of Mutualism: Ecology and Evolution*. Oxford University Press.

de Lacaze-Duthiers, H. 1874. Histoire des ascidies simples des cotes de France. *I. Arch. Zool. Exp. et Gén.* 3: 304–313.

Douglas, A. E., and Smith, D. C. 1983. The cost of symbionts to their host in green hydra. In: Schenk, H. E. A., and Schwemmler, W., eds., *Endocytobiology II*. Walter de Gruyter.

Friedman, T. B., Polanco, G. E., Appold, J. C., and Mayle, J. E. 1985. On the loss of uricolytic activity during primate evolution. I. Silencing of urate oxidase in a hominoid ancestor. *Comp. Biochem. Physiol.* 81B: 653–659.

Giard, A. 1888. Sur les *Nephromyces*, genre nouveau de champignons parasites du rein des Molgulidées. *Compt. Rend. Hebd. Séanc. Acad. Sci. Paris* 106: 1180–1182.

Hanks, J., and Tolbert, N. E. 1982. Localization of purine degradation in animal and plant organelles. *Ann. N.Y. Acad. Sci.* 386: 420–421.

Harant, H. 1931. Contribution a l'histoire naturelle des ascides et de leurs parasites. 2. Chytridinées. *Ann. Inst. Ocean. Monaco* 8: 349–352.

Heithaus, E. R., Culver, D. C., and Beattie, A. J. 1980. Models of some ant-plant mutualisms. *Amer. Nat.* 116: 347–361.

Hochberg, F. G. 1983. The parasites of cephalopods: A review. *Mem. Nat. Mus. Victoria* 44: 109–145.

Hoegh-Guldberg, O., McCloskey, L. R., and Muscatine, L. 1987. Expulsion of zooxanthellae by symbiotic cnidarians from the Red Sea. *Coral Reefs* 5: 201–204.

Huang, A. H. C., Trelease, R. N., and Moore, T. S., Jr. 1983. *Plant Peroxisomes*. Academic Press.

Janzen, D. H. 1985. Natural history of mutualisms. In: Boucher, D. H., ed., *The Biology of Mutualism: Ecology and Evolution*. Oxford University Press.

Keeler, K. H. 1985. Cost:benefit models of mutualism. In: Boucher, D. H., ed., *The Biology of Mutualism: Ecology and Evolution*. Oxford University Press.

Kott, P. 1969. Antarctic ascidiacea. *Antarctic Research Series* (*Amer. Geophys. Union*) 13: 145–171.

Kott, P. 1972. Some sublittoral ascidians in Moreton Bay, and their seasonal occurrence. *Mem. Qd. Mus.* 16: 233–260.

Lapan, E. A., and Morowitz, H. J. 1972. The mesozoa. *Sci. Amer.* 222(1): 94–101.

Law, R. 1985. Evolution in a mutualistic environment. In: Boucher, D. H., ed., *The Biology of Mutualism: Ecology and Evolution*. Oxford University Press.

Law, R. 1988. Some ecological properties of intimate mutualisms involving plants. In: Davy, A. J., Hutchings, M. J., and Watkinson, A. R., eds., *Plant Population Ecology*. Blackwell.

Law, R., and Lewis, D. H. 1983. Biotic environments and the maintenance of sex: Some evidence from mutualistic symbioses. *Biol. J. Linn. Soc.* 20: 249–276.

Law, R., Hutson, V., and Lewis, D. H. 1984. Mutualism disagreement. *Nature* 310: 104.

Lewis, D. H. 1973. The relevance of symbiosis to taxonomy and ecology, with particular reference to mutualistic symbioses and the exploitation of marginal habitats. In: Heywood, V. H., ed., *Taxonomy and Ecology*. Academic Press.

Lewis, D. H. 1982. Mutualistic lives. *Nature* 297: 176.

May, R. 1973. *Stability and Complexity in Model Ecosystems*. Princeton University Press.

May, R. 1976. *Theoretical Ecology. Principles and Applications*. W. B. Saunders.

May, R. 1982. Mutualistic interactions among species. *Nature* 296: 803–804.

May, R. 1984. A test of ideas about mutualism. *Nature* 307: 410–411.

McGee, R. D., III, Drake, J. F., Fredrickson, A. G., and Tsuchiya, H. M. 1972. Studies in intermicrobial symbiosis. *Saccharomyces cerevisiae* and *Lactobacillus casei*. *Can. J. Microbiol*. 18: 1733–1742.

Monniot, C. 1969. Les molgulidae des mers européenes. *Mém. Mus. Nat. d'Hist. Nat*. 60 (sér. A): 171–272.

Paine, R. T. 1979. Disaster, catastrophe and local persistence of the sea palm *Postelsia palmaeformis*. Science 205: 685–687.

Plehn, M. 1916. Pathogen Schimmelpilze in der fischniere. *Z. Fisch. Deren Hilfswiss*. 18: 51–54.

Roughgarden, J. 1975. Evolution of marine symbiosis: A simple cost-benefit model. *Ecology* 56: 1201–1208.

Saffo, M. B. 1977. Studies on the Renal Sac of *Molgula manhattensis* De Kay (Ascidiacea, Tunicata, Phylum Chordata). Ph.D. thesis, Stanford University.

Saffo, M. B. 1978. Studies on the renal sac of the ascidian *Molgula manhattensis*. I. Development of the renal sac. *J. Morph*. 155: 287–310.

Saffo, M. B. 1981. The enigmatic protist *Nephromyces*. *BioSystems* 14: 487–490.

Saffo, M. B. 1982. Distribution of the endosymbiont *Nephromyces Giard within the ascidian family Molgulidae*. *Biol. Bull*. 162: 95–104.

Saffo, M. B. 1983. A new mutualism? The symbiosis of molgulid tunicates with the protist *Nephromyces*. *Amer. Zool*. 24: 122A.

Saffo, M. B. 1986. Renal sac function and the Nephromyces-molgulid symbiosis: Does uric acid play the central metabolic role? *Amer. Zool*. 26: 22A.

Saffo, M. B. 1987. Symbiosis within a symbiosis: The renal sac endosymbiont *Nephromyces* contains intracellular bacteria. *Amer. Zool.* 27: 15A.

Saffo, M. B. 1988. Nitrogen waste or nitrogen source? Urate degradation in the renal sac of molgulid tunicates. *Biol. Bull.* 175: 403–409.

Saffo, M. B. 1990. Symbiosis within a symbiosis: Intracellular bacteria within the endosymbiotic protist *Nephromyces. Mar. Biol.* (in press)

Saffo, M. B., and Davis, W. 1982. Modes of infection of the ascidian *Molgula manhattensis* by its endosymbiont *Nephromyces* Giard. *Biol. Bull.* 162: 105–112.

Saffo, M. B., and Fultz, S. 1986. Chitin in the symbiotic protist *Nephromyces. Can. J. Bot.* 64: 1306–1310.

Saffo, M. B., and Lowenstam, H. A. 1978. Calcareous deposits in the renal sac of a molgulid tunicate. *Science* 200:1166–1168.

Saffo, M. B., and Nelson, R. 1983. The cells of *Nephromyces*: Developmental stages of a single life cycle. *Can. J. Bot.* 61: 3230–3239.

Scott, P. J., Visentin, L. P., and Allen, J. M. 1969. The enzymatic characteristics of peroxisomes of amphibian and avian liver and kidney. *Ann. N.Y. Acad. Sci.* 168: 244–264.

Smith, D. C., and Douglas, A. E. 1987. *The Biology of Symbiosis.* Edward Arnold.

Templeton, A. R., and Gilbert, L. E. 1985. Population genetics and the coevolution of mutualism. In: Boucher, D. H., ed., *The Biology of Mutualism: Ecology and Evolution.* Oxford University Press.

Tokioka, T., and Kado, Y. 1972. The occurrence of *Molgula manhattensis* (De Kay) in brackish water near Hiroshima, Japan. *Publ. Seto. Mar. Biol. Lab.* 21: 121–129.

Vandermeer, J. H., and Boucher, D. H. 1978. Varieties of mutualistic interactions in population models. *J. Theor. Biol.* 74: 549–558.

van Name, W. G. 1945. The North and South American ascidians. *Bull. Amer. Mus. Nat. Hist.* 84: 372–443.

Vermeij, G. J. 1983. Intimate associations and coevolution in the sea. In: Futuyma, D. J., and Slatkin, M., eds., *Coevolution.* Sinauer.

Williams, G. C. 1966. *Adaptation and Natural Selection: A Critique of Some Current Evolutionary Thought.* Princeton University Press.

Williamson, M. 1972. *The Analysis of Biological Populations.* Edward Arnold.

Wilson, D. S. 1983. The effect of population structure on the evolution of mutualism: A field test involving burying beetles and their phoretic mites. *Amer. Nat.* 121: 851–870.

About the Authors

Peter Atsatt is a professor in the Department of Ecology and Evolutionary Biology at the University of California, Irvine. His interests include the ecology and evolution of parasitic plants, plant-insect interactions, symbiosis, and the evolution of plants.

David Bermudes is a postdoctoral research associate at the University of Wisconsin's Center for Great Lakes Studies. His research on symbiosis has included studies of terrestrial and epiphytic mycorrhizae, nitrogen fixation in association with bromeliads, spirochetes, and the origin of the undulipodium (cilium, flagellum).

Richard C. Back is a graduate fellow studying plankton ecology at the University of Wisconsin's Center for Great Lakes Studies. His doctoral research focuses on the relationship between the biochemical composition of phytoplankton and the grazing, assimilation, and growth of zooplankton.

Paola Bonfante-Fasolo is a professor of botany at the University of Torino. Her major contributions deal with the biology of mycorrhizal symbiosis, mostly in the fields of the cellular mechanisms regulating partner development. She is a co-editor of *Cell-to-Cell Signals in Plant, Animal and Microbial Symbiosis* (Springer-Verlag, 1988).

René Fester, a graduate student at Northern Arizona University, is a co-editor of *Global Ecology: Towards a Science of the Biosphere* (Academic Press, 1989).

Lynda J. Goff is a professor of biology at the University of California at Santa Cruz. She has pioneered modern molecular-biological studies of red algae and their symbionts.

Anne-Marie Grenier, an engineer with the National Institute for Agronomic Research (INRA), is working on the symbiosis of *Sitophilus* (Insecta, Coleoptera) with Paul Nardon at the National Institute of Applied Sciences (INSA) in Lyon, France.

Ricardo Guerrero is a *Catedrático* (full professor) in the Department of Microbiology at the University of Barcelona. He is president of the Sociedad Catalana de Biologica.

Robert H. Haynes is Distinguished Research Professor of Biology at York University, in Toronto, and a Fellow of the Royal Society of Canada. He has worked on problems of DNA repair and mutagenesis in microorganisms for the past 30 years.

Gregory Hinkle, a postdoctoral researcher at the University of Massachusetts at Amherst, co-directs the Planetary Biology Program of NASA's Life Sciences Division.

Rosmarie Honegger is a *Privatdozentin* (lecturer) in the Institute of Plant Biology at the University of Zürich. Her main interest is in mutualistic symbioses, especially lichen-forming fungi and their photobionts.

Kwang W. Jeon is a professor of cell biology in the Department of Zoology at the University of Tennessee, Knoxville. He is the senior editor of the *International Review of Cytology—A Survey of Cell Biology* and has worked on endosymbiosis using the amoeba-bacteria symbiosis system.

Bryce Kendrick is a professor of mycology in the Department of Biology at the University of Waterloo. A former Guggenheim Fellow, he is a fellow of the Royal Society of Canada and the secretary of the Academy of Science. Over the past 30 years, he has contributed to many areas of mycology, especially ecology, systematics, and development, and has authored or edited many books, notably *The Fifth Kingdom* and *The Whole Fungus*.

Richard Law is a lecturer in the Department of Biology at the University of York. His research lies at the interface of ecology and evolutionary biology, and includes contributions to the evolution of symbioses.

David Lewis holds a personal chair at the University of Sheffield, and is currently chairman of the Department of Animal and Plant Sciences. His research interests center on the carbohydrate metabolism of plants and fungi, and on the interactions—mutualistic and antagonistic—between species from these two kingdoms. He has also contributed to theoretical considerations concerning symbiosis. He is executive editor of the *New Phytologist*, a journal which continues to publish many papers on fungus-plant mutualisms. He was president of the British Mycological Society in 1989.

Lynn Margulis is a Distinguished University Professor in the Department of Botany at the University of Massachusetts. A member of the National Academy of Sciences, she has contributed to the scientific literature, primarily in the fields of cell evolution and microbial ecology. She works with James E. Lovelock on the Gaia hypothesis.

John Maynard Smith is Emeritus Professor of Biology at the University of Sussex. He is a fellow of the Royal Society and a foreign member of the National Academy of Sciences. He has worked on many aspects of evolutionary biology, including the evolution of sex and recombination, and on the evolution of behavior.

Margaret McFall-Ngai is an assistant professor in the Department of Biological Sciences at the University of Southern California. Her research work focuses on the morphological and biochemical characteristics of photogenic and photoreceptive tissues of vertebrates.

Paul Nardon is a professor of biology, and the head of the Department of Biochemistry, at the National Institute of Applied Sciences (Villeurbanne, France). He has contributed to the study of bacterial symbioses in insects, especially beetles.

Kenneth Nealson is the Shaw Distinguished Professor of Biology at the University of Wisconsin–Milwaukee and the Center for Great Lakes Studies. A microbial physiologist by training, he has been actively involved with the study of bioluminescence and symbiosis for over 20 years.

Kris Pirozynski is a paleomycologist at the National Museum of Natural Sciences in Ottawa. His main interest is the evolutionary interaction between fungi and plants through the ages.

Peter Price, professor of biology at Northern Arizona University, has authored the books *Insect Ecology* and *Evolutionary Biology of Parasites*. His research involves principally the relationships among plants, insect herbivores, and natural enemies, and their population dynamics. He teaches courses in ecology and evolution.

Mary Beth Saffo is an associate research marine biologist at the Institute of Marine Sciences of the University of California at Santa Cruz. A Fellow of the American Association for the Advancement of Science, she has published papers on symbiosis, physiological ecology, biomineralization, functional morphology, and protist phylogeny.

Jan Sapp is an associate professor in the Department of History and Philosophy of Science at the University of Melbourne. He has written widely in the history of the modern life sciences, and is the author of *Beyond the Gene* (Oxford University Press, 1987) and *Where the Truth Lies* (Cambridge University Press, 1990).

Silvano Scannerini is a professor of botany, the director of the postgraduate course Biology and Biotechnology of Fungi and the head of the Electron Microscopy Center at the University of Torino. A member of the Council of the Academy of Agriculture and a fellow of the Academy of Science of Torino, he has contributed to the scientific literature, primarily in the fields of functional morphology and morphogenesis of fungi and mycorrhizae.

Werner Schwemmler is a university professor in the Department of Biology at the Free University of Berlin. He has contributed to the scientific literature, primarily in the fields of insect symbiosis (with Georg Gassner: *Insect Endocytobiosis*, CRC Press, 1989) and cell evolution (*Symbiogenesis as Macro-Mechanism of Evolution*, Walter De Gruyter, 1989). He initiated (with Hainfried Schenk) the interdisciplinary research area of endocytobiology and the international journal *Endocytobiosis and Cell Research*.

David Smith is Principal and Vice-Chancellor of the University of Edinburgh. Before taking up this office in 1987, he had spent a third of a century researching various symbiotic associations, ranging from lichens to green hydra. He is a co-author of *The Biology of Symbiosis* (Edward Arnold, 1987).

Sorin Sonea is a professor in the Department of Microbiology and Immunology at the Université de Montréal and a member of the Academy of Sciences of the Royal Society of Canada. He has presented a new approach to the bacteria world as a partially unified global organism.

Toomas Tiivel is a researcher in the Laboratory of Molecular Genetics at the Institute of Chemical Physics and Biophysics of the Estonian

Academy of Sciences. He has contributed to the scientific literature in the fields of cellular ultrastructure, endocytobiosis of insects, and cell evolution. He is a co-editor of *Lectures in Theoretical Biology* (Valgus, Tallinn, 1988).

Robert Trench is a professor of biology at the University of California at Santa Barbara. His interests include symbiosis, algal phylogenies, and the evolution of symbiotic interactions.

Russell D. Vetter is an assistant research biologist at the University of California's Scripps Institution of Oceanography, in San Diego. Previously, he was a Mellon Foundation Fellow at the same institution. He is a comparative physiologist interested in the adaptations of marine organisms to extreme environments. He has studied adaptations to temperature, salinity, low oxygen, and anthropogenic toxicants as well as adaptations to hydrogen sulfide. He has devoted the past seven years to study of hydrothermal-vent ecosystems.

Index

Printed in the United States
by Baker & Taylor Publisher Services